ATF

Federal Firearms Regulations

Reference Guide

2014

By

U.S. Department of Justice

Bureau of Alcohol, Tobacco, Firearms, and Explosives

PrepperPress

Post-Apocalyptic Fiction & Survival Nonfiction

www.PrepperPress.com

ATF
Federal Firearms Regulations
Reference Guide
2014

ISBN: 978-1-939473-58-5

Prepper Press
Printed in the United States of America

Bureau of Alcohol, Tobacco,
Firearms and Explosives

Office of the Director

Washington, DC 20226

Dear Federal Firearms Licensees:

The Bureau of Alcohol, Tobacco, Firearms and Explosives (ATF), a component of the United States Department of Justice, is a law enforcement agency charged with protecting our communities from violent criminals, criminal organizations, the illegal possession, use and trafficking of firearms, the illegal possession, use and storage of explosives, acts of arson and bombings, and the illegal diversion of alcohol and tobacco products. We are proud to partner with industries, law enforcement, and the community to protect the public we serve.

Federal firearms licensees play a key role in safeguarding the public from violent crime by maintaining accurate records, instituting internal controls, and performing background checks on potential firearms purchasers. These practices have saved lives, prohibited violent criminals from obtaining firearms, and prevented firearms-related crimes.

The 2014 edition of the Federal Firearms Regulations Reference Guide contains information that will help you comply with Federal laws and regulations governing the manufacture, importation and distribution of firearms and ammunition. This edition contains new and amended statutes enacted since publication of the 2005 edition, as well as updated regulations and rulings issued by ATF. In addition to these updated materials, in response to inquiries received from industry members, the public, and partner agencies, the 2014 edition contains additional and amended Questions and Answers to assist with compliance.

Laws may change over time, as will information in this guide. The ATF website at www.ATF.gov is a good source of updated information and is one method ATF uses to communicate with licensees about changes to laws and regulations. Additionally, as always, you are welcome to contact your local ATF office for information or assistance.

Thank you for working with us to make our communities safer.

Sincerely,

B. Todd Jones
Director

On January 15, 2015, ATF posted the 2014 Federal Firearms Regulations Reference Guide (the "2014 Guide") to the ATF website (www.atf.gov). ATF last published this guide in 2005; the 2014 Guide incorporates almost a decade of updates to the federal firearms regulations.

Since posting the 2014 Guide, ATF became aware that it contained some inadvertent omissions and editing errors. ATF has corrected these omissions and errors in an update to 2014 Guide posted on March 7, 2014.

The corrected omissions and editing errors are:

(1) At page 90, within Section III, Laws and Regulations, Sub-Part B., National Firearms Act, the citation and title for one of the regulations was not included with the text of the regulation. That citation and title, "§ 479.105 Transfer and possession of machine guns," is now included.

(2) At page 190, within the General Information section, the last two subparts of Item 10, Armor Piercing Ammunition, were not included. These subparts consist of: (a) the text of the 1994 amendment Congress made to the Gun Control Act's (GCA) definition of armor piercing ammunition and, (b) a listing of the projectiles that have been granted exemptions to the GCA's prohibition on armor piercing ammunition. These subparts are now included.

(3) At page 102, within Section III, Laws and Regulations, Sub-part C, U.S. Munitions Import List, Category XIV, Toxicological Agents and Equipment and Radiological Equipment, the text of 27 CFR 447.21(c) did not reflect recent regulatory changes. Those language changes are now included.

(4) At page 198, Question & Answer Item B8, grammatical errors have been corrected.

(5) At page 206, Question & Answer Item I3, grammatical errors have been corrected.

The 2014 Guide functions as a convenient reference source for the public and industry. The omissions and errors in the original posting in no way altered the legal force and effect of these provisions.

ATF continues to review the 2014 Guide to ensure no other typographical or editing errors occurred in the publishing process. The public is welcome to provide input regarding the identification of any potential error in the 2014 Guide, or any other suggestions on improving the 2014 Guide. Please submit to: ORA@atf.gov.

Any future corrections, additions or modifications to the 2014 Guide will be identified on this page. ATF apologizes for any confusion or inconvenience caused by publishing errors, and looks forward to continued dialogue with our regulated community and the general public.

FEDERAL FIREARMS REGULATIONS
REFERENCE GUIDE
2014

TABLE OF CONTENTS

EDITOR'S NOTE

The cross references, bracketed notes, and Editor's notes seen in the laws and regulations are for guidance and assistance only and do not appear in the official United States Code and Code of Federal Regulations.

THE GUN CONTROL ACT OF 1968

TITLE 18, UNITED STATE CODE, CHAPTER 44

TITLE I: STATE FIREARMS CONTROL ASSISTANCE

PURPOSE

Sec. 101. The Congress hereby declares that the purpose of this title is to provide support to Federal, State, and local law enforcement officials in their fight against crime and violence, and it is not the purpose of this title to place any undue or unnecessary Federal restrictions or burdens on law–abiding citizens with respect to the acquisition, possession, or use of firearms appropriate to the purpose of hunting, trapshooting, target shooting, personal protection, or any other lawful activity, and that this title is not intended to discourage or eliminate the private ownership or use of firearms by law–abiding citizens for lawful purposes, or provide for the imposition by Federal regulations of any procedures or requirements other than those reasonably necessary to implement and effectuate the provisions of this title.

Chapter 44—Firearms

§ 921 Definitions.

(a) As used in this chapter—

(1) The term **"person"** and the term **"whoever"** include any individual, corporation, company, association, firm, partnership, society, or joint stock company.

(2) The term **"interstate or foreign commerce"** includes commerce between any place in a State and any place outside of that State, or within any possession of the United States (not including the Canal Zone) or the District of Columbia, but such term does not include commerce between places within the same State but through any place outside of that State. The term **"State"** includes the District of Columbia, the Commonwealth of Puerto Rico, and the possessions of the United States (not including the Canal Zone).

(3) The term **"firearm"** means (A) any weapon (including a starter gun) which will or is designed to or may readily be converted to expel a projectile by the action of an explosive; (B) the frame or receiver of any such weapon; (C) any firearm muffler or firearm silencer; or (D) any destructive device. Such term does not include an antique firearm.

(4) The term **"destructive device"** means—

(A) any explosive, incendiary, or poison gas—

    (i) bomb,

    (ii) grenade,

    (iii) rocket having a propellant charge of more than four ounces,

    (iv) missile having an explosive or incendiary charge of more than one–quarter ounce,

    (v) mine, or

    (vi) device similar to any of the devices described in the preceding clauses;

(B) any type of weapon (other than a shotgun or a shotgun shell which the Attorney General finds is generally recognized as particularly suitable for sporting purposes) by whatever name known which will, or which may be readily converted to, expel a projectile by the action of an explosive or other propellant, and which has any barrel with a bore of more than one–half inch in diameter; and

(C) any combination of parts either designed or intended for use in converting any device into any destructive device described in subparagraph (A) or (B) and from which a destructive device may be readily assembled.

The term **"destructive device"** shall not include any device which is neither designed nor redesigned for use as a weapon; any device, although originally designed for use as a weapon, which is redesigned for use as a signaling, pyrotechnic, line throwing, safety, or similar device; surplus ordnance sold, loaned, or given by the Secretary of the Army pursuant to the provisions of section 4684(2), 4685, or 4686 of title 10; or any other device which the Attorney General finds is not likely to be used as a weapon, is an antique, or is a rifle which the owner intends to use solely for sporting, recreational or cultural purposes.

(5) The term **"shotgun"** means a weapon designed or redesigned, made or remade, and intended to be fired from the shoulder and designed or redesigned and made or remade to use the energy of an explosive to fire through a smooth bore either a number of ball shot or a single projectile for each single pull of the trigger.

(6) The term **"short–barreled shotgun"** means a shotgun having one or more barrels less than eighteen inches in length and any weapon made from a shotgun (whether by alteration, modification or otherwise) if such a weapon as modified has an overall length of less than twenty–six inches.

(7) The term **"rifle"** means a weapon designed or redesigned, made or remade, and intended to be fired from the shoulder and designed or redesigned and made or remade to use the energy of an explosive to fire only a single projectile through a rifled bore for each single pull of the trigger.

(8) The term **"short–barreled rifle"** means a rifle having one or more barrels less than sixteen inches in length and any weapon made from a rifle (whether by alteration, modification, or otherwise) if such weapon, as modified, has an overall length of less than twenty–six inches.

(9) The term **"importer"** means any person engaged in the business of importing or bringing firearms or ammunition into the United States for purposes of sale or distribution; and the term **"licensed importer"** means any such person licensed under the provisions of this chapter.

(10) The term **"manufacturer"** means any person engaged in the business of manufacturing firearms or ammunition for purposes of sale or distribution; and the term **"licensed manufacturer"** means any such person licensed under the provisions of this chapter.

(11) The term **"dealer"** means (A) any person engaged in the business of selling firearms at wholesale or retail, (B) any person engaged in the business of repairing firearms or of making or fitting special barrels, stocks, or trigger mechanisms to firearms, or (C) any person who is a pawnbroker. The term **"licensed dealer"** means any dealer who is licensed under the provisions of this chapter.

(12) The term **"pawnbroker"** means any person whose business or occupation includes the taking or receiving, by way of pledge or pawn, of any firearm as security for the payment or repayment of money.

(13) The term **"collector"** means any person who acquires, holds, or disposes of firearms as curios or relics, as the Attorney General shall by regulation define, and the term **"licensed collector"** means any such person licensed under the provisions of this chapter.

(14) The term **"indictment"** includes an indictment or information in any court under which a crime punishable by imprisonment for a term exceeding one year may be prosecuted.

(15) The term **"fugitive from justice"** means any person who has fled from any State to avoid prosecution for a crime or to avoid giving testimony in any criminal proceeding.

(16) The term **"antique firearm"** means—

(A) any firearm (including any firearm with a matchlock, flintlock, percussion cap, or similar type of ignition system) manufactured in or before 1898; or

(B) any replica of any firearm described in subparagraph (A) if such replica–

(i) is not designed or redesigned for using rimfire or conventional centerfire fixed ammunition, or

(ii) uses rimfire or conventional centerfire fixed ammunition which is no longer manufactured in the United States and which is not readily available in the ordinary channels of commercial trade; or

(C) any muzzle loading rifle, muzzle loading shotgun, or muzzle loading pistol, which is designed to use black powder, or a black powder substitute, and which cannot use fixed ammunition. For purposes of this subparagraph, the term **"antique firearm"** shall not include any weapon which incorporates a firearm frame or receiver, any firearm which is converted into a muzzle loading weapon, or any muzzle loading weapon which can be readily converted to fire fixed ammunition by replacing the barrel, bolt, breechblock, or any combination thereof.

(17)(A) The term **"ammunition"** means ammunition or cartridge cases, primers, bullets, or propellent powder designed for use in any firearm.

(B) The term **"armor piercing ammunition"** means—

(i) a projectile or projectile core which may be used in a handgun and which is constructed entirely (excluding the presence of traces of other substances) from one or a combination of tungsten alloys, steel, iron, brass, bronze, beryllium copper, or depleted uranium; or

(ii) a full jacketed projectile larger than .22 caliber designed and intended for use in a handgun and whose jacket has a weight of more than 25 percent of the total weight of the projectile.

(C) The term **"armor piercing ammunition"** does not include shotgun shot required by Federal or State environmental or game regulations for hunting purposes, a frangible projectile designed for target shooting, a projectile which the Attorney General finds is primarily intended to be used for sporting purposes, or any other projectile or projectile core which the Attorney General finds is intended to be used for industrial purposes, including a charge used in an oil and gas well perforating device.

(18) The term **"Attorney General"** means the Attorney General of the United States[1]

(19) The term **"published ordinance"** means a published law of any political subdivision of a State which the Attorney General determines to be relevant to the enforcement of this chapter and which is contained on a list compiled by the Attorney General, which list shall be published in the Federal Register, revised annually, and furnished to each licensee under this chapter.

(20) The term **"crime punishable**

by imprisonment for a term exceeding one year" does not include—

(A) any Federal or State offenses pertaining to antitrust violations, unfair trade practices, restraints of trade, or other similar offenses relating to the regulation of business practices, or

(B) any State offense classified by the laws of the State as a misdemeanor and punishable by a term of imprisonment of two years or less.

What constitutes a conviction of such a crime shall be determined in accordance with the law of the jurisdiction in which the proceedings were held. Any conviction which has been expunged, or set aside or for which a person has been pardoned or has had civil rights restored shall not be considered a conviction for purposes of this chapter, unless such pardon, expungement, or restoration of civil rights expressly provides that the person may not ship, transport, possess, or receive firearms.

(21) The term **"engaged in the business"** means—

(A) as applied to a manufacturer of firearms, a person who devotes time, attention, and labor to manufacturing firearms as a regular course of trade or business with the principal objective of livelihood and profit through the sale or distribution of the firearms manufactured;

(B) as applied to a manufacturer of ammunition, a person who devotes time, attention, and labor to manufacturing ammunition as a regular course of trade or business with the principal objective of livelihood and profit through the sale or distribution of the ammunition manufactured;

(C) as applied to a dealer in firearms, as defined in section 921(a) (11)(A), a person who devotes time, attention, and labor to dealing in firearms as a regular course of trade or business with the principal objective of livelihood and profit through the repetitive purchase and resale of firearms, but such term shall not include a person who makes occasional sales, exchanges, or purchases of firearms for the enhancement of a personal collection or for a hobby, or who sells all or part of his personal collection of firearms;

(D) as applied to a dealer in firearms, as defined in section 921(a) (11)(B), a person who devotes time, attention, and labor to engaging in such activity as a regular course of trade or business with the principal objective of livelihood and profit, but such term shall not include a person who makes occasional repairs of firearms, or who occasionally fits special barrels, stocks, or trigger mechanisms to firearms;

(E) as applied to an importer of firearms, a person who devotes time, attention, and labor to importing firearms as a regular course of trade or business with the principal objective of livelihood and profit through the sale or distribution of the firearms imported; and

(F) as applied to an importer of ammunition, a person who devotes time, attention, and labor to importing ammunition as a regular course of trade or business with the principal objective of livelihood and profit through the sale or distribution of the ammunition imported.

(22) The term **"with the principal objective of livelihood and profit"** means that the intent underlying the sale or disposition of firearms is predominantly one of obtaining livelihood and pecuniary gain, as opposed to other intents, such as improving or liquidating a personal firearms collection: **Provided,** That proof of profit shall not be required as to a person who engages in the regular and repetitive purchase and disposition of firearms for criminal purposes or terrorism. For purposes of this paragraph, the term **"terrorism"** means activity, directed against United States persons, which—

(A) is committed by an individual who is not a national or permanent resident alien of the United States;

(B) involves violent acts or acts dangerous to human life which would be a criminal violation if committed within the jurisdiction of the United States; and

(C) is intended—

(i) to intimidate or coerce a civilian population;

(ii) to influence the policy of a government by intimidation or coercion; or

(iii) to affect the conduct of a government by assassination or kidnapping.

(23) The term **"machinegun"** has the meaning given such term in section 5845(b) of the National Firearms Act (26 U.S.C. 5845(b)).

(24) The terms **"firearm silencer"** and **"firearm muffler"** mean any device for silencing, muffling, or diminishing the report of a portable firearm, including any combination of parts, designed or redesigned, and intended for use in assembling or fabricating a firearm silencer or firearm muffler, and any part intended only for use in such assembly or fabrication.

(25) The term **"school zone"** means—

(A) in, or on the grounds of, a public, parochial or private school; or

(B) within a distance of 1,000 feet from the grounds of a public, parochial or private school.

(26) The term **"school"** means a school which provides elementary or secondary education, as determined under State law.

(27) The term **"motor vehicle"** has the meaning given such term in section 13102 of title 49, United States Code.

(28) The term **"semiautomatic rifle"** means any repeating rifle which utilizes a portion of the energy of a firing cartridge to extract the fired cartridge case and chamber the next round, and which requires a separate pull of the trigger to fire each cartridge.

(29) The term **"handgun"** means—

(A) a firearm which has a short stock and is designed to be held and fired by the use of a single hand; and

(B) any combination of parts from which a firearm described in subparagraph (A) can be assembled.

[(30), (31) Repealed. Pub. L. 103–322, title XI, §110105(2), Sept. 13, 1994, 108 Stat. 2000.]

(32) The term **"intimate partner"** means, with respect to a person, the spouse of the person, a former spouse of the person, an individual who is a parent of a child of the person, and an individual who cohabitates or has cohabited with the person.

(33)(A) Except as provided in subparagraph (C), the term **"misdemeanor crime of domestic violence"** means an offense that—

   (i) is a misdemeanor under Federal, State, or Tribal law; and

   (ii) has, as an element, the use or attempted use of physical force, or the threatened use of a deadly weapon, committed by a current or former spouse, parent, or guardian of the victim, by a person with whom the victim shares a child in common, by a person who is cohabiting with or has cohabited with the victim as a spouse, parent, or guardian, or by a person similarly situated to a spouse, parent, or guardian of the victim.

(B)(i) A person shall not be considered to have been convicted of such an offense for purposes of this chapter, unless—

   (I) the person was represented by counsel in the case, or knowingly and intelligently waived the right to counsel in the case; and

   (II) in the case of a prosecution for an offense described in this paragraph for which a person was entitled to a jury trial in the jurisdiction in which the case was tried, either

      (aa) the case was tried by a jury, or

      (bb) the person knowingly and intelligently waived the right to have the case tried by a jury, by guilty plea or otherwise.

(ii) A person shall not be considered to have been convicted of such an offense for purposes of this chapter if the conviction has been expunged or set aside, or is an offense for which the person has been pardoned or has had civil rights restored (if the law of the applicable jurisdiction provides for the loss of civil rights under such an offense) unless the pardon, expungement, or restoration of civil rights expressly provides that the person may not ship, transport, possess, or receive firearms.

Editor's Note:

Subparagraph (C) referenced in 921(a)(33)(A) never was enacted. We presume the reference should have been to subparagraph (B).

(34) The term **"secure gun storage or safety device"** means—

   (A) a device that, when installed on a firearm, is designed to prevent the firearm from being operated without first deactivating the device;

   (B) a device incorporated into the design of the firearm that is designed to prevent the operation of the firearm by anyone not having access to the device; or

   (C) a safe, gun safe, gun case, lock box, or other device that is designed to be or can be used to store a firearm and that is designed to be unlocked only by means of a key, a combination, or other similar means.

(35) The term **"body armor"** means any product sold or offered for sale, in interstate or foreign commerce, as personal protective body covering intended to protect against gunfire, regardless of whether the product is to be worn alone or is sold as a complement to another product or garment.

(b) For the purposes of this chapter, a member of the Armed Forces on active duty is a resident of the State in which his permanent duty station is located.

§ 922 Unlawful acts.

(a) It shall be unlawful—

   (1) for any person—

      (A) except a licensed importer, licensed manufacturer, or licensed dealer, to engage in the business of importing, manufacturing, or dealing in firearms, or in the course of such business to ship, transport, or receive any firearm in interstate or foreign commerce; or

      (B) except a licensed importer or licensed manufacturer, to engage in the business of importing or manufacturing ammunition, or in the course of such business, to ship, transport, or receive any ammunition in interstate or foreign commerce;

   (2) for any importer, manufacturer, dealer, or collector licensed under the provisions of this chapter to ship or transport in interstate or foreign commerce any firearm to any person other than a licensed importer, licensed manufacturer, licensed dealer, or licensed collector, except that—

      (A) this paragraph and subsection (b)(3) shall not be held to preclude a licensed importer, licensed manufacturer, licensed dealer, or licensed collector from returning a firearm or replacement firearm of the same kind and type to a person from whom it was received; and this paragraph shall not be held to preclude an individual from mailing a firearm owned in compliance with Federal, State, and local law to a licensed importer, licensed manufacturer, licensed dealer, or licensed collector;

      (B) this paragraph shall not be held to preclude a licensed importer, licensed manufacturer, or licensed dealer from depositing a firearm for conveyance in the mails to any officer, employee, agent, or watchman who, pursuant to the provisions of section 1715 of this title, is eligible to receive through the mails pistols, revolvers, and other firearms capable of being concealed on the person, for use in connection with his official duty; and

      (C) nothing in this paragraph shall be construed as applying in any manner in the District of Columbia, the Commonwealth of Puerto Rico, or any possession of the United States differently than it would apply if the District of Columbia, the Commonwealth of Puerto Rico, or the possession were in fact a State of the United States;

(3) for any person, other than a licensed importer, licensed manufacturer, licensed dealer, or licensed collector to transport into or receive in the State where he resides (or if the person is a corporation or other business entity, the State where it maintains a place of business) any firearm purchased or otherwise obtained by such person outside that State, except that this paragraph **(A)** shall not preclude any person who lawfully acquires a firearm by bequest or intestate succession in a State other than his State of residence from transporting the firearm into or receiving it in that State, if it is lawful for such person to purchase or possess such firearm in that State, **(B)** shall not apply to the transportation or receipt of a firearm obtained in conformity with subsection (b)(3) of this section, and **(C)** shall not apply to the transportation of any firearm acquired in any State prior to the effective date of this chapter;

(4) for any person, other than a licensed importer, licensed manufacturer, licensed dealer, or licensed collector, to transport in interstate or foreign commerce any destructive device, machinegun (as defined in section 5845 of the Internal Revenue Code of 1986), short–barreled shotgun, or short–barreled rifle, except as specifically authorized by the Attorney General consistent with public safety and necessity;

(5) for any person (other than a licensed importer, licensed manufacturer, licensed dealer, or licensed collector) to transfer, sell, trade, give, transport, or deliver any firearm to any person (other than a licensed importer, licensed manufacturer, licensed dealer, or licensed collector) who the transferor knows or has reasonable cause to believe does not reside in (or if the person is a corporation or other business entity, does not maintain a place of business in) the State in which the transferor resides; except that this paragraph shall not apply to **(A)** the transfer, transportation, or delivery of a firearm made to carry out a bequest of a firearm to, or an acquisition by intestate succession of a firearm by, a person who is permitted to acquire or possess a firearm under the laws of the State of his residence, and **(B)** the loan or rental of a firearm to any person for temporary use for lawful sporting purposes;

(6) for any person in connection with the acquisition or attempted acquisition of any firearm or ammunition from a licensed importer, licensed manufacturer, licensed dealer, or licensed collector, knowingly to make any false or fictitious oral or written statement or to furnish or exhibit any false, fictitious, or misrepresented identification, intended or likely to deceive such importer, manufacturer, dealer, or collector with respect to any fact material to the lawfulness of the sale or other disposition of such firearm or ammunition under the provisions of this chapter;

(7) for any person to manufacture or import armor piercing ammunition, unless—

(A) the manufacture of such ammunition is for the use of the United States, any department or agency of the United States, any State, or any department, agency, or political subdivision of a State;

(B) the manufacture of such ammunition is for the purpose of exportation; or

(C) the manufacture or importation of such ammunition is for the purpose of testing or experimentation and has been authorized by the Attorney General;

(8) for any manufacturer or importer to sell or deliver armor piercing ammunition, unless such sale or delivery—

(A) is for the use of the United States, any department or agency of the United States, any State, or any department, agency, or political subdivision of a State;

(B) is for the purpose of exportation; or

(C) is for the purpose of testing or experimentation and has been authorized by the Attorney General;[1]

(9) for any person, other than a licensed importer, licensed manufacturer, licensed dealer, or licensed collector, who does not reside in any State to receive any firearms unless such receipt is for lawful sporting purposes.

(b) It shall be unlawful for any licensed importer, licensed manufacturer, licensed dealer, or licensed collector to sell or deliver—

(1) any firearm or ammunition to any individual who the licensee knows or has reasonable cause to believe is less than eighteen years of age, and, if the firearm, or ammunition is other than a shotgun or rifle, or ammunition for a shotgun or rifle, to any individual who the licensee knows or has reasonable cause to believe is less than twenty–one years of age;

(2) any firearm to any person in any State where the purchase or possession by such person of such firearm would be in violation of any State law or any published ordinance applicable at the place of sale, delivery or other disposition, unless the licensee knows or has reasonable cause to believe that the purchase or possession would not be in violation of such State law or such published ordinance;

(3) any firearm to any person who the licensee knows or has reasonable cause to believe does not reside in (or if the person is a corporation or other business entity, does not maintain a place of business in) the State in which the licensee's place of business is located, except that this paragraph **(A)** shall not apply to the sale or delivery of any rifle or shotgun to a resident of a State other than a State in which the licensee's place of business is located if the transferee meets in person with the transferor to accomplish the transfer, and the sale, delivery, and receipt fully comply with the legal conditions of sale in both such States (and any licensed manufacturer, importer or dealer shall be presumed, for purposes of this subparagraph, in the absence of evidence to the contrary, to have had actual knowledge of the State laws and published ordinances of both States), and **(B)** shall not apply to the loan or rental of a firearm to any person for temporary use for lawful sporting purposes;

(4) to any person any destructive device, machinegun (as defined in section 5845 of the Internal Revenue Code of 1986), short–barreled shotgun, or short–barreled rifle, except as specifically authorized by the Attorney General consistent with public safety and necessity; and

(5) any firearm or armor–piercing ammunition to any person unless the licensee notes in his records, required to be kept pursuant to section 923 of this chapter, the name, age, and place of residence of such person if the person is an individual, or the identity and principal and local places of business of such person if the person is a corporation or other business entity.

Paragraphs (1), (2), (3), and (4) of this subsection shall not apply to transactions between licensed importers, licensed manufacturers, licensed dealers, and licensed collectors. Paragraph (4) of this subsection shall not apply to a sale or delivery to any research organization designated by the Attorney General.

(c) In any case not otherwise prohibited by this chapter, a licensed importer, licensed manufacturer, or licensed dealer may sell a firearm to a person who does not appear in person at the licensee's business premises (other than another licensed importer, manufacturer, or dealer) only if—

(1) the transferee submits to the transferor a sworn statement in the following form:

"Subject to penalties provided by law, I swear that, in the case of any firearm other than a shotgun or a rifle, I am twenty-one years or more of age, or that, in the case of a shotgun or a rifle, I am eighteen years or more of age; that I am not prohibited by the provisions of chapter 44 of title 18, United States Code, from receiving a firearm in interstate or foreign commerce; and that my receipt of this firearm will not be in violation of any statute of the State and published ordinance applicable to the locality in which I reside. Further, the true title, name, and address of the principal law enforcement officer of the locality to which the firearm will be delivered are _____

Signature _____.

Date_____."

and containing blank spaces for the attachment of a true copy of any permit or other information required pursuant to such statute or published ordinance;

(2) the transferor has, prior to the shipment or delivery of the firearm, forwarded by registered or certified mail (return receipt requested) a copy of the sworn statement, together with a description of the firearm, in a form prescribed by the Attorney General, to the chief law enforcement officer of the transferee's place of residence, and has received a return receipt evidencing delivery of the statement or has had the statement returned due to the refusal of the named addressee to accept such letter in accordance with United States Post Office Department regulations; and

(3) the transferor has delayed shipment or delivery for a period of at least seven days following receipt of the notification of the acceptance or refusal of delivery of the statement.

A copy of the sworn statement and a copy of the notification to the local law enforcement officer, together with evidence of receipt or rejection of that notification shall be retained by the licensee as a part of the records required to be kept under section 923(g).

(d) It shall be unlawful for any person to sell or otherwise dispose of any firearm or ammunition to any person knowing or having reasonable cause to believe that such person—

(1) is under indictment for, or has been convicted in any court of, a crime punishable by imprisonment for a term exceeding one year;

(2) is a fugitive from justice;

(3) is an unlawful user of or addicted to any controlled substance (as defined in section 102 of the Controlled Substances Act (21 U.S.C. 802));

(4) has been adjudicated as a mental defective or has been committed to any mental institution;

(5) who, being an alien—

(A) is illegally or unlawfully in the United States; or

(B) except as provided in subsection (y)(2), has been admitted to the United States under a nonimmigrant visa (as that term is defined in section 101(a)(26) of the Immigration and Nationality Act (8 U.S.C. 1101(a)(26)));

(6) who has been discharged from the Armed Forces under dishonorable conditions;

(7) who, having been a citizen of the United States, has renounced his citizenship;

(8) is subject to a court order that restrains such person from harassing, stalking, or threatening an intimate partner of such person or child of such intimate partner or person, or engaging in other conduct that would place an intimate partner in reasonable fear of bodily injury to the partner or child, except that this paragraph shall only apply to a court order that—

(A) was issued after a hearing of which such person received actual notice, and at which such person had the opportunity to participate; and

(B)(i) includes a finding that such person represents a credible threat to the physical safety of such intimate partner or child; or

(ii) by its terms explicitly prohibits the use, attempted use, or threatened use of physical force against such intimate partner or child that would reasonably be expected to cause bodily injury; or

(9) has been convicted in any court of a misdemeanor crime of domestic violence.

This subsection shall not apply with respect to the sale or disposition of a firearm or ammunition to a licensed importer, licensed manufacturer, licensed dealer, or licensed collector who pursuant to subsection (b) of section 925 of this chapter is not precluded from dealing in firearms or ammunition, or to a person who has been granted relief from disabilities pursuant to subsection (c) of section 925 of this chapter.

(e) It shall be unlawful for any person knowingly to deliver or cause to be delivered to any common or contract carrier for transportation or shipment in interstate or foreign commerce, to persons other than licensed importers, licensed manufacturers, licensed dealers, or licensed collectors, any package or other container in which there is any firearm or ammunition without written notice to the carrier that such firearm or ammunition is being transported or shipped; except that any passenger who owns or legally possesses a firearm or ammunition being transported aboard any common or contract carrier for movement with the passenger in interstate or foreign commerce may deliver said firearm or ammunition into the custody of the pilot, captain, conductor or operator of such common or contract carrier for the duration of the trip without violating any of the provisions of this chapter. No common or contract carrier shall require or cause any label, tag, or other written notice to be placed on the outside of any package, luggage, or other container that such package, luggage, or other container contains a firearm.

(f) (1) It shall be unlawful for any common or contract carrier to transport or deliver in interstate or foreign commerce any firearm or ammunition with knowledge or reasonable cause to believe that the shipment, transportation, or receipt thereof would be in violation of the provisions of this chapter.

(2) It shall be unlawful for any common or contract carrier to deliver in interstate or foreign commerce any firearm without obtaining written acknowledgement of receipt from the recipient of the package or other container in which there is a firearm.

(g) It shall be unlawful for any person—

(1) who has been convicted in any court of, a crime punishable by imprisonment for a term exceeding one year;

(2) who is a fugitive from justice;

(3) who is an unlawful user of or addicted to any controlled substance (as defined in section 102 of the Controlled Substances Act (21 U.S.C. 802));

(4) who has been adjudicated as a mental defective or who has been committed to a mental institution;

(5) who, being an alien—

(A) is illegally or unlawfully in the United States; or

(B) except as provided in subsection (y)(2), has been admitted to the United States under a nonimmigrant visa (as that term is defined in section 101(a)(26) of the Immigration and Nationality Act (8 U.S.C. 1101(a)(26)));

(6) who has been discharged from the Armed Forces under dishonorable conditions;

(7) who, having been a citizen of the United States, has renounced his citizenship;

(8) who is subject to a court order that—

(A) was issued after a hearing of which such person received actual notice, and at which such person had an opportunity to participate;

(B) restrains such person from harassing, stalking, or threatening an intimate partner of such person or child of such intimate partner or person, or engaging in other conduct that would place an intimate partner in reasonable fear of bodily injury to the partner or child; and

(C)(i) includes a finding that such person represents a credible threat to the physical safety of such intimate partner or child; or

(ii) by its terms explicitly prohibits the use, attempted use, or threatened use of physical force against such intimate partner or child that would reasonably be expected to cause bodily injury; or

(9) who has been convicted in any court of a misdemeanor crime of domestic violence, to ship or transport in interstate or foreign commerce, or possess in or affecting commerce, any firearm or ammunition; or to receive any firearm or ammunition which has been shipped or transported in interstate or foreign commerce.

(h) It shall be unlawful for any individual, who to that individual's knowledge and while being employed for any person described in any paragraph of subsection (g) of this section, in the course of such employment—

(1) to receive, possess, or transport any firearm or ammunition in or affecting interstate or foreign commerce; or

(2) to receive any firearm or ammunition which has been shipped or transported in interstate or foreign commerce.

(i) It shall be unlawful for any person to transport or ship in interstate or foreign commerce, any stolen firearm or stolen ammunition, knowing or having reasonable cause to believe that the firearm or ammunition was stolen.

(j) It shall be unlawful for any person to receive, possess, conceal, store, barter, sell, or dispose of any stolen firearm or stolen ammunition, or pledge or accept as security for a loan any stolen firearm or stolen ammunition, which is moving as, which is a part of, which constitutes, or which has been shipped or transported in, interstate or foreign commerce, either before or after it was stolen, knowing or having reasonable cause to believe that the firearm or ammunition was stolen.

(k) It shall be unlawful for any person knowingly to transport, ship, or receive, in interstate or foreign commerce, any firearm which has had the importer's or manufacturer's serial number removed, obliterated, or altered or to possess or receive any firearm which has had the importer's or manufacturer's serial number removed, obliterated, or altered and has, at any time, been shipped or transported in interstate or foreign commerce.

(l) Except as provided in section 925(d) of this chapter, it shall be unlawful for any person knowingly to import or bring into the United States or any possession thereof any firearm or ammunition; and it shall be unlawful for any person knowingly to receive any firearm or ammunition which has been imported or brought into the United States or any possession thereof in violation of the provisions of this chapter.

(m) It shall be unlawful for any licensed importer, licensed manufacturer, licensed dealer, or licensed collector knowingly to make any false entry in, to fail to make appropriate entry in, or to fail to properly maintain, any record which he is required to keep pursuant to section 923 of this chapter or regulations promulgated thereunder.

(n) It shall be unlawful for any person who is under indictment for a crime punishable by imprisonment for a term exceeding one year to ship or transport in interstate or foreign commerce any firearm or ammunition or receive any firearm or ammunition which has been shipped or transported in interstate or foreign commerce.

(o)(1) Except as provided in paragraph (2), it shall be unlawful for any person to transfer or possess a machinegun.

(2) This subsection does not apply with respect to—

(A) a transfer to or by, or possession by or under the authority of, the United States or any department or agency thereof or a State, or a department, agency, or political subdivision thereof; or

(B) any lawful transfer or lawful possession of a machinegun that was lawfully possessed before the date this subsection takes effect.

(p)(1) It shall be unlawful for any person to manufacture, import, sell, ship, deliver, possess, transfer, or receive any firearm—

(A) that, after removal of grips, stocks, and magazines, is not as detectable as the Security Exemplar, by walk–through metal detectors calibrated and operated to detect the Security Exemplar; or

(B) any major component of which, when subjected to inspection by the types of x–ray machines commonly used at airports, does not generate an image that accurately depicts the shape of the component. Barium sulfate or other compounds may be used in the fabrication of the component.

(2) For purposes of this subsection—

(A) the term **"firearm"** does not include the frame or receiver of any such weapon;

(B) the term **"major component"** means, with respect to a firearm, the barrel, the slide or cylinder, or the frame or receiver of the firearm; and

(C) the term **"Security Exemplar"** means an object, to be fabricated at the direction of the Attorney General, that is—

(i) constructed of, during the 12–month period beginning on the date of the enactment of this subsection, 3.7 ounces of material type 17–4 PH stainless steel in a shape resembling a handgun; and

(ii) suitable for testing and calibrating metal detectors:

Provided, however, That at the close of such 12–month period, and at appropriate times thereafter the Attorney General shall promulgate regulations to permit the manufacture, importation, sale, shipment, delivery, possession, transfer, or receipt of firearms previously prohibited under this subparagraph that are as detectable as a **"Security Exemplar"** which contains 3.7 ounces of material type 17–4 PH stainless steel, in a shape resembling a handgun, or such lesser amount as is detectable in view of advances in state–of–the–art developments in weapons detection technology.

(3) Under such rules and regulations as the Attorney General shall prescribe, this subsection shall not apply to the manufacture, possession, transfer, receipt, shipment, or delivery of a firearm by a licensed manufacturer or any person acting pursuant to a contract with a licensed manufacturer, for the purpose of examining and testing such firearm to determine whether paragraph (1) applies to such firearm. The Attorney General shall ensure that rules and regulations adopted pursuant to this paragraph do not impair the manufacture of prototype firearms or the development of new technology.

(4) The Attorney General shall permit the conditional importation of a firearm by a licensed importer or licensed manufacturer, for examination and testing to determine whether or not the unconditional importation of such firearm would violate this subsection.

(5) This subsection shall not apply to any firearm which—

(A) has been certified by the Secretary of Defense or the Director of Central Intelligence, after consultation with the Attorney General and the Administrator of the Federal Aviation Administration, as necessary for military or intelligence applications; and

(B) is manufactured for and sold exclusively to military or intelligence agencies of the United States.

(6) This subsection shall not apply with respect to any firearm manufactured in, imported into, or possessed in the United States before the date of the enactment of the Undetectable Firearms Act of 1988.

(q)(1) The Congress finds and declares that—

(A) crime, particularly crime involving drugs and guns, is a pervasive, nationwide problem;

(B) crime at the local level is exacerbated by the interstate movement of drugs, guns, and criminal gangs;

(C) firearms and ammunition move easily in interstate commerce and have been found in increasing numbers in and around schools, as documented in numerous hearings in both the Committee on the Judiciary[3] the House of Representatives and the Committee on the Judiciary of the Senate;

(D) in fact, even before the sale of a firearm, the gun, its component parts, ammunition, and the raw materials from which they are made have considerably moved in interstate commerce;

(E) while criminals freely move from State to State, ordinary citizens and foreign visitors may fear to travel to or through certain parts of the country due to concern about violent crime and gun violence, and parents may decline to send their children to school for the same reason;

(F) the occurrence of violent crime in school zones has resulted in a decline in the quality of education in our country;

(G) this decline in the quality of education has an adverse impact on interstate commerce and the foreign commerce of the United States;

(H) States, localities, and school systems find it almost impossible to handle gun–related crime by themselves–even States, localities, and school systems that have made strong efforts to prevent, detect, and punish gun–related crime find their efforts unavailing due in part to the failure or inability of other States or localities to take strong measures; and

(I) the Congress has the power, under the interstate commerce clause and other provisions of the Constitution, to enact measures to ensure the integrity and safety of the Nation's schools by enactment of this subsection.

(2)(A) It shall be unlawful for any individual knowingly to possess a firearm that has moved in or that otherwise affects interstate or foreign commerce at a place that the individual knows, or has reasonable cause to believe, is a school zone.

(B) Subparagraph (A) does not apply to the possession of a firearm—

(i) on private property not part of school grounds;

(ii) if the individual possessing the firearm is licensed to do so by the State in which the school zone is located or a political subdivision of the State, and the law of the State or political subdivision requires that, before an individual obtains such a license, the law enforcement authorities of the State or political subdivision verify that the individual is qualified under law to receive the license;

(iii) that is—

(I) not loaded; and

(II) in a locked container, or a locked firearms rack that is on a motor vehicle;

(iv) by an individual for use in a program approved by a school in the school zone;

(v) by an individual in accordance with a contract entered into between a school in the school zone and the individual or an employer of the individual;

(vi) by a law enforcement officer acting in his or her official capacity; or

(vii) that is unloaded and is possessed by an individual while traversing school premises for the purpose of gaining access to public or private lands open to hunting, if the entry on school premises is authorized by school authorities.

(3)(A) Except as provided in subparagraph (B), it shall be unlawful for any person, knowingly or with reckless disregard for the safety of another, to discharge or attempt to discharge a firearm that has moved in or that otherwise affects interstate or foreign commerce at a place that the person knows is a school zone.

(B) Subparagraph (A) does not apply to the discharge of a firearm—

(i) on private property not part of school grounds;

(ii) as part of a program approved by a school in the school zone, by an individual who is participating in the program;

(iii) by an individual in accordance with a contract entered into between a school in a school zone and the individual or an employer of the individual; or

(iv) by a law enforcement officer acting in his or her official capacity.

(4) Nothing in this subsection shall be construed as preempting or preventing a State or local government from enacting a statute establishing gun free school zones as provided in this subsection.

(r) It shall be unlawful for any person to assemble from imported parts any semiautomatic rifle or any shotgun which is identical to any rifle or shotgun prohibited from importation under section 925(d)(3) of this chapter as not being particularly suitable for or readily adaptable to sporting purposes except that this subsection shall not apply to—

(1) the assembly of any such rifle or shotgun for sale or distribution by a licensed manufacturer to the United States or any department or agency thereof or to any State or any department, agency, or political subdivision thereof; or

(2) the assembly of any such rifle or shotgun for the purposes of testing or experimentation authorized by the Attorney General.

(s) (1) Beginning on the date that is 90 days after the date of enactment of this subsection and ending on the day

before the date that is 60 months after such date of enactment, it shall be unlawful for any licensed importer, licensed manufacturer, or licensed dealer to sell, deliver, or transfer a handgun (other than the return of a handgun to the person from whom it was received) to an individual who is not licensed under section 923, unless—

(A) after the most recent proposal of such transfer by the transferee—

(i) the transferor has—

(I) received from the transferee a statement of the transferee containing the information described in paragraph (3);

(II) verified the identity of the transferee by examining the identification document presented;

(III) within 1 day after the transferee furnishes the statement, provided notice of the contents of the statement to the chief law enforcement officer of the place of residence of the transferee; and

(IV) within 1 day after the transferee furnishes the statement, transmitted a copy of the statement to the chief law enforcement officer of the place of residence of the transferee; and

(ii)(I) 5 business days (meaning days on which State offices are open) have elapsed from the date the transferor furnished notice of the contents of the statement to the chief law enforcement officer, during which period the transferor has not received information from the chief law enforcement officer that receipt or possession of the handgun by the transferee would be in violation of Federal, State, or local law; or

(II) the transferor has received notice from the chief law enforcement officer that the officer has no information indicating that receipt or possession of the handgun by the transferee would violate Federal, State, or local law;

(B) the transferee has presented to the transferor a written statement, issued by the chief law enforcement officer of the place of residence of the transferee during the 10–day period ending on the date of the most recent proposal of such transfer by the transferee, stating that the transferee requires access to a handgun because of a threat to the life of the transferee or of any member of the household of the transferee;

(C)(i) the transferee has presented to the transferor a permit that—

(I) allows the transferee to possess or acquire a handgun; and

(II) was issued not more than 5 years earlier by the State in which the transfer is to take place; and

(ii) the law of the State provides that such a permit is to be issued only after an authorized government official has verified that the information available to such official does not indicate that possession of a handgun by the transferee would be in violation of the law;

(D) the law of the State requires that, before any licensed importer, licensed manufacturer, or licensed dealer completes the transfer of a handgun to an individual who is not licensed under section 923, an authorized government official verify that the information available to such official does not indicate that possession of a handgun by the transferee would be in violation of law;

(E) the Attorney General has approved the transfer under section 5812 of the Internal Revenue Code of 1986; or

(F) on application of the transferor, the Attorney General has certified that compliance with subparagraph (A)(i)(III) is impracticable because—

(i) the ratio of the number of law enforcement officers of the State in which the transfer is to occur to the number of square miles of land area of the State does not exceed 0.0025;

(ii) the business premises of the transferor at which the transfer is to occur are extremely remote in relation to the chief law enforcement officer; and

(iii) there is an absence of telecommunications facilities in the geographical area in which the business premises are located.

(2) A chief law enforcement officer to whom a transferor has provided notice pursuant to paragraph (1)(A)(i)(III) shall make a reasonable effort to ascertain within 5 business days whether receipt or possession would be in violation of the law, including research in whatever State and local recordkeeping systems are available and in a national system designated by the Attorney General.

(3) The statement referred to in paragraph (1)(A)(i)(I) shall contain only—

(A) the name, address, and date of birth appearing on a valid identification document (as defined in section 1028(d)(1)[4]) of the transferee containing a photograph of the transferee and a description of the identification used;

(B) a statement that the transferee—

(i) is not under indictment for, and has not been convicted in any court of, a crime punishable by imprisonment for a term exceeding 1 year, and has not been convicted in any court of a misdemeanor crime of domestic violence;

(ii) is not a fugitive from justice;

(iii) is not an unlawful user of or addicted to any controlled substance (as defined in section 102 of the Controlled Substances Act);

(iv) has not been adjudicated as a mental defective or been committed to a mental institution;

(v) is not an alien who—

(I) is illegally or unlawfully in the United States; or

(II) subject to subsection (y)(2), has been admitted to the United States under a nonimmigrant visa (as that term is defined in section 101(a)(26) of the Immigration and Nationality Act (8 U.S.C. 1101(a)(26)));

(vi) has not been discharged from the Armed Forces under dishonorable conditions; and

(vii) is not a person who, having been a citizen of the United States, has renounced such citizenship;

(C) the date the statement is made; and

(D) notice that the transferee intends to obtain a handgun from the transferor.

(4) Any transferor of a handgun who, after such transfer, receives a report from a chief law enforcement officer containing information that receipt or possession of the handgun by the transferee violates Federal, State, or local law shall, within 1 business day after receipt of such request, communicate any information related to the transfer that the transferor has about the transfer and the transferee to—

(A) the chief law enforcement officer of the place of business of the transferor; and

(B) the chief law enforcement officer of the place of residence of the transferee.

(5) Any transferor who receives information, not otherwise available to the public, in a report under this subsection shall not disclose such information except to the transferee, to law enforcement authorities, or pursuant to the direction of a court of law.

(6)(A) Any transferor who sells, delivers, or otherwise transfers a handgun to a transferee shall retain the copy of the statement of the transferee with respect to the handgun transaction, and shall retain evidence that the transferor has complied with subclauses (III) and (IV) of paragraph (1)(A)(i) with respect to the statement.

(B) Unless the chief law enforcement officer to whom a statement is transmitted under paragraph (1)(A)(i)(IV) determines that a transaction would violate Federal, State, or local law—

(i) the officer shall, within 20 business days after the date the transferee made the statement on the basis of which the notice was provided, destroy the statement, any record containing information derived from the statement, and any record created as a result of the notice required by paragraph (1)(A)(i)(III);

(ii) the information contained in the statement shall not be conveyed to any person except a person who has a need to know in order to carry out this subsection; and

(iii) the information contained in the statement shall not be used for any purpose other than to carry out this subsection.

(C) If a chief law enforcement officer determines that an individual is ineligible to receive a handgun and the individual requests the officer to provide the reason for such determination, the officer shall provide such reasons to the individual in writing within 20 business days after receipt of the request.

(7) A chief law enforcement officer or other person responsible for providing criminal history background information pursuant to this subsection shall not be liable in an action at law for damages—

(A) for failure to prevent the sale or transfer of a handgun to a person whose receipt or possession of the handgun is unlawful under this section; or

(B) for preventing such a sale or transfer to a person who may lawfully receive or possess a handgun.

(8) For purposes of this subsection, the term **"chief law enforcement officer"** means the chief of police, the sheriff, or an equivalent officer or the designee of any such individual.

(9) The Attorney General shall take necessary actions to ensure that the provisions of this subsection are published and disseminated to licensed dealers, law enforcement officials, and the public.

(t) (1) Beginning on the date that is 30 days after the Attorney General notifies licensees under section 103(d) of the Brady Handgun Violence Prevention Act that the national instant criminal background check system is established, a licensed importer, licensed manufacturer, or licensed dealer shall not transfer a firearm to any other person who is not licensed under this chapter, unless—

(A) before the completion of the transfer, the licensee contacts the national instant criminal background check system established under section 103 of that Act;

(B)(i) the system provides the licensee with a unique identification number; or

(ii) 3 business days (meaning a day on which State offices are open) have elapsed since the licensee contacted the system, and the system has not notified the licensee that the receipt of a firearm by such other person would violate subsection (g) or (n) of this section; and

(C) the transferor has verified the identity of the transferee by examining a valid identification document (as defined in section 1028(d) of this title) of the transferee containing a photograph of the transferee.

(2) If receipt of a firearm would not violate subsection (g) or (n) or State law, the system shall—

(A) assign a unique identification number to the transfer;

(B) provide the licensee with the number; and

(C) destroy all records of the system with respect to the call (other than the identifying number and the date the number was assigned) and all records of the system relating to the person or the transfer.

(3) Paragraph (1) shall not apply to a firearm transfer between a licensee and another person if—

(A)(i) such other person has presented to the licensee a permit that—

(I) allows such other person to possess or acquire a firearm; and

(II) was issued not more than 5 years earlier by the State in which the transfer is to take place; and

(ii) the law of the State provides that such a permit is to be issued only after an authorized government official has verified that the information available to such official does not indicate that possession of a firearm by such other person would be in violation of law;

(B) the Attorney General has approved the transfer under section 5812 of the Internal Revenue Code of 1986; or

(C) on application of the transferor, the Attorney General has certified that compliance with paragraph (1)(A) is impracticable because—

(i) the ratio of the number of law enforcement officers of the State in which the transfer is to occur to the number of square miles of land area of the State does not exceed 0.0025;

(ii) the business premises of the licensee at which the transfer is to occur are extremely remote in relation to the chief law enforcement officer (as defined in subsection (s)(8)); and

(iii) there is an absence of telecommunications facilities in the geographical area in which the business premises are located.

(4) If the national instant criminal background check system notifies the licensee that the information available to the system does not demonstrate that the receipt of a firearm by such other person would violate subsection (g) or (n) or State law, and the licensee transfers a firearm to such other person, the licensee shall include in the record of the transfer the unique identification number provided by the system with respect to the transfer.

(5) If the licensee knowingly transfers a firearm to such other person and knowingly fails to comply with paragraph (1) of this subsection with respect to the transfer and, at the time such other person most recently proposed the transfer, the national instant criminal background check sys-

tem was operating and information was available to the system demonstrating that receipt of a firearm by such other person would violate subsection (g) or (n) of this section or State law, the Attorney General may, after notice and opportunity for a hearing, suspend for not more than 6 months or revoke any license issued to the licensee under section 923, and may impose on the licensee a civil fine of not more than $5,000.

(6) Neither a local government nor an employee of the Federal Government or of any State or local government, responsible for providing information to the national instant criminal background check system shall be liable in an action at law for damages—

(A) for failure to prevent the sale or transfer of a firearm to a person whose receipt or possession of the firearm is unlawful under this section; or

(B) for preventing such a sale or transfer to a person who may lawfully receive or possess a firearm.

(u) It shall be unlawful for a person to steal or unlawfully take or carry away from the person or the premises of a person who is licensed to engage in the business of importing, manufacturing, or dealing in firearms, any firearm in the licensee's business inventory that has been shipped or transported in interstate or foreign commerce.

[(v), (w) Repealed. Pub. L. 103–322, title XI, §110105(2), Sept. 13, 1994, 108 Stat. 2000.]

(x) (1) It shall be unlawful for a person to sell, deliver, or otherwise transfer to a person who the transferor knows or has reasonable cause to believe is a juvenile—

(A) a handgun; or

(B) ammunition that is suitable for use only in a handgun.

(2) It shall be unlawful for any person who is a juvenile to knowingly possess—

(A) a handgun; or

(B) ammunition that is suitable for use only in a handgun.

(3) This subsection does not apply to—

(A) a temporary transfer of a handgun or ammunition to a juvenile or to the possession or use of a handgun or ammunition by a juvenile if the handgun and ammunition are possessed and used by the juvenile—

(i) in the course of employment, in the course of ranching or farming related to activities at the residence of the juvenile (or on property used for ranching or farming at which the juvenile, with the permission of the property owner or lessee, is performing activities related to the operation of the farm or ranch), target practice, hunting, or a course of instruction in the safe and lawful use of a handgun;

(ii) with the prior written consent of the juvenile's parent or guardian who is not prohibited by Federal, State, or local law from possessing a firearm, except—

(I) during transportation by the juvenile of an unloaded handgun in a locked container directly from the place of transfer to a place at which an activity described in clause (i) is to take place and transportation by the juvenile of that handgun, unloaded and in a locked container, directly from the place at which such an activity took place to the transferor; or

(II) with respect to ranching or farming activities as described in clause (i), a juvenile may possess and use a handgun or ammunition with the prior written approval of the juvenile's parent or legal guardian and at the direction of an adult who is not prohibited by Federal, State or local law from possessing a firearm;

(iii) the juvenile has the prior written consent in the juvenile's possession at all times when a handgun is in the possession of the juvenile; and

(iv) in accordance with State and local law;

(B) a juvenile who is a member of the Armed Forces of the United States or the National Guard who possesses or is armed with a handgun in the line of duty;

(C) a transfer by inheritance of title (but not possession) of a handgun or ammunition to a juvenile; or

(D) the possession of a handgun or ammunition by a juvenile taken in defense of the juvenile or other persons against an intruder into the residence of the juvenile or a residence in which the juvenile is an invited guest.

(4) A handgun or ammunition, the possession of which is transferred to a juvenile in circumstances in which the transferor is not in violation of this subsection shall not be subject to permanent confiscation by the Government if its possession by the juvenile subsequently becomes unlawful because of the conduct of the juvenile, but shall be returned to the lawful owner when such handgun or ammunition is no longer required by the Government for the purposes of investigation or prosecution.

(5) For purposes of this subsection, the term **"juvenile"** means a person who is less than 18 years of age.

(6)(A) In a prosecution of a violation of this subsection, the court shall require the presence of a juvenile defendant's parent or legal guardian at all proceedings.

(B) The court may use the contempt power to enforce subparagraph (A).

(C) The court may excuse attendance of a parent or legal guardian of a juvenile defendant at a proceeding in a prosecution of a violation of this subsection for good cause shown.

(y) Provisions Relating to Aliens Admitted Under Nonimmigrant Visas.—

(1) Definitions. In this subsection—

(A) the term **"alien"** has the same meaning as in section 101(a)(3) of the Immigration and Nationality Act (8 U.S.C. 1101(a)(3)); and

(B) the term **"nonimmigrant visa"** has the same meaning as in section 101(a)(26) of the Immigration and Nationality Act (8 U.S.C. 1101(a)(26)).

(2) Exceptions. Subsections (d)(5)(B), (g)(5)(B), and (s)(3)(B)(v)(II) do not apply to any alien who has been lawfully admitted to the United States under a nonimmigrant visa, if that alien is—

(A) admitted to the United States for lawful hunting or sporting purposes or is in possession of a hunting license or permit lawfully issued in the United States;

(B) an official representative of a foreign government who is—

(i) accredited to the United States Government or the Government's mission to an international organization having its headquarters in the United States; or

(ii) en route to or from another country to which that alien is accredited;

(C) an official of a foreign government or a distinguished foreign visitor who has been so designated by the Department of State; or

(D) a foreign law enforcement officer of a friendly foreign government entering the United States on official law enforcement business.

(3) Waiver.—

(A) Conditions for waiver.–Any individual who has been admitted to the United States under a nonimmigrant visa may receive a waiver from the requirements of subsection (g)(5), if—

(i) the individual submits to the Attorney General a petition that meets the requirements of subparagraph (C); and

(ii) the Attorney General approves the petition.

(B) Petition. Each petition under subparagraph (B) shall—

(i) demonstrate that the petitioner has resided in the United States for a continuous period of not less than 180 days before the date on which the petition is submitted under this paragraph; and

(ii) include a written statement from the embassy or consulate of the petitioner, authorizing the petitioner to acquire a firearm or ammunition and certifying that the alien would not, absent the application of subsection (g)(5)(B), otherwise be prohibited from such acquisition under subsection (g).

(C) Approval of petition. The Attorney General shall approve a petition submitted in accordance with this paragraph, if the Attorney General determines that waiving the requirements of subsection (g)(5)(B) with respect to the petitioner—

(i) would be in the interests of justice; and

(ii) would not jeopardize the public safety.

(z) Secure Gun Storage or Safety Device.—

(1) In general. Except as provided under paragraph (2), it shall be unlawful for any licensed importer, licensed manufacturer, or licensed dealer to sell, deliver, or transfer any handgun to any person other than any person licensed under this chapter, unless the transferee is provided with a secure gun storage or safety device (as defined in section 921(a)(34)) for that handgun.

(2) Exceptions. Paragraph (1) shall not apply to—

(A)(i) the manufacture for, transfer to, or possession by, the United States, a department or agency of the United States, a State, or a department, agency, or political subdivision of a State, of a handgun; or

(ii) the transfer to, or possession by, a law enforcement officer employed by an entity referred to in clause (i) of a handgun for law enforcement purposes (whether on or off duty); or

(B) the transfer to, or possession by, a rail police officer employed by a rail carrier and certified or commissioned as a police officer under the laws of a State of a handgun for purposes of law enforcement (whether on or off duty);

(C) the transfer to any person of a handgun listed as a curio or relic by the Secretary pursuant to section 921(a)(13); or

(D) the transfer to any person of a handgun for which a secure gun storage or safety device is temporarily unavailable for the reasons described in the exceptions stated in section 923(e), if the licensed manufacturer, licensed importer, or licensed dealer delivers to the transferee within 10 calendar days from the date of the delivery of the handgun to the transferee a secure gun storage or safety device for the handgun.

(3) Liability for use.—

(A) In general. Notwithstanding any other provision of law, a person who has lawful possession and control of a handgun, and who uses a secure gun storage or safety device with the handgun, shall be entitled to immunity from a qualified civil liability action.

(B) Prospective actions. A qualified civil liability action may not be brought in any Federal or State court.

(C) Defined term. As used in this paragraph, the term **"qualified civil liability action"**—

(i) means a civil action brought by any person against a person described in subparagraph (A) for damages resulting from the criminal or unlawful misuse of the handgun by a third party, if—

(I) the handgun was accessed by another person who did not have the permission or authorization of the person having lawful possession and control of the handgun to have access to it; and

(II) at the time access was gained by the person not so authorized, the handgun had been made inoperable by use of a secure gun storage or safety device; and

(ii) shall not include an action brought against the person having lawful possession and control of the handgun for negligent entrustment or negligence per se.

§ 923 Licensing.

(a) No person shall engage in the business of importing, manufacturing, or dealing in firearms, or importing or manufacturing ammunition, until he has filed an application with and received a license to do so from the Attorney General. The application shall be in such form and contain only that information necessary to determine eligibility for licensing as the Attorney General shall by regulation prescribe and shall include a photograph and fingerprints of the applicant. Each applicant shall pay a fee for obtaining such a license, a separate fee being required for each place in which the applicant is to do business, as follows:

(1) If the applicant is a manufacturer—

(A) of destructive devices, ammunition for destructive devices or armor piercing ammunition, a fee of $1,000 per year;

(B) of firearms other than destructive devices, a fee of $50 per year; or

(C) of ammunition for firearms, other than ammunition for destructive devices or armor piercing ammunition, a fee of $10 per year.

(2) If the applicant is an importer—

(A) of destructive devices, ammunition for destructive devices or armor piercing ammunition, a fee of $1,000 per year; or

(B) of firearms other than destructive devices or ammunition for firearms other than destructive devices, or ammunition other than armor piercing ammunition, a fee of $50 per year.

(3) If the applicant is a dealer—

(A) in destructive devices or ammunition for destructive devices, a fee of $1,000 per year; or

(B) who is not a dealer in destructive devices, a fee of $200 for 3 years, except that the fee for renewal of a valid license shall be $90 for 3 years.

(b) Any person desiring to be licensed as a collector shall file an application for such license with the Attorney General. The application shall be in such form and contain only that information necessary to determine eligibility as the Attorney General shall by regulation prescribe. The fee for such license shall be $10 per year. Any license granted under this subsection shall only apply to transactions in curios and relics.

(c) Upon the filing of a proper application and payment of the prescribed fee, the Attorney General shall issue to a qualified applicant the appropriate license which, subject to the provisions of this chapter and other applicable provisions of law, shall entitle the licensee to transport, ship, and receive firearms and ammunition covered by such license in interstate or foreign commerce during the period stated in the license. Nothing in this chapter shall be construed to prohibit a licensed manufacturer, importer, or dealer from maintaining and disposing of a personal collection of firearms, subject only to such restrictions as apply in this chapter to dispositions by a person other than a licensed manufacturer, importer, or dealer. If any firearm is so disposed of by a licensee within one year after its transfer from his business inventory into such licensee's personal collection or if such disposition or any other acquisition is made for the purpose of willfully evading the restrictions placed upon licensees by this chapter, then such firearm shall be deemed part of such licensee's business inventory, except that any licensed manufacturer, importer, or dealer who has maintained a firearm as part of a personal collection for one year and who sells or otherwise disposes of such firearm shall record the description of the firearm in a bound volume, containing the name and place of residence and date of birth of the transferee if the transferee is an individual, or the identity and principal and local places of business of the transferee if the transferee is a corporation or other business entity: **Provided,** That no other recordkeeping shall be required.

(d) (1) Any application submitted under subsection (a) or (b) of this section shall be approved if—

(A) the applicant is twenty–one years of age or over;

(B) the applicant (including, in the case of a corporation, partnership, or association, any individual possessing, directly or indirectly, the power to direct or cause the direction of the management and policies of the corporation, partnership, or association) is not prohibited from transporting, shipping, or receiving firearms or ammunition in interstate or foreign commerce under section 922(g) and (n) of this chapter;

(C) the applicant has not willfully violated any of the provisions of this chapter or regulations issued thereunder;

(D) the applicant has not willfully failed to disclose any material information required, or has not made any false statement as to any material fact, in connection with his application;

(E) the applicant has in a State (i) premises from which he conducts business subject to license under this chapter or from which he intends to conduct such business within a reasonable period of time, or (ii) in the case of a collector, premises from which he conducts his collecting subject to license under this chapter or from which he intends to conduct such collecting within a reasonable period of time;

(F) the applicant certifies that—

(i) the business to be conducted under the license is not prohibited by State or local law in the place where the licensed premise is located;

(ii)(I) within 30 days after the application is approved the business will comply with the requirements of State and local law applicable to the conduct of the business; and

(II) the business will not be conducted under the license until the requirements of State and local law applicable to the business have been met; and

(iii) that the applicant has sent or delivered a form to be prescribed by the Attorney General, to the chief law enforcement officer of the locality in which the premises are located, which indicates that the applicant intends to apply for a Federal firearms license; and

(G) in the case of an application to be licensed as a dealer, the applicant certifies that secure gun storage or safety devices will be available at any place in which firearms are sold under the license to persons who are not licensees (subject to the exception that in any case in which a secure gun storage or safety device is temporarily unavailable because of theft, casualty loss, consumer sales, backorders from a manufacturer, or any other similar reason beyond the control of the licensee, the dealer shall not be considered to be in violation of the requirement under this subparagraph to make available such a device).

(2) The Attorney General must approve or deny an application for a license within the 60–day period beginning on the date it is received. If the Attorney General fails to act within such period, the applicant may file an action under section 1361 of title 28 to compel the Attorney General to act. If the Attorney General approves an applicant's application, such applicant shall be issued a license upon the payment of the prescribed fee.

(e) The Attorney General may, after notice and opportunity for hearing, revoke any license issued under this section if the holder of such license has willfully violated any provision of this chapter or any rule or regulation prescribed by the Attorney General under this chapter or fails to have secure gun storage or safety devices available at any place in which firearms are sold under the license to persons who are not licensees (except that in any case in which a secure gun storage or safety device is temporarily unavailable because of theft, casualty loss, consumer sales, backorders from a manufacturer, or any other similar reason beyond the control of the licensee, the dealer shall not be considered to be in violation of the requirement to make available such a device). The Attorney General may, after notice and opportunity for hearing, revoke the license of a dealer who willfully transfers armor piercing ammunition. The Secretary's[1] action under this subsection may be reviewed only as provided in subsection (f) of this section.

(f) (1) Any person whose application for a license is denied and any holder of a license which is revoked shall receive a written notice from the Attorney General stating specifically the grounds upon which the application was denied or upon which the license was revoked. Any notice of a re-vocation of a license shall be given to the holder of such license before the effective date of the revocation.

(2) If the Attorney General denies an application for, or revokes, a license, he shall, upon request by the aggrieved party, promptly hold a hearing to review his denial or revocation. In the case of a revocation of a license, the Attorney General shall upon the request of the holder of the license stay the effective date of the revocation. A hearing held under this paragraph shall be held at a location convenient to the aggrieved party.

(3) If after a hearing held under paragraph (2) the Attorney General decides not to reverse his decision to deny an application or revoke a license, the Attorney General shall give notice of his decision to the aggrieved party. The aggrieved party may at any time within sixty days after the date notice was given under this paragraph file a petition with the United States district court for the district in which he resides or has his principal place of business for a de novo judicial review of such denial or revocation. In a proceeding conducted under this subsection, the court may consider any evidence submitted by the parties to the proceeding whether or not such evidence was considered at the hearing held under paragraph (2). If the court decides that the Attorney General was not authorized to deny the application or to revoke the license, the court shall order the Attorney General to take such action as may be necessary to comply with the judgment of the court.

(4) If criminal proceedings are instituted against a licensee alleging any violation of this chapter or of rules or regulations prescribed under this chapter, and the licensee is acquitted of such charges, or such proceedings are terminated, other than upon motion of the Government before trial upon such charges, the Attorney General shall be absolutely barred from denying or revoking any license granted under this chapter where such denial or revocation is based in whole or in part on the facts which form the basis of such criminal charges. No proceedings for the revocation of a license shall be instituted by the Attorney General more than one year after the filing of the indictment or information.

(g) (1) (A) Each licensed importer, licensed manufacturer, and licensed dealer shall maintain such records of importation, production, shipment, receipt, sale, or other disposition of firearms at his place of business for such period, and in such form, as the Attorney General may by regulations prescribe. Such importers, manufacturers, and dealers shall not be required to submit to the Attorney General reports and information with respect to such records and the contents thereof, except as expressly required by this section. The Attorney General, when he has reasonable cause to believe a violation of this chapter has occurred and that evidence thereof may be found on such premises, may, upon demonstrating such cause before a Federal magistrate judge and securing from such magistrate judge a warrant authorizing entry, enter during business hours the premises (including places of storage) of any licensed firearms importer, licensed manufacturer, licensed dealer, licensed collector, or any licensed importer or manufacturer of ammunition, for the purpose of inspecting or examining—

> **(i)** any records or documents required to be kept by such licensed importer, licensed manufacturer, licensed dealer, or licensed collector under this chapter or rules or regulations under this chapter, and

> **(ii)** any firearms or ammunition kept or stored by such licensed importer, licensed manufacturer, licensed dealer, or licensed collector, at such premises.

(B) The Attorney General may inspect or examine the inventory and records of a licensed importer, licensed manufacturer, or licensed dealer without such reasonable cause or warrant—

> **(i)** in the course of a reasonable inquiry during the course of a criminal investigation of a person or persons other than the licensee;

> **(ii)** for ensuring compliance with the record keeping requirements of this chapter—

> > **(I)** not more than once during any 12–month period; or

(II) at any time with respect to records relating to a firearm involved in a criminal investigation that is traced to the licensee; or

(iii) when such inspection or examination may be required for determining the disposition of one or more particular firearms in the course of a bona fide criminal investigation.

(C) The Attorney General may inspect the inventory and records of a licensed collector without such reasonable cause or warrant—

(i) for ensuring compliance with the record keeping requirements of this chapter not more than once during any twelve–month period; or

(ii) when such inspection or examination may be required for determining the disposition of one or more particular firearms in the course of a bona fide criminal investigation.

(D) At the election of a licensed collector, the annual inspection of records and inventory permitted under this paragraph shall be performed at the office of the Attorney General designated for such inspections which is located in closest proximity to the premises where the inventory and records of such licensed collector are maintained. The inspection and examination authorized by this paragraph shall not be construed as authorizing the Attorney General to seize any records or other documents other than those records or documents constituting material evidence of a violation of law. If the Attorney General seizes such records or documents, copies shall be provided the licensee within a reasonable time. The Attorney General may make available to any Federal, State, or local law enforcement agency any information which he may obtain by reason of this chapter with respect to the identification of persons prohibited from purchasing or receiving firearms or ammunition who have purchased or received firearms or ammunition, together with a description of such firearms or ammunition, and he may provide information to the extent such information may be contained in the records required to be maintained by this chapter, when so requested by any Federal, State, or local law enforcement agency.

(2) Each licensed collector shall maintain in a bound volume the nature of which the Attorney General may by regulations prescribe, records of the receipt, sale, or other disposition of firearms. Such records shall include the name and address of any person to whom the collector sells or otherwise disposes of a firearm. Such collector shall not be required to submit to the Attorney General reports and information with respect to such records and the contents thereof, except as expressly required by this section.

(3)(A) Each licensee shall prepare a report of multiple sales or other dispositions whenever the licensee sells or otherwise disposes of, at one time or during any five consecutive business days, two or more pistols, or revolvers, or any combination of pistols and revolvers totalling two or more, to an unlicensed person. The report shall be prepared on a form specified by the Attorney General and forwarded to the office specified thereon and to the department of State police or State law enforcement agency of the State or local law enforcement agency of the local jurisdiction in which the sale or other disposition took place, not later than the close of business on the day that the multiple sale or other disposition occurs.

(B) Except in the case of forms and contents thereof regarding a purchaser who is prohibited by subsection (g) or (n) of section 922 of this title from receipt of a firearm, the department of State police or State law enforcement agency or local law enforcement agency of the local jurisdiction shall not disclose any such form or the contents thereof to any person or entity, and shall destroy each such form and any record of the contents thereof no more than 20 days from the date such form is received. No later than the date that is 6 months after the effective date of this subparagraph, and at the end of each 6–month period thereafter, the department of State police or State law enforcement agency or local law enforcement agency of the local jurisdiction shall certify to the Attorney General of the United States that no disclosure contrary to this subparagraph has been made and that all forms and any record of the contents thereof have been destroyed as provided in this subparagraph.

(4) Where a firearms or ammunition business is discontinued and succeeded by a new licensee, the records required to be kept by this chapter shall appropriately reflect such facts and shall be delivered to the successor. Where discontinuance of the business is absolute, such records shall be delivered within thirty days after the business discontinuance to the Attorney General. However, where State law or local ordinance requires the delivery of records to other responsible authority, the Attorney General may arrange for the delivery of such records to such other responsible authority.

(5)(A) Each licensee shall, when required by letter issued by the Attorney General, and until notified to the contrary in writing by the Attorney General, submit on a form specified by the Attorney General, for periods and at the times specified in such letter, all record information required to be kept by this chapter or such lesser record information as the Attorney General in such letter may specify.

(B) The Attorney General may authorize such record information to be submitted in a manner other than that prescribed in subparagraph (A) of this paragraph when it is shown by a licensee that an alternate method of reporting is reasonably necessary and will not unduly hinder the effective administration of this chapter. A licensee may use an alternate method of reporting if the licensee describes the proposed alternate method of reporting and the need therefor in a letter application submitted to the Attorney General, and the Attorney General approves such alternate method of reporting.

(6) Each licensee shall report the theft or loss of a firearm from the licensee's inventory or collection, within 48 hours after the theft or loss is discovered, to the Attorney General and to the appropriate local authorities.

(7) Each licensee shall respond immediately to, and in no event later than 24 hours after the receipt of, a request by the Attorney General for information contained in the records required to be kept by this chapter as may be required for determining the disposition of 1 or more firearms in the course of a bona fide criminal investigation. The requested information shall be provided orally or in writing, as the Attorney General may require. The Attorney General shall implement a system whereby the licensee can positively identify and establish that an individual requesting information via telephone is employed by and authorized by the agency to request such information.

(h) Licenses issued under the provisions of subsection (c) of this section shall be kept posted and kept available for inspection on the premises covered by the license.

(i) Licensed importers and licensed manufacturers shall identify by means of a serial number engraved or cast on the receiver or frame of the weapon, in such manner as the Attorney General shall by regulations prescribe, each firearm imported or manufactured by such importer or manufacturer.

(j) A licensed importer, licensed manufacturer, or licensed dealer may, under rules or regulations prescribed by the Attorney General, conduct business temporarily at a location other than the location specified on the license if such temporary location is the location for a gun show or event sponsored by any national, State, or local organization, or any affiliate of any such organization devoted to the collection, competitive use, or other sporting use of firearms in the community, and such location is in the State which is specified on the license. Records of receipt and disposition of firearms transactions conducted at such temporary location shall include the location of the sale or other disposition and shall be entered in the permanent records of the licensee and retained on the location specified on the license. Nothing in this subsection shall authorize any licensee to conduct business in or from any motorized or towed vehicle. Notwithstanding the provisions of subsection (a) of this section, a separate fee shall not be required of a licensee with respect to business conducted under this subsection. Any inspection or examination of inventory or records under this chapter by the Attorney General at such temporary location shall be limited to inventory consisting of, or records relating to, firearms held or disposed at such temporary location. Nothing in this subsection shall be construed to authorize the Attorney General to inspect or examine the inventory or records of a licensed importer, licensed manufacturer, or licensed dealer at any location other than the location specified on the license. Nothing in this subsection shall be construed to diminish in any manner any right to display, sell, or otherwise dispose of firearms or ammunition, which is in effect before the date of the enactment of the Firearms Owners' Protection Act, including the right of a licensee to conduct **"curios or relics"** firearms transfers and business away from their business premises with another licensee without regard as to whether the location of where the business is conducted is located in the State specified on the license of either licensee.

(k) Licensed importers and licensed manufacturers shall mark all armor piercing projectiles and packages containing such projectiles for distribution in the manner prescribed by the Attorney General by regulation. The Attorney General shall furnish information to each dealer licensed under this chapter defining which projectiles are considered armor piercing ammunition as defined in section 921(a)(17)(B).

(l) The Attorney General shall notify the chief law enforcement officer in the appropriate State and local jurisdictions of the names and addresses of all persons in the State to whom a firearms license is issued.

§ 924 Penalties.

(a) (1) Except as otherwise provided in this subsection, subsection (b), (c), (f), or (p) of this section, or in section 929, whoever—

    (A) knowingly makes any false statement or representation with respect to the information required by this chapter to be kept in the records of a person licensed under this chapter or in applying for any license or exemption or relief from disability under the provisions of this chapter;

    (B) knowingly violates subsection (a)(4), (f), (k), or (q) of section 922;

    (C) knowingly imports or brings into the United States or any possession thereof any firearm or ammunition in violation of section 922(l); or

    (D) willfully violates any other provision of this chapter, shall be fined under this title, imprisoned not more than five years, or both.

(2) Whoever knowingly violates subsection (a)(6), (d), (g), (h), (i), (j), or (o) of section 922 shall be fined as provided in this title, imprisoned not more than 10 years, or both.

(3) Any licensed dealer, licensed importer, licensed manufacturer, or licensed collector who knowingly—

    (A) makes any false statement or representation with respect to the information required by the provisions of this chapter to be kept in the records of a person licensed under this chapter, or

    (B) violates subsection (m) of section 922, shall be fined under this title, imprisoned not more than one year, or both.

(4) Whoever violates section 922(q) shall be fined under this title, imprisoned for not more than 5 years, or both. Notwithstanding any other provision of law, the term of imprisonment imposed under this paragraph shall not run concurrently with any other term of imprisonment imposed under any other provision of law. Except for the authorization of a term of imprisonment of not more than 5 years made in this paragraph, for the purpose of any other law a violation of section 922(q) shall be deemed to be a misdemeanor.

(5) Whoever knowingly violates subsection (s) or (t) of section 922 shall be fined under this title, imprisoned for not more than 1 year, or both.

(6)(A)(i) A juvenile who violates section 922(x) shall be fined under this title, imprisoned not more than 1 year, or both, except that a juvenile described in clause (ii) shall be sentenced to probation on appropriate conditions and shall not be incarcerated unless the juvenile fails to comply with a condition of probation.

(ii) A juvenile is described in this clause if—

(I) the offense of which the juvenile is charged is possession of a handgun or ammunition in violation of section 922(x)(2); and

(II) the juvenile has not been convicted in any court of an offense (including an offense under section 922(x) or a similar State law, but not including any other offense consisting of conduct that if engaged in by an adult would not constitute an offense) or adjudicated as a juvenile delinquent for conduct that if engaged in by an adult would constitute an offense.

(B) A person other than a juvenile who knowingly violates section 922(x)—

(i) shall be fined under this title, imprisoned not more than 1 year, or both; and

(ii) if the person sold, delivered, or otherwise transferred a handgun or ammunition to a juvenile knowing or having reasonable cause to know that the juvenile intended to carry or otherwise possess or discharge or otherwise use the handgun or ammunition in the commission of a crime of violence, shall be fined under this title, imprisoned not more than 10 years, or both.

(7) Whoever knowingly violates section 931 shall be fined under this title, imprisoned not more than 3 years, or both.

(b) Whoever, with intent to commit therewith an offense punishable by imprisonment for a term exceeding one year, or with knowledge or reasonable cause to believe that an offense punishable by imprisonment for a term exceeding one year is to be committed therewith, ships, transports, or receives a firearm or any ammunition in interstate or foreign commerce shall be fined under this title, or imprisoned not more than ten years, or both.

(c) (1) (A) Except to the extent that a greater minimum sentence is otherwise provided by this subsection or by any other provision of law, any person who,

during and in relation to any crime of violence or drug trafficking crime (including a crime of violence or drug trafficking crime that provides for an enhanced punishment if committed by the use of a deadly or dangerous weapon or device) for which the person may be prosecuted in a court of the United States, uses or carries a firearm, or who, in furtherance of any such crime, possesses a firearm, shall, in addition to the punishment provided for such crime of violence or drug trafficking crime—

(i) be sentenced to a term of imprisonment of not less than 5 years;

(ii) if the firearm is brandished, be sentenced to a term of imprisonment of not less than 7 years; and

(iii) if the firearm is discharged, be sentenced to a term of imprisonment of not less than 10 years.

(B) If the firearm possessed by a person convicted of a violation of this subsection—

(i) is a short–barreled rifle, short–barreled shotgun, or semiautomatic assault weapon, the person shall be sentenced to a term of imprisonment of not less than 10 years; or

(ii) is a machinegun or a destructive device, or is equipped with a firearm silencer or firearm muffler, the person shall be sentenced to a term of imprisonment of not less than 30 years.

(C) In the case of a second or subsequent conviction under this subsection, the person shall—

(i) be sentenced to a term of imprisonment of not less than 25 years; and

(ii) if the firearm involved is a machinegun or a destructive device, or is equipped with a firearm silencer or firearm muffler, be sentenced to imprisonment for life.

(D) Notwithstanding any other provision of law—

(i) a court shall not place on probation any person convicted of a violation of this subsection; and

(ii) no term of imprisonment imposed on a person under this subsection shall run concurrently with any other term of imprisonment imposed on the person, including any term of imprisonment imposed for the crime of violence or drug trafficking crime during which the firearm was used, carried, or possessed.

(2) For purposes of this subsection, the term **"drug trafficking crime"** means any felony punishable under the Controlled Substances Act (21 U.S.C. 801 et seq.), the Controlled Substances Import and Export Act (21 U.S.C. 951 et seq.), or chapter 705 of title 46.

(3) For purposes of this subsection the term **"crime of violence"** means an offense that is a felony and—

(A) has as an element the use, attempted use, or threatened use of physical force against the person or property of another, or

(B) that by its nature, involves a substantial risk that physical force against the person or property of another may be used in the course of committing the offense.

(4) For purposes of this subsection, the term **"brandish"** means, with respect to a firearm, to display all or part of the firearm, or otherwise make the presence of the firearm known to another person, in order to intimidate that person, regardless of whether the firearm is directly visible to that person.

(5) Except to the extent that a greater minimum sentence is otherwise provided under this subsection, or by any other provision of law, any person who, during and in relation to any crime of violence or drug trafficking crime (including a crime of violence or drug trafficking crime that provides for an enhanced punishment if committed by the use of a deadly or dangerous weapon or device) for which the person may be prosecuted in a court of the United States, uses or carries armor piercing ammunition, or who, in furtherance of any such crime, possesses armor piercing ammunition, shall, in addition to the punishment provided for such crime of violence or drug trafficking crime or conviction under this section—

(A) be sentenced to a term of imprisonment of not less than 15 years; and

(B) if death results from the use of such ammunition—

(i) if the killing is murder (as defined in section 1111), be punished by death or sentenced to a term of imprisonment for any term of years or for life; and

(ii) if the killing is manslaughter (as defined in section 1112), be punished as provided in section 1112.

(d) **(1)** Any firearm or ammunition involved in or used in any knowing violation of subsection (a)(4), (a)(6), (f), (g), (h), (i), (j), or (k) of section 922, or knowing importation or bringing into the United States or any possession thereof any firearm or ammunition in violation of section 922(l), or knowing violation of section 924, or willful violation of any other provision of this chapter or any rule or regulation promulgated thereunder, or any violation of any other criminal law of the United States, or any firearm or ammunition intended to be used in any offense referred to in paragraph (3) of this subsection, where such intent is demonstrated by clear and convincing evidence, shall be subject to seizure and forfeiture, and all provisions of the Internal Revenue Code of 1986 relating to the seizure, forfeiture, and disposition of firearms, as defined in section 5845(a) of that Code, shall, so far as applicable, extend to seizures and forfeitures under the provisions of this chapter: **Provided,** That upon acquittal of the owner or possessor, or dismissal of the charges against him other than upon motion of the Government prior to trial, or lapse of or court termination of the restraining order to which he is subject, the seized or relinquished firearms or ammunition shall be returned forthwith to the owner or possessor or to a person delegated by the owner or possessor unless the return of the firearms or ammunition would place the owner or possessor or his delegate in violation of law. Any action or proceeding for the forfeiture of firearms or ammunition shall be commenced within one hundred and twenty days of such seizure.

(2)(A) In any action or proceeding for the return of firearms or ammunition seized under the provisions of this chapter, the court shall allow the prevailing party, other than the United States, a reasonable attorney's fee, and the United States shall be

liable therefor.

(B) In any other action or proceeding under the provisions of this chapter, the court, when it finds that such action was without foundation, or was initiated vexatiously, frivolously, or in bad faith, shall allow the prevailing party, other than the United States, a reasonable attorney's fee, and the United States shall be liable therefor.

(C) Only those firearms or quantities of ammunition particularly named and individually identified as involved in or used in any violation of the provisions of this chapter or any rule or regulation issued thereunder, or any other criminal law of the United States or as intended to be used in any offense referred to in paragraph (3) of this subsection, where such intent is demonstrated by clear and convincing evidence, shall be subject to seizure, forfeiture, and disposition.

(D) The United States shall be liable for attorneys' fees under this paragraph only to the extent provided in advance by appropriation Acts.

(3) The offenses referred to in paragraphs (1) and (2)(C) of this subsection are—

(A) any crime of violence, as that term is defined in section 924(c)(3) of this title;

(B) any offense punishable under the Controlled Substances Act (21 U.S.C. 801 et seq.) or the Controlled Substances Import and Export Act (21 U.S.C. 951 et seq.);

(C) any offense described in section 922(a)(1), 922(a)(3), 922(a)(5), or 922(b)(3) of this title, where the firearm or ammunition intended to be used in any such offense is involved in a pattern of activities which includes a violation of any offense described in section 922(a)(1), 922(a)(3), 922(a)(5), or 922(b)(3) of this title;

(D) any offense described in section 922(d) of this title where the firearm or ammunition is intended to be used in such offense by the transferor of such firearm or ammunition;

(E) any offense described in section 922(i), 922(j), 922(l), 922(n), or 924(b) of this title; and

(F) any offense which may be prosecuted in a court of the United States which involves the exportation of firearms or ammunition.

(e) **(1)** In the case of a person who violates section 922(g) of this title and has three previous convictions by any court referred to in section 922(g)(1) of this title for a violent felony or a serious drug offense, or both, committed on occasions different from one another, such person shall be fined under this title and imprisoned not less than fifteen years, and, notwithstanding any other provision of law, the court shall not suspend the sentence of, or grant a probationary sentence to, such person with respect to the conviction under section 922(g).

(2) As used in this subsection—

(A) the term **"serious drug offense"** means—

(i) an offense under the Controlled Substances Act (21 U.S.C. 801 et seq.), the Controlled Substances Import and Export Act (21 U.S.C. 951 et seq.), or chapter 705 of title 46 for which a maximum term of imprisonment of ten years or more is prescribed by law; or

(ii) an offense under State law, involving manufacturing, distributing, or possessing with intent to manufacture or distribute, a controlled substance (as defined in section 102 of the Controlled Substances Act (21 U.S.C. 802)), for which a maximum term of imprisonment of ten years or more is prescribed by law;

(B) the term **"violent felony"** means any crime punishable by imprisonment for a term exceeding one year, or any act of juvenile delinquency involving the use or carrying of a firearm, knife, or destructive device that would be punishable by imprisonment for such term if committed by an adult, that—

(i) has as an element the use, attempted use, or threatened use of physical force against the person of another; or

(ii) is burglary, arson, or extortion, involves use of explosives, or otherwise involves conduct that presents a serious potential risk of physical injury to another; and

(C) the term **"conviction"** includes a finding that a person has committed an act of juvenile delinquency involving a violent felony.

(f) In the case of a person who knowingly violates section 922(p), such person shall be fined under this title, or imprisoned not more than 5 years, or both.

(g) Whoever, with the intent to engage in conduct which—

(1) constitutes an offense listed in section 1961(1),

(2) is punishable under the Controlled Substances Act (21 U.S.C. 801 et seq.), the Controlled Substances Import and Export Act (21 U.S.C. 951 et seq.), or chapter 705 of title 46,

(3) violates any State law relating to any controlled substance (as defined in section 102(6) of the Controlled Substances Act (21 U.S.C. 802(6))), or

(4) constitutes a crime of violence (as defined in subsection (c)(3)),

travels from any State or foreign country into any other State and acquires, transfers, or attempts to acquire or transfer, a firearm in such other State in furtherance of such purpose, shall be imprisoned not more than 10 years, fined in accordance with this title, or both.

(h) Whoever knowingly transfers a firearm, knowing that such firearm will be used to commit a crime of violence (as defined in subsection (c)(3)) or drug trafficking crime (as defined in subsection (c)(2)) shall be imprisoned not more than 10 years, fined in accordance with this title, or both.

(i) (1) A person who knowingly violates section 922(u) shall be fined under this title, imprisoned not more than 10 years, or both.

(2) Nothing contained in this subsection shall be construed as indicating an intent on the part of Congress to occupy the field in which provisions of this subsection operate to the exclusion of State laws on the same subject

matter, nor shall any provision of this subsection be construed as invalidating any provision of State law unless such provision is inconsistent with any of the purposes of this subsection.

(j) A person who, in the course of a violation of subsection (c), causes the death of a person through the use of a firearm, shall—

(1) if the killing is a murder (as defined in section 1111), be punished by death or by imprisonment for any term of years or for life; and

(2) if the killing is manslaughter (as defined in section 1112), be punished as provided in that section.

(k) A person who, with intent to engage in or to promote conduct that—

(1) is punishable under the Controlled Substances Act (21 U.S.C. 801 et seq.), the Controlled Substances Import and Export Act (21 U.S.C. 951 et seq.), or chapter 705 of title 46;

(2) violates any law of a State relating to any controlled substance (as defined in section 102 of the Controlled Substances Act, 21 U.S.C. 802); or

(3) constitutes a crime of violence (as defined in subsection (c)(3)), smuggles or knowingly brings into the United States a firearm, or attempts to do so, shall be imprisoned not more than 10 years, fined under this title, or both.

(l) A person who steals any firearm which is moving as, or is a part of, or which has moved in, interstate or foreign commerce shall be imprisoned for not more than 10 years, fined under this title, or both.

(m) A person who steals any firearm from a licensed importer, licensed manufacturer, licensed dealer, or licensed collector shall be fined under this title, imprisoned not more than 10 years, or both.

(n) A person who, with the intent to engage in conduct that constitutes a violation of section 922(a)(1)(A), travels from any State or foreign country into any other State and acquires, or attempts to acquire, a firearm in such other State in furtherance of such purpose shall be imprisoned for not more than 10 years.

(o) A person who conspires to commit an offense under subsection (c) shall be imprisoned for not more than 20 years, fined under this title, or both; and if the firearm is a machinegun or destructive device, or is equipped with a firearm silencer or muffler, shall be imprisoned for any term of years or life.

(p) Penalties Relating To Secure Gun Storage or Safety Device.—

(1) In general.—

(A) Suspension or revocation of license; civil penalties. With respect to each violation of section 922(z)(1) by a licensed manufacturer, licensed importer, or licensed dealer, the Secretary may, after notice and opportunity for hearing—

(i) suspend for not more than 6 months, or revoke, the license issued to the licensee under this chapter that was used to conduct the firearms transfer; or

(ii) subject the licensee to a civil penalty in an amount equal to not more than $2,500.

(B) Review. An action of the Secretary under this paragraph may be reviewed only as provided under section 923(f).

(2) Administrative remedies. The suspension or revocation of a license or the imposition of a civil penalty under paragraph (1) shall not preclude any administrative remedy that is otherwise available to the Secretary.

§ 925 Exceptions: Relief from disabilities.

(a) (1) The provisions of this chapter, except for sections 922(d)(9) and 922(g)(9) and provisions relating to firearms subject to the prohibitions of section 922(p), shall not apply with respect to the transportation, shipment, receipt, possession, or importation of any firearm or ammunition imported for, sold or shipped to, or issued for the use of, the United States or any department or agency thereof or any State or any department, agency, or political subdivision thereof.

(2) The provisions of this chapter, except for provisions relating to firearms subject to the prohibitions of section 922(p), shall not apply with respect to (A) the shipment or receipt of firearms or ammunition when sold or issued by the Secretary of the Army pursuant to section 4308 of title 10 before the repeal of such section by section 1624(a) of the Corporation for the Promotion of Rifle Practice and Firearms Safety Act, and (B) the transportation of any such firearm or ammunition carried out to enable a person, who lawfully received such firearm or ammunition from the Secretary of the Army, to engage in military training or in competitions.

(3) Unless otherwise prohibited by this chapter, except for provisions relating to firearms subject to the prohibitions of section 922(p), or any other Federal law, a licensed importer, licensed manufacturer, or licensed dealer may ship to a member of the United States Armed Forces on active duty outside the United States or to clubs, recognized by the Department of Defense, whose entire membership is composed of such members, and such members or clubs may receive a firearm or ammunition determined by the Attorney General to be generally recognized as particularly suitable for sporting purposes and intended for the personal use of such member or club.

(4) When established to the satisfaction of the Attorney General to be consistent with the provisions of this chapter, except for provisions relating to firearms subject to the prohibitions of section 922(p), and other applicable Federal and State laws and published ordinances, the Attorney General may authorize the transportation, shipment, receipt, or importation into the United States to the place of residence of any member of the United States Armed Forces who is on active duty outside the United States (or who has been on active duty outside the United States within the sixty day period immediately preceding the transportation, shipment, receipt, or importation), of any firearm or ammunition which is (A) determined by the Attorney General to be generally recognized as particularly suitable for sporting purposes, or determined by the Department of Defense to be a type of firearm normally classified as a war souvenir, and (B) intended for the personal use of such member.

(5) For the purpose of paragraph (3) of this subsection, the term **"United States"** means each of the several States and the District of Columbia.

(b) A licensed importer, licensed manufacturer, licensed dealer, or licensed collector who is indicted for a crime punishable by imprisonment for a term exceeding one year, may, notwithstanding any other provision of this chapter, continue operation pursuant to his existing license (if prior to the expiration of the term of the existing license timely application is made for a new license) during the term of such indictment and until any conviction pursuant to the indictment becomes final.

(c) A person who is prohibited from possessing, shipping, transporting, or receiving firearms or ammunition may make application to the Attorney General for relief from the disabilities imposed by Federal laws with respect to the acquisition, receipt, transfer, shipment, transportation, or possession of firearms, and the Attorney General may grant such relief if it is established to his satisfaction that the circumstances regarding the disability, and the applicant's record and reputation, are such that the applicant will not be likely to act in a manner dangerous to public safety and that the granting of the relief would not be contrary to the public interest. Any person whose application for relief from disabilities is denied by the Attorney General may file a petition with the United States district court for the district in which he resides for a judicial review of such denial. The court may in its discretion admit additional evidence where failure to do so would result in a miscarriage of justice. A licensed importer, licensed manufacturer, licensed dealer, or licensed collector conducting operations under this chapter, who makes application for relief from the disabilities incurred under this chapter, shall not be barred by such disability from further operations under his license pending final action on an application for relief filed pursuant to this section. Whenever the Attorney General grants relief to any person pursuant to this section he shall promptly publish in the Federal Register notice of such action, together with the reasons therefor.

(d) The Attorney General shall authorize a firearm or ammunition to be imported or brought into the United States or any possession thereof if the firearm or ammunition—

(1) is being imported or brought in for scientific or research purposes, or is for use in connection with competition or training pursuant to chapter 401 of title 10;

(2) is an unserviceable firearm, other than a machinegun as defined in section 5845(b) of the Internal Revenue Code of 1986 (not readily restorable to firing condition), imported or brought in as a curio or museum piece;

(3) is of a type that does not fall within the definition of a firearm as defined in section 5845(a) of the Internal Revenue Code of 1986 and is generally recognized as particularly suitable for or readily adaptable to sporting purposes, excluding surplus military firearms, except in any case where the Attorney General has not authorized the importation of the firearm pursuant to this paragraph, it shall be unlawful to import any frame, receiver, or barrel of such firearm which would be prohibited if assembled; or

(4) was previously taken out of the United States or a possession by the person who is bringing in the firearm or ammunition.

The Attorney General shall permit the conditional importation or bringing in of a firearm or ammunition for examination and testing in connection with the making of a determination as to whether the importation or bringing in of such firearm or ammunition will be allowed under this subsection.

(e) Notwithstanding any other provision of this title, the Attorney General shall authorize the importation of, by any licensed importer, the following:

(1) All rifles and shotguns listed as curios or relics by the Attorney General pursuant to section 921(a)(13), and

(2) All handguns, listed as curios or relics by the Attorney General pursuant to section 921(a)(13), provided that such handguns are generally recognized as particularly suitable for or readily adaptable to sporting purposes.

(f) The Attorney General shall not authorize, under subsection (d), the importation of any firearm the importation of which is prohibited by section 922(p).

§ 925A Remedy for erroneous denial of firearm.

Any person denied a firearm pursuant to subsection (s) or (t) of section 922—

(1) due to the provision of erroneous information relating to the person by any State or political subdivision thereof, or by the national instant criminal background check system established under section 103 of the Brady Handgun Violence Prevention Act; or

(2) who was not prohibited from receipt of a firearm pursuant to subsection (g) or (n) of section 922,

may bring an action against the State or political subdivision responsible for providing the erroneous information, or responsible for denying the transfer, or against the United States, as the case may be, for an order directing that the erroneous information be corrected or that the transfer be approved, as the case may be. In any action under this section, the court, in its discretion, may allow the prevailing party a reasonable attorney's fee as part of the costs.

§ 926 Rules and regulations.

(a) The Attorney General may prescribe only such rules and regulations as are necessary to carry out the provisions of this chapter, including—

(1) regulations providing that a person licensed under this chapter, when dealing with another person so licensed, shall provide such other licensed person a certified copy of this license;

(2) regulations providing for the issuance, at a reasonable cost, to a person licensed under this chapter, of certified copies of his license for use as provided under regulations issued under paragraph (1) of this subsection; and

(3) regulations providing for effective receipt and secure storage of firearms relinquished by or seized from persons described in subsection (d)(8) or (g)(8) of section 922.

No such rule or regulation prescribed after the date of the enactment of the Firearms Owners' Protection Act may require that records required to be maintained under this chapter or any portion of the contents of such records, be recorded at or transferred to a facility owned, managed, or controlled by the United States or any State or any political subdivision thereof, nor that any

system of registration of firearms, firearms owners, or firearms transactions or dispositions be established. Nothing in this section expands or restricts the Secretary's[1] authority to inquire into the disposition of any firearm in the course of a criminal investigation.

(b) The Attorney General shall give not less than ninety days public notice, and shall afford interested parties opportunity for hearing, before prescribing such rules and regulations.

(c) The Attorney General shall not prescribe rules or regulations that require purchasers of black powder under the exemption provided in section 845(a)(5) of this title to complete affidavits or forms attesting to that exemption.

§ 926A Interstate transportation of firearms.

Notwithstanding any other provision of any law or any rule or regulation of a State or any political subdivision thereof, any person who is not otherwise prohibited by this chapter from transporting, shipping, or receiving a firearm shall be entitled to transport a firearm for any lawful purpose from any place where he may lawfully possess and carry such firearm to any other place where he may lawfully possess and carry such firearm if, during such transportation the firearm is unloaded, and neither the firearm nor any ammunition being transported is readily accessible or is directly accessible from the passenger compartment of such transporting vehicle: *Provided,* That in the case of a vehicle without a compartment separate from the driver's compartment the firearm or ammunition shall be contained in a locked container other than the glove compartment or console.

§ 926B Carrying of concealed firearms by qualified law enforcement officers.

(a) Notwithstanding any other provision of the law of any State or any political subdivision thereof, an individual who is a qualified law enforcement officer and who is carrying the identification required by subsection (d) may carry a concealed firearm that has been shipped or transported in interstate or foreign commerce, subject to subsection (b).

(b) This section shall not be construed to supersede or limit the laws of any State that—

(1) permit private persons or entities to prohibit or restrict the possession of concealed firearms on their property; or

(2) prohibit or restrict the possession of firearms on any State or local government property, installation, building, base, or park.

(c) As used in this section, the term **"qualified law enforcement officer"** means an employee of a governmental agency who—

(1) is authorized by law to engage in or supervise the prevention, detection, investigation, or prosecution of, or the incarceration of any person for, any violation of law, and has statutory powers of arrest or apprehension under section 807(b) of title 10, United States Code (article 7(b) of the Uniform Code of Military Justice);

(2) is authorized by the agency to carry a firearm;

(3) is not the subject of any disciplinary action by the agency which could result in suspension or loss of police powers;

(4) meets standards, if any, established by the agency which require the employee to regularly qualify in the use of a firearm;

(5) is not under the influence of alcohol or another intoxicating or hallucinatory drug or substance; and

(6) is not prohibited by Federal law from receiving a firearm.

(d) The identification required by this subsection is the photographic identification issued by the governmental agency for which the individual is employed that identifies the employee as a police officer or law enforcement officer of the agency.

(e) As used in this section, the term **"firearm"**—

(1) except as provided in this subsection, has the same meaning as in section 921 of this title;

(2) includes ammunition not expressly prohibited by Federal law or subject to the provisions of the National Firearms Act; and

(3) does not include—

(A) any machinegun (as defined in section 5845 of the National Firearms Act);

(B) any firearm silencer (as defined in section 921 of this title); and

(C) any destructive device (as defined in section 921 of this title).

(f) For the purposes of this section, a law enforcement officer of the Amtrak Police Department, a law enforcement officer of the Federal Reserve, or a law enforcement or police officer of the executive branch of the Federal Government qualifies as an employee of a governmental agency who is authorized by law to engage in or supervise the prevention, detection, investigation, or prosecution of, or the incarceration of any person for, any violation of law, and has statutory powers of arrest or apprehension under section 807(b) of title 10, United States Code (article 7(b) of the Uniform Code of Military Justice).

§ 926C Carrying of concealed firearms by qualified retired law enforcement officers.

(a) Notwithstanding any other provision of the law of any State or any political subdivision thereof, an individual who is a qualified retired law enforcement officer and who is carrying the identification required by subsection (d) may carry a concealed firearm that has been shipped or transported in interstate or foreign commerce, subject to subsection (b).

(b) This section shall not be construed to supersede or limit the laws of any State that—

(1) permit private persons or entities to prohibit or restrict the possession of concealed firearms on their property; or

(2) prohibit or restrict the possession of firearms on any State or local government property, installation, building, base, or park.

(c) As used in this section, the term **"qualified retired law enforcement officer"** means an individual who—

(1) separated from service in good standing from service with a public agency as a law enforcement officer;

(2) before such separation, was authorized by law to engage in or supervise the prevention, detection, investigation, or prosecution of, or the incarceration of any person for, any violation of law, and had statutory powers of arrest or apprehension under section 807(b) of title 10, United States Code (article 7(b) of the Uniform Code of Military Justice);

(3)(A) before such separation, served as a law enforcement officer for an aggregate of 10 years or more; or

(B) separated from service with such agency, after completing any applicable probationary period of such service, due to a service–connected disability, as determined by such agency;

(4) during the most recent 12–month period, has met, at the expense of the individual, the standards for qualification in firearms training for active law enforcement officers, as determined by the former agency of the individual, the State in which the individual resides or, if the State has not established such standards, either a law enforcement agency within the State in which the individual resides or the standards used by a certified firearms instructor that is qualified to conduct a firearms qualification test for active duty officers within that State;

(5)(A) has not been officially found by a qualified medical professional employed by the agency to be unqualified for reasons relating to mental health and as a result of this finding will not be issued the photographic identification as described in subsection (d)(1); or

(B) has not entered into an agreement with the agency from which the individual is separating from service in which that individual acknowledges he or she is not qualified under this section for reasons relating to mental health and for those reasons will not receive or accept the photographic identification as described in subsection (d)(1);

(6) is not under the influence of alcohol or another intoxicating or hallucinatory drug or substance; and

(7) is not prohibited by Federal law from receiving a firearm.

(d) The identification required by this subsection is—

(1) a photographic identification issued by the agency from which the individual separated from service as a law enforcement officer that identifies the person as having been employed as a police officer or law enforcement officer and indicates that the individual has, not less recently than one year before the date the individual is carrying the concealed firearm, been tested or otherwise found by the agency to meet the active duty standards for qualification in firearms training as established by the agency to carry a firearm of the same type as the concealed firearm; or

(2)(A) a photographic identification issued by the agency from which the individual separated from service as a law enforcement officer that identifies the person as having been employed as a police officer or law enforcement officer; and

(B) a certification issued by the State in which the individual resides or by a certified firearms instructor that is qualified to conduct a firearms qualification test for active duty officers within that State that indicates that the individual has, not less than 1 year before the date the individual is carrying the concealed firearm, been tested or otherwise found by the State or a certified firearms instructor that is qualified to conduct a firearms qualification test for active duty officers within that State to have met—

(I) the active duty standards for qualification in firearms training, as established by the State, to carry a firearm of the same type as the concealed firearm; or

(II) if the State has not established such standards, standards set by any law enforcement agency within that State to carry a firearm of the same type as the concealed firearm.

(e) As used in this section—

(1) the term **"firearm"**—

(A) except as provided in this paragraph, has the same meaning as in section 921 of this title;

(B) includes ammunition not expressly prohibited by Federal law or subject to the provisions of the National Firearms Act; and

(C) does not include—

(i) any machinegun (as defined in section 5845 of the National Firearms Act);

(ii) any firearm silencer (as defined in section 921 of this title); and

(iii) any destructive device (as defined in section 921 of this title); and

(2) the term **"service with a public agency as a law enforcement officer"** includes service as a law enforcement officer of the Amtrak Police Department, service as a law enforcement officer of the Federal Reserve, or service as a law enforcement or police officer of the executive branch of the Federal Government.

§ 927 Effect on State law.

No provision of this chapter shall be construed as indicating an intent on the part of the Congress to occupy the field in which such provision operates to the exclusion of the law of any State on the same subject matter, unless there is a direct and positive conflict between such provision and the law of the State so that the two cannot be reconciled or consistently stand together.

§ 928 Separability.

If any provision of this chapter or the application thereof to any person or circumstance is held invalid, the remainder of the chapter and the application of such provision to other persons not similarly situated or to other circumstances shall not be affected thereby.

§ 929 Use of restricted ammunition.

(a) (1) Whoever, during and in relation to the commission of a crime of violence or drug trafficking crime (including a crime of violence or drug trafficking crime which provides for an enhanced punishment if committed by the use of a dead-ly or dangerous weapon or device) for which he may be prosecuted in a court of the United States, uses or carries a firearm and is in possession of armor piercing ammunition capable of being fired in that firearm, shall, in addition to the punishment provided for the commission of such crime of violence or drug trafficking crime be sentenced to a term of imprisonment for not less than five years.

(2) For purposes of this subsection, the term **"drug trafficking crime"** means any felony punishable under the Controlled Substances Act (21 U.S.C. 801 et seq.), the Controlled Substances Import and Export Act (21 U.S.C. 951 et seq.), or chapter 705 of title 46.

(b) Notwithstanding any other provision of law, the court shall not suspend the sentence of any person convicted of a violation of this section, nor place the person on probation, nor shall the terms of imprisonment run concurrently with any other terms of imprisonment, including that imposed for the crime in which the armor piercing ammunition was used or possessed.

§ 930 Possession of firearms and dangerous weapons in Federal facilities.

(a) Except as provided in subsection (d), whoever knowingly possesses or causes to be present a firearm or other dangerous weapon in a Federal facility (other than a Federal court facility), or attempts to do so, shall be fined under this title or imprisoned not more than 1 year, or both.

(b) Whoever, with intent that a firearm or other dangerous weapon be used in the commission of a crime, knowingly possesses or causes to be present such firearm or dangerous weapon in a Federal facility, or attempts to do so, shall be fined under this title or imprisoned not more than 5 years, or both.

(c) A person who kills any person in the course of a violation of subsection (a) or (b), or in the course of an attack on a Federal facility involving the use of a firearm or other dangerous weapon, or attempts or conspires to do such an act, shall be punished as provided in sections 1111, 1112, 1113, and 1117.

(d) Subsection (a) shall not apply to—

(1) the lawful performance of official duties by an officer, agent, or employee of the United States, a State, or a political subdivision thereof, who is authorized by law to engage in or supervise the prevention, detection, investigation, or prosecution of any violation of law;

(2) the possession of a firearm or other dangerous weapon by a Federal official or a member of the Armed Forces if such possession is authorized by law; or

(3) the lawful carrying of firearms or other dangerous weapons in a Federal facility incident to hunting or other lawful purposes.

(e) (1) Except as provided in paragraph (2), whoever knowingly possesses or causes to be present a firearm or other dangerous weapon in a Federal court facility, or attempts to do so, shall be fined under this title, imprisoned not more than 2 years, or both.

(2) Paragraph (1) shall not apply to conduct which is described in paragraph (1) or (2) of subsection (d).

(f) Nothing in this section limits the power of a court of the United States to punish for contempt or to promulgate rules or orders regulating, restricting, or prohibiting the possession of weapons within any building housing such court or any of its proceedings, or upon any grounds appurtenant to such building.

(g) As used in this section:

(1) The term **"Federal facility"** means a building or part thereof owned or leased by the Federal Government, where Federal employees are regularly present for the purpose of performing their official duties.

(2) The term **"dangerous weapon"** means a weapon, device, instrument, material, or substance, animate or inanimate, that is used for, or is readily capable of, causing death or serious bodily injury, except that such term does not include a pocket knife with a blade of less than 2½ inches in length.

(3) The term **"Federal court facility"** means the courtroom, judges' chambers, witness rooms, jury deliberation rooms, attorney conference rooms, prisoner holding cells, offices of the court clerks, the United States attorney, and the United States marshal, probation and parole offices, and adjoining corridors of any court of the United States.

(h) Notice of the provisions of subsections (a) and (b) shall be posted conspicuously at each public entrance to each Federal facility, and notice of subsection (e) shall be posted conspicuously at each public entrance to each Federal court facility, and no person shall be convicted of an offense under subsection (a) or (e) with respect to a Federal facility if such notice is not so posted at such facility, unless such person had actual notice of subsection (a) or (e), as the case may be.

§ 931 Prohibition on purchase, ownership, or possession of body armor by violent felons.

(a) In General. Except as provided in subsection (b), it shall be unlawful for a person to purchase, own, or possess body armor, if that person has been convicted of a felony that is—

(1) a crime of violence (as defined in section 16); or

(2) an offense under State law that would constitute a crime of violence under paragraph (1) if it occurred within the special maritime and territorial jurisdiction of the United States.

(b) Affirmative Defense.

(1) In general. It shall be an affirmative defense under this section that—

(A) the defendant obtained prior written certification from his or her employer that the defendant's purchase, use, or possession of body armor was necessary for the safe performance of lawful business activity; and

(B) the use and possession by the defendant were limited to the course of such performance.

(2) Employer. In this subsection, the term **"employer"** means any other individual employed by the defendant's business that supervises defendant's activity. If that defendant has no supervisor, prior written certification is acceptable from any other employee of the business.

TITLE 27 CFR CHAPTER II

PART 478—COMMERCE IN FIREARMS AND AMMUNITION
(This Part was formerly designated as Part 178)

33

NOTICE: ASSAULT WEAPONS BAN

On September 13, 1994, Congress passed the Violent Crime Control and Law Enforcement Act of 1994, Public Law 103–322. Title IX, Subtitle A, Section 110105 of this Act generally made it unlawful to manufacture, transfer, and possess semiautomatic assault weapons (SAWs)and to transfer and possess large capacity ammunition feeding devices (LCAFDs). The law also required importers and manufacturers to place certain markings on SAWs and LCAFDs, designating they were for export or law enforcement/government use. Significantly, the law provided that it would expire 10 years from the date of enactment. Accordingly, effective 12:00 am on September 13, 2004, these provisions of the law ceased to apply and the following provisions of the regulations in Part 478 no longer apply:

Section 478.11	—	Definitions of the terms "semiautomatic assault weapon" and "large capacity ammunition feeding device;"
Section 478.40	—	Entire section;
Section 478.40a	—	Entire section;
Section 478.57	—	Paragraphs (b) and (c);
Section 478.92	—	Paragraphs (a)(3) and (c);
Section 478.119	—	Entire section (Note: an import permit is still needed pursuant to the Arms Export Control Act—see 27 CFR 447.41(a));
Section 478.132	—	Entire section; and
Section 478.153	—	Entire section.

References to "ammunition feeding device" in section 478.116 are not applicable. References to "semiautomatic assault weapons" in section 478.171 are not applicable.

Subpart A—Introduction

§ 478.1 Scope of regulations.

(a) **General.** The regulations contained in this part relate to commerce in firearms and ammunition and are promulgated to implement Title I, State Firearms Control Assistance.

(b) **Procedural and substantive requirements.** This part contains the procedural and substantive requirements relative to:

(1) The interstate or foreign commerce in firearms and ammunition;

(2) The licensing of manufacturers and importers of firearms and ammunition, collectors of firearms, and dealers in firearms;

(3) The conduct of business or activity by licensees;

(4) The importation of firearms and ammunition;

(5) The records and reports required of licensees;

(6) Relief from disabilities under this part;

(7) Exempt interstate and foreign commerce in firearms and ammunition; and

(8) Restrictions on armor piercing ammunition.

§ 478.2 Relation to other provisions of law.

The provisions in this part are in addition to, and are not in lieu of, any other provision of law, or regulations, respecting commerce in firearms or ammunition. For regulations applicable to traffic in machine guns, destructive devices, and certain other firearms, see Part 479 of this chapter. For statutes applicable to the registration and licensing of persons engaged in the business of manufacturing, importing or exporting arms, ammunition, or implements of war, see section 38 of the Arms Export Control Act (22 U.S.C. 2778) and regulations thereunder and Part 447 of this chapter. For statutes applicable to nonmailable firearms, see 18 U.S.C. 1715 and regulations thereunder.

Subpart B—Definitions

§ 478.11 Meaning of terms.

When used in this part and in forms prescribed under this part, where not otherwise distinctly expressed or manifestly incompatible with the intent thereof, terms shall have the meanings ascribed in this section. Words in the plural form shall include the singular, and vice versa, and words importing the masculine gender shall include the feminine. The terms **"includes"** and **"including"** do not exclude other things not enumerated which are in the same general class or are otherwise within the scope thereof.

Act. 18 U.S.C. Chapter 44.

Adjudicated as a mental defective.

(a) A determination by a court, board, commission, or other lawful authority that a person, as a result of marked subnormal intelligence, or mental illness, incompetency, condition, or disease:

(1) Is a danger to himself or to others; or

(2) Lacks the mental capacity to contract or manage his own affairs.

(b) The term shall include–

(1) A finding of insanity by a court in a criminal case; and

(2) Those persons found incompetent to stand trial or found not guilty by reason of lack of mental responsibility pursuant to articles 50a and 72b of the Uniform Code of Military Justice, 10 U.S.C. 850a, 876b.

Admitted to the United States for lawful hunting or sporting purposes. (a) Is entering the United States to participate in a competitive target shooting event sponsored by a national, State, or local organization, devoted to the competitive use or other sporting use of firearms; or **(b)** Is entering the United States to display firearms at a sports or hunting trade show sponsored by a national, State, or local firearms trade organization, devoted to the competitive use or other sporting use of firearms.

Alien. Any person not a citizen or national of the United States.

Alien illegally or unlawfully in the United States. Aliens who are unlawfully in the United States are not in valid immigrant, nonimmigrant or parole status. The term includes any alien—

(a) Who unlawfully entered the United States without inspection and authorization by an immigration officer and who has not been paroled into the United States under section 212(d)(5) of the Immigration and Nationality Act (INA);

(b) Who is a nonimmigrant and whose authorized period of stay has expired or who has violated the terms of the nonimmigrant category in which he or she was admitted;

(c) Paroled under INA section 212(d)(5) whose authorized period of parole has expired or whose parole status has been terminated; or

(d) Under an order of deportation, exclusion, or removal, or under an order to depart the United States voluntarily, whether or not he or she has left the United States.

Ammunition. Ammunition or cartridge cases, primers, bullets, or propellent powder designed for use in any firearm other than an antique firearm. The term shall not include (a) any shotgun shot or pellet not designed for use as the single, complete projectile load for one shotgun hull or casing, nor (b) any unloaded, non–metallic shotgun hull or casing not having a primer.

Antique firearm. (a) Any firearm (including any firearm with a matchlock, flintlock, percussion cap, or similar type of ignition system) manufactured in or before 1898; and **(b)** any replica of any firearm described in paragraph **(a)** of this definition if such replica **(1)** is not designed or redesigned for using rimfire or conventional centerfire fixed ammunition, or **(2)** uses rimfire or conventional centerfire fixed ammunition which is no longer manufactured in the United States and which is not readily available in the ordinary channels of commercial trade.

Armor piercing ammunition. Projectiles or projectile cores which may be used in a handgun and which are constructed entirely (excluding the presence of traces of other substances) from one or a combination of tungsten alloys, steel, iron, brass, bronze, beryllium copper, or depleted uranium; or full jacketed projectiles larger than .22 caliber designed and intended for use in a handgun and whose jacket has a weight of more than 25 percent of the total weight of the projectile. The term does not include shotgun shot required by Federal or State environmental or game regulations for hunting purposes, frangible projectiles designed for target shooting, projectiles which the Director finds are primarily intended to be used for sporting purposes, or any other projectiles or projectile cores which the Director finds are intended to be used for industrial purposes, including charges used in oil and gas well perforating devices.

ATF officer. An officer or employee of the Bureau of Alcohol, Tobacco, Firearms and Explosives (ATF) authorized to perform any function relating to the administration or enforcement of this part.

Business premises. The property on which the manufacturing or importing of firearms or ammunition or the dealing in firearms is or will be conducted. A private dwelling, no part of which is open to the public, shall not be recognized as coming within the meaning of the term.

Chief, Federal Firearms Licensing Center (FFLC). The ATF official responsible for the issuance and renewal of licenses under this part.

Collector. Any person who acquires, holds, or disposes of firearms as curios or relics.

Collection premises. The premises described on the license of a collector as the location at which he maintains his collection of curios and relics.

Commerce. Travel, trade, traffic, commerce, transportation, or communication among the several States, or between the District of Columbia and any State, or between any foreign country or any territory or possession and any State or the District of Columbia, or between points in the same State but through any other State or the District of Columbia or a foreign country.

Committed to a mental institution. A formal commitment of a person to a mental institution by a court, board, commission, or other lawful authority. The term includes a commitment to a mental institution involuntarily. The term includes commitment for mental defectiveness or mental illness. It also includes commitments for other reasons, such as for drug use. The term does not include a person in a mental institution for observation or a voluntary admission to a mental institution.

Controlled substance. A drug or other substance, or immediate precursor, as defined in section 102 of the Controlled Substances Act, 21 U.S.C. 802. The term includes, but is not limited to, marijuana, depressants, stimulants, and narcotic drugs. The term does not include distilled spirits, wine, malt beverages, or tobacco, as those terms are defined or used in Subtitle E of the Internal Revenue Code of 1986, as amended.

Crime punishable by imprisonment for a term exceeding 1 year. Any Federal, State or foreign offense for which the maximum penalty, whether or not imposed, is capital punishment or imprisonment in excess of 1 year. The term shall not include **(a)** any Federal or State offenses pertaining to antitrust violations, unfair trade practices, restraints of trade, or other similar offenses relating to the regulation of business practices or **(b)** any State offense classified by the laws of the State as a misdemeanor and punishable by a term of imprisonment of 2 years or less. What constitutes a conviction of such a crime shall be determined in accordance with the law of the jurisdiction in which the proceedings were held. Any conviction which has been expunged or set aside or for which a person has been pardoned or has had civil rights restored shall not be considered a conviction for the purposes of the Act or this part, unless such pardon, expunction, or restoration of civil rights expressly provides that the person may not ship, transport, possess, or receive firearms, or unless the person is prohibited by the law of the jurisdiction in which the proceedings were held from receiving or possessing any firearms.

Curios or relics. Firearms which are of special interest to collectors by reason of some quality other than is associated with firearms intended for sporting use or as offensive or defensive weapons. To be recognized as curios or relics, firearms must fall within one of the following categories:

(a) Firearms which were manufactured at least 50 years prior to the current date, but not including replicas thereof;

(b) Firearms which are certified by the curator of a municipal, State, or Federal museum which exhibits firearms to be curios or relics of museum interest; and

(c) Any other firearms which derive a substantial part of their monetary value from the fact that they are novel, rare, bizarre, or because of their association with some historical figure, period, or event. Proof of qualification of a particular firearm under this category may be established by evidence of present value and evidence that like firearms are not available except as collector's items, or that the value of like firearms available in ordinary commercial channels is substantially less.

ATF Publication 5300.11, *Firearms Curios and Relics List,* **consists of lists of those firearms determined to be curios or relics from 1972 to the present.**

Customs officer. Any officer of the U.S. Customs and Border Protection, any commissioned, warrant, or petty officer of the Coast Guard, or any agent or other person authorized by law to perform the duties of a customs officer.

Dealer. Any person engaged in the business of selling firearms at wholesale or retail; any person engaged in the business of repairing firearms or of making or fitting special barrels, stocks, or trigger mechanisms to firearms; or any person who is a pawnbroker. The term shall include any person who engages in such business or occupation on a part–time basis.

Destructive device. (a) Any explosive, incendiary, or poison gas **(1)** bomb, **(2)** grenade, **(3)** rocket having a propellant charge of more than 4 ounces, **(4)** missile having an explosive or incendiary charge of more than one–quarter ounce, **(5)** mine, or **(6)** device similar to any of the devices described in the preceding paragraphs of this definition; **(b)** any type of weapon (other than a shotgun or a shotgun shell which the Director finds is generally recognized as particularly suitable for sporting purposes) by whatever name known which will, or which may be readily converted to, expel a projectile by the action of an explosive or other propellant, and which has any barrel with a bore of more than one–half inch in diameter; and **(c)** any combination of parts either designed or intended for use in converting any device into any destructive device described in paragraph (a) or (b) of this section and from which a destructive device may be readily assembled. The term shall not include any device which is neither designed nor redesigned for use as a weapon; any device, although originally designed for use as a weapon, which is redesigned for use as a signalling, pyrotechnic, line throwing, safety, or similar device; surplus ordnance sold, loaned, or given by the Secretary of the Army pursuant to the provisions of section 4684(2), 4685, or 4686 of title 10, United States Code; or any other device which the Director finds is not likely to be used as a weapon, is an antique, or is a rifle which the owner intends to use solely for sporting, recreational, or cultural purposes.

Director. The Director, Bureau of Alcohol, Tobacco, Firearms and Explosives, the Department of Justice, Washington, DC.

Director of Industry Operations. The principal ATF official in a Field Operations division responsible for administering regulations in this part.

Discharged under dishonorable conditions. Separation from the U.S. Armed Forces resulting from a dishonorable discharge or dismissal adjudged by a general court–martial. The term does not include any separation from the Armed Forces resulting from any other discharge, e.g., a bad conduct discharge.

Division. A Bureau of Alcohol, Tobacco, Firearms, and Explosives.

Engaged in the business—(a) Manufacturer of firearms. A person who devotes time, attention, and labor to manufacturing firearms as a regular course of trade or business with the principal objective of livelihood and profit through the sale or distribution of the firearms manufactured;

(b) Manufacturer of ammunition. A person who devotes time, attention, and labor to manufacturing ammunition as a regular course of trade or business with the principal objective of livelihood and profit through the sale or distribution of the ammunition manufactured;

(c) Dealer in firearms other than a gunsmith or a pawnbroker. A person who devotes time, attention, and labor to dealing in firearms as a regular course of trade or business with the principal objective of livelihood and profit through the repetitive purchase and resale of firearms, but such a term shall not include a person who makes occasional sales, exchanges, or purchases of firearms for the enhancement of a personal collection or for a hobby, or who sells all or part of his personal collection of firearms;

(d) Gunsmith. A person who devotes time, attention, and labor to engaging in such activity as a regular course of trade or business with the principal objective of livelihood and profit, but such a term shall not include a person who makes occasional repairs of firearms or who occasionally fits special barrels, stocks, or trigger mechanisms to firearms;

(e) Importer of firearms. A person who devotes time, attention, and labor to importing firearms as a regular course of trade or business with the principal objective of livelihood and profit through the sale or distribution of the firearms imported; and,

(f) Importer of ammunition. A person who devotes time, attention, and labor to importing ammunition as a regular course of trade or business with the principal objective of livelihood and profit through the sale or distribution of the ammunition imported.

Executed under penalties of perjury. Signed with the prescribed declaration under the penalties of perjury as provided on or with respect to the return form, or other document or, where no form of declaration is prescribed, with the declaration:

"I declare under the penalties of perjury that this–(insert type of document, such as, statement, application, request, certificate), including the documents submit-

ted in support thereof, has been examined by me and, to the best of my knowledge and belief, is true, correct, and complete."

Federal Firearms Act. 15 U.S.C. Chapter 18.

Firearm. Any weapon, including a starter gun, which will or is designed to or may readily be converted to expel a projectile by the action of an explosive; the frame or receiver of any such weapon; any firearm muffler or firearm silencer; or any destructive device; but the term shall not include an antique firearm. In the case of a licensed collector, the term shall mean only curios and relics.

Firearm frame or receiver. That part of a firearm which provides housing for the hammer, bolt or breechblock, and firing mechanism, and which is usually threaded at its forward portion to receive the barrel.

Firearm muffler or firearm silencer. Any device for silencing, muffling, or diminishing the report of a portable firearm, including any combination of parts, designed or redesigned, and intended for use in assembling or fabricating a firearm silencer or firearm muffler, and any part intended only for use in such assembly or fabrication.

Friendly foreign government. Any government with whom the United States has diplomatic relations and whom the United States has not identified as a State sponsor of terrorism.

Fugitive from justice. Any person who has fled from any State to avoid prosecution for a felony or a misdemeanor; or any person who leaves the State to avoid giving testimony in any criminal proceeding. The term also includes any person who knows that misdemeanor or felony charges are pending against such person and who leaves the State of prosecution.

Handgun. (a) Any firearm which has a short stock and is designed to be held and fired by the use of a single hand; and

(b) Any combination of parts from which a firearm described in paragraph (a) can be assembled.

Hunting license or permit lawfully issued in the United States. A license or permit issued by a State for hunting which is valid and unexpired.

Identification document. A document containing the name, residence address, date of birth, and photograph of the holder and which was made or issued by or under the authority of the United States Government, a State, political subdivision of a State, a foreign government, political subdivision of a foreign government, an international governmental or an international quasi–governmental organization which, when completed with information concerning a particular individual, is of a type intended or commonly accepted for the purpose of identification of individuals.

Importation. The bringing of a firearm or ammunition into the United States; except that the bringing of a firearm or ammunition from outside the United States into a foreign–trade zone for storage pending shipment to a foreign country or subsequent importation into this country, pursuant to this part, shall not be deemed importation.

Importer. Any person engaged in the business of importing or bringing firearms or ammunition into the United States. The term shall include any person who engages in such business on a part–time basis.

Indictment. Includes an indictment or information in any court, under which a crime punishable by imprisonment for a term exceeding 1 year (as defined in this section) may be prosecuted, or in military cases to any offense punishable by imprisonment for a term exceeding 1 year which has been referred to a general court–martial. An information is a formal accusation of a crime, differing from an indictment in that it is made by a prosecuting attorney and not a grand jury.

Interstate or foreign commerce. Includes commerce between any place in a State and any place outside of that State, or within any possession of the United States (not including the Canal Zone) or the District of Columbia. The term shall not include commerce between places within the same State but through any place outside of that State.

Intimate partner. With respect to a person, the spouse of the person, a former spouse of the person, an individual who is a parent of a child of the person, and an individual who cohabitates or has cohabitated with the person.

Large capacity ammunition feeding device.

Editor's Note:

The term large capacity ammunition feeding device is not applicable on or after September 13, 2004.

Licensed dealer. A dealer licensed under the provisions of this part.

Licensed importer. An importer licensed under the provisions of this part.

Licensed manufacturer. A manufacturer licensed under the provisions of this part.

Machine gun. Any weapon which shoots, is designed to shoot, or can be readily restored to shoot, automatically more than one shot, without manual reloading, by a single function of the trigger. The term shall also include the frame or receiver of any such weapon, any part designed and intended solely and exclusively, or combination of parts designed and intended, for use in converting a weapon into a machine gun, and any combination of parts from which a machine gun can be assembled if such parts are in the possession or under the control of a person.

Manufacturer. Any person engaged in the business of manufacturing firearms or ammunition. The term shall include any person who engages in such business on a part–time basis.

Mental institution. Includes mental health facilities, mental hospitals, sanitariums, psychiatric facilities, and other facilities that provide diagnoses by licensed professionals of mental retardation or mental illness, including a psychiatric ward in a general hospital.

Misdemeanor crime of domestic violence. (a) Is a Federal, State or local offense that:

(1) Is a misdemeanor under Federal or State law or, in States which do not classify offenses as misdemeanors, is an offense punishable by imprisonment for a term of one year or less, and includes offenses that are punishable only by a fine. (This is true whether or not the State statute specifically defines the offense as a **"misdemeanor"** or as a **"misdemeanor crime of domestic violence."** The term includes all such misdemeanor convictions in Indian Courts established pursuant to 25 CFR Part 11.);

(2) Has, as an element, the use or attempted use of physical force (e.g., assault and battery), or the threatened use of a deadly weapon; and

(3) Was committed by a current or former spouse, parent, or guardian of the victim, by a person with whom the victim shares a child in common, by a person who is cohabiting with or has cohabited with the victim as a spouse, parent, or guardian, (e.g., the equivalent of a **"common law"** marriage even if such relationship is not recognized under the law), or a person similarly situated to a spouse, parent, or guardian of the victim (e.g., two persons who are residing at the same location in an intimate relationship with the intent to make that place their home would be similarly situated to a spouse).

(b) A person shall not be considered to have been convicted of such an offense for purposes of this part unless:

(1) The person is considered to have been convicted by the jurisdiction in which the proceedings were held.

(2) The person was represented by counsel in the case, or knowingly and intelligently waived the right to counsel in the case; and

(3) In the case of a prosecution for which a person was entitled to a jury trial in the jurisdiction in which the case was tried, either

(i) The case was tried by a jury, or

(ii) The person knowingly and intelligently waived the right to have the case tried by a jury, by guilty plea or otherwise.

(c) A person shall not be considered to have been convicted of such an offense for purposes of this part if the conviction has been expunged or set aside, or is an offense for which the person has been pardoned or has had civil rights restored (if the law of the jurisdiction in which the proceedings were held provides for the loss of civil rights upon conviction for such an offense) unless the pardon, expunction, or restoration of civil rights expressly provides that the person may not ship, transport, possess, or receive firearms, and the person is not otherwise prohibited by the law of the jurisdiction in which the proceedings were held from receiving or possessing any firearms.

National Firearms Act. 26 U.S.C. Chapter 53.

NICS. The National Instant Criminal Background Check System established by the Attorney General pursuant to 18 U.S.C. 922(t).

Nonimmigrant alien. An alien in the United States in a nonimmigrant classification as defined by section 101(a)(15) of the Immigration and Nationality Act (8 U.S.C. 1101(a)(15)).

Nonimmigrant visa. A visa properly issued to an alien as an eligible nonimmigrant by a competent officer as provided in the Immigration and Nationality Act, 8 U.S.C. 1101 *et seq.*

Pawnbroker. Any person whose business or occupation includes the taking or receiving, by way of pledge or pawn, of any firearm as security for the payment or repayment of money. The term shall include any person who engages in such business on a part–time basis.

Permanently inoperable. A firearm which is incapable of discharging a shot by means of an explosive and incapable of being readily restored to a firing condition. An acceptable method of rendering most firearms permanently inoperable is to fusion weld the chamber closed and fusion weld the barrel solidly to the frame. Certain unusual firearms require other methods to render the firearm permanently inoperable. Contact ATF for instructions.

Person. Any individual, corporation, company, association, firm, partnership, society, or joint stock company.

Pistol. A weapon originally designed, made, and intended to fire a projectile (bullet) from one or more barrels when held in one hand, and having (a) a chamber(s) as an integral part(s) of, or permanently aligned with, the bore(s); and (b) a short stock designed to be gripped by one hand and at an angle to and extending below the line of the bore(s).

Principal objective of livelihood and profit. The intent underlying the sale or disposition of firearms is predominantly one of obtaining livelihood and pecuniary gain, as opposed to other intents such as improving or liquidating a personal firearms collection: Provided, That proof of profit shall not be required as to a person who engages in the regular and repetitive purchase and disposition of firearms for criminal purposes or terrorism. For purposes of this part, the term **"terrorism"** means activity, directed against United States persons, which—

(a) Is committed by an individual who is not a national or permanent resident alien of the United States;

(b) Involves violent acts or acts dangerous to human life which would be a criminal violation if committed within the jurisdiction of the United States; and

(c) Is intended—

(1) To intimidate or coerce a civilian population;

(2) To influence the policy of a government by intimidation or coercion; or

(3) To affect the conduct of a government by assassination or kidnapping.

Published ordinance. A published law of any political subdivision of a State which the Director determines to be relevant to the enforcement of this part and which is contained on a list compiled by the Director, which list is incorporated by reference in the FEDERAL REGISTER, revised annually, and furnished to licensees under this part.

Renounced U.S. citizenship. (a) A person has renounced his U.S. citizenship if the person, having been a citizen of the United States, has renounced citizenship either—

(1) Before a diplomatic or consular officer of the United States in a foreign state pursuant to 8 U.S.C. 1481(a)(5); or

(2) Before an officer designated by the Attorney General when the United States is in a state of war pursuant to 8 U.S.C. 1481(a)(6).

(b) The term shall not include any renunciation of citizenship that has been reversed as a result of administrative or judicial appeal.

Revolver. A projectile weapon, of the pistol type, having a breechloading chambered cylinder so arranged that the cocking of the hammer or movement of the trigger rotates it and brings the next cartridge in line with the barrel for firing.

Rifle. A weapon designed or redesigned, made or remade, and intended to be fired from the shoulder, and designed or redesigned and made or remade to use the energy of the explosive in a fixed metallic cartridge to fire only a single projectile through a rifled bore for each single pull of the trigger.

Semiautomatic assault weapon.

Editor's Note:

The term semiautomatic assault weapon is not applicable on or after September 13, 2004.

Semiautomatic pistol. Any repeating pistol which utilizes a portion of the energy of a firing cartridge to extract the fired cartridge case and chamber the next round, and which requires a separate pull of the trigger to fire each cartridge.

Semiautomatic rifle. Any repeating rifle which utilizes a portion of the energy of a firing cartridge to extract the fired cartridge case and chamber the next round, and which requires a separate pull of the trigger to fire each cartridge.

Semiautomatic shotgun. Any repeating shotgun which utilizes a portion of the energy of a firing cartridge to extract the fired cartridge case and chamber the next round, and which requires a separate pull of the trigger to fire each cartridge.

Short–barreled rifle. A rifle having one or more barrels less than 16 inches in length, and any weapon made from a rifle, whether by alteration, modification, or otherwise, if such weapon, as modified, has an overall length of less than 26 inches.

Short–barreled shotgun. A shotgun having one or more barrels less than 18 inches in length, and any weapon made from a shotgun, whether by alteration, modification, or otherwise, if such weapon as modified has an overall length of less than 26 inches.

Shotgun. A weapon designed or redesigned, made or remade, and intended to be fired from the shoulder, and designed or redesigned and made or remade to use the energy of the explosive in a fixed shotgun shell to fire through a smooth bore either a number of ball shot or a single projectile for each single pull of the trigger.

State. A State of the United States. The term shall include the District of Columbia, the Commonwealth of Puerto Rico, and the possessions of the United States (not including the Canal Zone).

State of residence. The State in which an individual resides. An individual resides in a State if he or she is present in a State with the intention of making a home in that State. If an individual is on active duty as a member of the Armed Forces, the individual's State of residence is the State in which his or her permanent duty station is located, as stated in 18 U.S.C. 921(b). The following are examples that illustrate this definition:

Example 1. A maintains a home in State X. A travels to State Y on a hunting, fishing, business, or other type of trip. A does not become a resident of State Y by reason of such trip.

Example 2. A maintains a home in State X and a home in State Y. A resides in State X except for weekends or the summer months of the year and in State Y for the weekends or the summer months of the year. During the time that A actually resides in State X, A is a resident of State X, and during the time that A actually resides in State Y, A is a resident of State Y.

Example 3. A, an alien, travels to the United States on a three–week vacation to State X. A does not have a state of residence in State X because A does not have the intention of making a home in State X while on vacation. This is true regardless of the length of the vacation.

Example 4. A, an alien, travels to the United States to work for three years in State X. A rents a home in State X, moves his personal possessions into the home, and his family resides with him in the home. A intends to reside in State X during the 3–year period of his employment. A is a resident of State X.

Unlawful user of or addicted to any controlled substance. A person who uses a controlled substance and has lost the power of self–control with reference to the use of controlled substance; and any person who is a current user of a controlled substance in a manner other than as prescribed by a licensed physician. Such use is not limited to the use of drugs on a particular day, or within a matter of days or weeks before, but rather that the unlawful use has occurred recently enough to indicate that the individual is actively engaged in such conduct. A person may be an unlawful current user of a controlled substance even though the substance is not being used at the precise time the person seeks to acquire a firearm or receives or possesses a firearm. An inference of current use may be drawn from evidence of a recent use or possession of a controlled substance or a pattern of use or possession that reasonably covers the present time, e.g., a conviction for use or possession of a controlled substance within the past year; multiple arrests for such offenses within the past 5 years if the most recent arrest occurred within the past year; or persons found through a drug test to use a controlled substance unlawfully, provided that the test was administered within the past year. For a current or former member of the Armed Forces, an inference of current use may be drawn from recent disciplinary or other administrative action based on confirmed drug use, e.g., court–martial conviction, nonjudicial punishment, or an administrative discharge based on drug use or drug rehabilitation failure.

Unserviceable firearm. A firearm which is incapable of discharging a shot by means of an explosive and is incapable of being readily restored to a firing condition.

U.S.C. The United States Code.

Subpart C—Administrative and Miscellaneous Provisions

§ 478.21 Forms prescribed.

(a) The Director is authorized to prescribe all forms required by this part. All of the information called for in each form shall be furnished as indicated by the headings on the form and the instructions on or pertaining to the form. In addition, information called for in each form shall be furnished as required by this part.

(b) Requests for forms should be submitted to the ATF Distribution Center (http://www.atf.gov) or by calling (703) 870–7526 or (703) 870-7528

§ 478.22 Alternate methods or procedures; emergency variations from requirements.

(a) Alternate methods or procedures. The licensee, on specific approval by the Director as provided in this paragraph, may use an alternate method or procedure in lieu of a method or procedure specifically prescribed in this part. The Director may approve an alternate method or procedure, subject to stated conditions, when it is found that:

(1) Good cause is shown for the use of the alternate method or procedure;

(2) The alternate method or procedure is within the purpose of, and consistent with the effect intended by, the specifically prescribed method or procedure and that the alternate method or procedure is substantially equivalent to that specifically prescribed method or procedure; and

(3) The alternate method or procedure will not be contrary to any provision of law and will not result in an increase in cost to the Government or hinder the effective administration of this part. Where the licensee desires to employ an alternate method or procedure, a written application shall be submitted to the appropriate Director of Industry Operations, for transmittal to the Director. The application shall specifically describe the proposed alternate method or procedure and shall set forth the reasons for it. Alternate methods or procedures may not be employed until the application is approved by the Director. The licensee shall, during the period of authorization of an alternate method or procedure, comply with the terms of the approved application. Authorization of any alternate method or procedure may be withdrawn whenever, in the judgment of the Director, the effective administration of this part is hindered by the continuation of the authorization.

(b) Emergency variations from requirements. The Director may approve a method of operation other than as specified in this part, where it is found that an emergency exists and the proposed variation from the specified requirements are necessary and the proposed variations (1) will not hinder the effective administration of this part, and (2) will not be contrary to any provisions of law. Variations from requirements granted under this paragraph are conditioned on compliance with the procedures, conditions, and limitations set forth in the approval of the application. Failure to comply in good faith with the procedures, conditions, and limitations shall automatically terminate the authority for the variations, and the licensee shall fully comply with the prescribed requirements of regulations from which the variations were authorized. Authority for any variation may be withdrawn whenever, in the judgment of the Director, the effective administration of this part is hindered by the continuation of the variation. Where the licensee desires to employ an emergency variation, a written application shall be submitted to the appropriate Director of Industry Opera-

tions for transmittal to the Director. The application shall describe the proposed variation and set forth the reasons for it. Variations may not be employed until the application is approved.

(c) Retention of approved variations. The licensee shall retain, as part of the licensee's records, available for examination by ATF officers, any application approved by the Director under this section.

§ 478.23 Right of entry and examination.

(a) Except as provided in paragraph (b), any ATF officer, when there is reasonable cause to believe a violation of the Act has occurred and that evidence of the violation may be found on the premises of any licensed manufacturer, licensed importer, licensed dealer, or licensed collector, may, upon demonstrating such cause before a Federal magistrate and obtaining from the magistrate a warrant authorizing entry, enter during business hours (or, in the case of a licensed collector, the hours of operation) the premises, including places of storage, of any such licensee for the purpose of inspecting or examining:

(1) Any records or documents required to be kept by such licensee under this part and

(2) Any inventory of firearms or ammunition kept or stored by any licensed manufacturer, licensed importer, or licensed dealer at such premises or any firearms curios or relics or ammunition kept or stored by any licensed collector at such premises.

(b) Any ATF officer, without having reasonable cause to believe a violation of the Act has occurred or that evidence of the violation may be found and without demonstrating such cause before a Federal magistrate or obtaining from the magistrate a warrant authorizing entry, may enter during business hours the premises, including places of storage, of any licensed manufacturer, licensed importer, or licensed dealer for the purpose of inspecting or examining the records, documents, ammunition and firearms referred to in paragraph (a) of this section:

(1) In the course of a reasonable inquiry during the course of a criminal investigation of a person or persons other than the licensee,

(2) For insuring compliance with the recordkeeping requirements of this part:

(i) Not more than once during any 12–month period, or

(ii) At any time with respect to records relating to a firearm involved in a criminal investigation that is traced to the licensee, or

(3) When such inspection or examination may be required for determining the disposition of one or more particular firearms in the course of a bona fide criminal investigation.

(c) Any ATF officer, without having reasonable cause to believe a violation of the Act has occurred or that evidence of the violation may be found and without demonstrating such cause before a Federal magistrate or obtaining from the magistrate a warrant authorizing entry, may enter during hours of operation the premises, including places of storage, of any licensed collector for the purpose of inspecting or examining the records, documents, firearms, and ammunition referred to in paragraph (a) of this section

(1) for ensuring compliance with the recordkeeping requirements of this part not more than once during any 12–month period or (2) when such inspection or examination may be required for determining the disposition of one or more particular firearms in the course of a bona fide criminal investigation. At the election of the licensed collector, the annual inspection permitted by this paragraph shall be performed at the ATF office responsible for conducting such inspection in closest proximity to the collectors premises.

(d) The inspections and examinations provided by this section do not authorize an ATF officer to seize any records or documents other than those records or documents constituting material evidence of a violation of law. If an ATF officer seizes such records or documents, copies shall be provided the licensee within a reasonable time.

§ 478.24 Compilation of State laws and published ordinances.

(a) The Director shall annually revise and furnish Federal firearms licensees with a compilation of State laws and published ordinances which are relevant to the enforcement of this part.

The Director annually revises the compilation and publishes it as **"State Laws and Published Ordinances–Firearms"** which is furnished free of charge to licensees under this part. Where the compilation has previously been furnished to licensees, the Director need only furnish amendments of the relevant laws and ordinances to such licensees.

(b) **"State Laws and Published Ordinances–Firearms"** is incorporated by reference in this part. It is ATF Publication 5300.5, revised yearly. The current edition is available from the Superintendent of Documents, U.S. Government Printing Office, Washington, DC 20402. It is also available for inspection at the National Archives and Records Administration (NARA). For information on the availability of this material at NARA, call 202–741–6030, or go to: http://www.archives.gov/federal_register/code_of_federal_regulations/ibr_locations.html. This incorporation by reference was approved by the Director of the Federal Register.

§ 478.25 Disclosure of information.

The Director of Industry Operations may make available to any Federal, State or local law enforcement agency any information which is obtained by reason of the provisions of the Act with respect to the identification of persons prohibited from purchasing or receiving firearms or ammunition who have purchased or received firearms or ammunition, together with a description of such firearms or ammunition. Upon the request of any Federal, State or local law enforcement agency, the Director of Industry Operations may provide such agency any information contained in the records required to be maintained by the Act or this part.

§ 478.25a Responses to requests for information.

Each licensee shall respond immediately to, and in no event later than 24 hours after the receipt of, a request by an ATF officer at the National Tracing Center for information contained in the records required to be kept by this part for determining the disposition of one or more firearms in the course of a bona fide criminal investigation. The requested information shall be provided orally to the ATF officer within the 24–hour period. Verification of the identity and employment of National Tracing Center personnel requesting information may be established at the time the requested information is provided by telephoning the toll–free number 1–800–788–7132 or using the toll–free facsimile (FAX) number 1–800–788–7133.

(Approved by the Office of Management and Budget under control number 1140–0032)

§ 478.26 Curio and relic determination.

Any person who desires to obtain a determination whether a particular firearm is a curio or relic shall submit a written request, in duplicate, for a ruling thereon to the Director. Each such request shall be executed under the penalties of perjury and shall contain a complete and accurate description of the firearm, and such photographs, diagrams, or drawings as may be necessary to enable the Director to make a determination. The Director may require the submission of the firearm for examination and evaluation. If the submission of the firearm is impractical, the person requesting the determination shall so advise the Director and designate the place where the firearm will be available for examination and evaluation.

§ 478.27 Destructive device determination.

The Director shall determine in accordance with 18 U.S.C. 921(a)(4) whether a device is excluded from the definition of a destructive device. A person who desires to obtain a determination under that provision of law for any device which he believes is not likely to be used as a weapon shall submit a written request, in triplicate, for a ruling thereon to the Director. Each such request shall be executed under the penalties of perjury and contain a complete and accurate description of the device, the name and address of the manufacturer or importer thereof, the purpose of and use for which it is intended, and such photographs, diagrams, or drawings as may be necessary to enable the Director to make his determination. The Director may require the submission to him, of a sample of such device for examination and evaluation. If the submission of such device is impracticable, the person requesting the ruling shall so advise the Director and designate the place where the device will be available for examination and evaluation.

§ 478.28 Transportation of destructive devices and certain firearms.

(a) The Director may authorize a person to transport in interstate or foreign commerce any destructive device, machine gun, short–barreled shotgun, or short–barreled rifle, if he finds that such transportation is reasonably necessary and is consistent with public safety and applicable State and local law. A person who desires to transport in interstate or foreign commerce any such device or weapon shall submit a written request so to do, in duplicate, to the Director. The request shall contain:

(1) A complete description and identification of the device or weapon to be transported;

(2) A statement whether such transportation involves a transfer of title;

(3) The need for such transportation;

(4) The approximate date such transportation is to take place;

(5) The present location of such device or weapon and the place to which it is to be transported;

(6) The mode of transportation to be used (including, if by common or contract carrier, the name and address of such carrier); and

(7) Evidence that the transportation or possession of such device or weapon is not inconsistent with the laws at the place of destination.

(b) No person shall transport any destructive device, machine gun, short–barreled shotgun, or short–barreled rifle in interstate or foreign commerce under the provisions of this section until he has received specific authorization so to do from the Director. Authorization granted under this section does not carry or import relief from any other statutory or regulatory provision relating to firearms.

(c) This section shall not be construed as requiring licensees to obtain authorization to transport destructive devices, machine guns, short–barreled shotguns, and short–barreled rifles in interstate or foreign commerce: Provided, That in the case of a licensed importer, licensed manufacturer, or licensed dealer, such a licensee is qualified under the National Firearms Act (see also Part 479 of this chapter) and this part to engage in the business with respect to the device or weapon to be transported, and that in the case of a licensed collector, the device or weapon to be transported is a curio or relic.

§ 478.29 Out–of–State acquisition of firearms by nonlicensees.

No person, other than a licensed importer, licensed manufacturer, licensed dealer, or licensed collector, shall transport into or receive in the State where the person resides (or if a corporation or other business entity, where it maintains a place of business) any firearm purchased or otherwise obtained by such person outside that State: Provided, That the provisions of this section:

(a) Shall not preclude any person who lawfully acquires a firearm by bequest or intestate succession in a State other than his State of residence from transporting the firearm into or receiving it in that State, if it is lawful for such person to purchase or possess such firearm in that State,

(b) Shall not apply to the transportation or receipt of a rifle or shotgun obtained from a licensed manufacturer, licensed importer, licensed dealer, or licensed collector in a State other than the transferee's State of residence in an over–the–counter transaction at the licensee's premises obtained in conformity with the provisions of §478.96(c) and

(c) Shall not apply to the transportation or receipt of a firearm obtained in conformity with the provisions of §§478.30 and 478.97.

§ 478.29a Acquisition of firearms by nonresidents.

No person, other than a licensed importer, licensed manufacturer, licensed dealer, or licensed collector, who does not reside in any State shall receive any firearms unless such receipt is for lawful sporting purposes.

§ 478.30 Out–of–State disposition of firearms by nonlicensees.

No nonlicensee shall transfer, sell, trade, give, transport, or deliver any firearm to any other nonlicensee, who the transferor knows or has reasonable cause to believe does not reside in (or if the person is a corporation or other business entity, does not maintain a place of business in) the State in which the transferor resides: Provided, That the provisions of this section:

(a) shall not apply to the transfer, transportation, or delivery of a firearm made to carry out a bequest of a firearm to, or any acquisition by intestate succession of a firearm by, a person who is permitted to acquire or possess a firearm under the laws of the State of his residence; and

(b) shall not apply to the loan or rental of a firearm to any person for temporary use for lawful sporting purposes.

§ 478.31 Delivery by common or contract carrier.

(a) No person shall knowingly deliver or cause to be delivered to any common or contract carrier for transportation or shipment in interstate or foreign commerce to any person other than a licensed importer, licensed manufacturer, licensed dealer, or licensed collector, any package or other container in which there is any firearm or ammunition without written notice to the carrier that such firearm or ammunition is being transported or shipped: Provided, That any passenger who owns or legally possesses a firearm or ammunition being transported aboard any common or contract carrier for movement with the passenger in interstate or foreign commerce may deliver said firearm or ammunition into the custody of the pilot, captain, conductor or operator of such common or contract carrier for the duration of that trip without violating any provision of this part.

(b) No common or contract carrier shall require or cause any label, tag, or other written notice to be placed on the outside of any package, luggage, or other container indicating that such package, luggage, or other container contains a firearm.

(c) No common or contract carrier shall transport or deliver in interstate or foreign commerce any firearm or ammunition with knowledge or reasonable cause to believe that the shipment, transportation, or receipt thereof would be in violation of any provision of this part: Provided, however, That the provisions of this paragraph shall not apply in respect to the transportation of firearms or ammunition in in–bond shipment under Customs laws and regulations.

(d) No common or contract carrier shall knowingly deliver in interstate or foreign commerce any firearm without obtaining written acknowledgement of receipt from the recipient of the package or other container in which there is a firearm: Provided, That this paragraph shall not apply with respect to the return of a firearm to a passenger who places firearms in the carrier's custody for the duration of the trip.

§ 478.32 Prohibited shipment, transportation, possession, or receipt of firearms and ammunition by certain persons.

(a) No person may ship or transport any firearm or ammunition in interstate or foreign commerce, or receive any firearm or ammunition which has been shipped or transported in interstate or foreign commerce, or possess any firearm or ammunition in or affecting commerce, who:

(1) Has been convicted in any court of a crime punishable by imprisonment for a term exceeding 1 year,

(2) Is a fugitive from justice,

(3) Is an unlawful user of or addicted to any controlled substance (as defined in section 102 of the Controlled Substances Act, 21 U.S.C. 802),

(4) Has been adjudicated as a mental defective or has been committed to a mental institution,

(5) Being an alien—

(i) Is illegally or unlawfully in the United States; or

(ii) Except as provided in paragraph (f) of this section, has been admitted to the United States under a nonimmigrant visa: Provided, That the provisions of this paragraph

(a) (5) (ii) do not apply to any alien who has been lawfully admitted to the United States under a nonimmigrant visa, if that alien is—

(A) Admitted to the United States for lawful hunting or sporting purposes or is in possession of a hunting license or permit lawfully issued in the United States;

(B) An official representative of a foreign government who is either accredited to the United States Government or the Government's mission to an international organization having its headquarters in the United States or is en route to or from another country to which that alien is accredited. This exception only applies if the firearm or ammunition is shipped, transported, possessed, or received in the representative's official capacity;

(C) An official of a foreign government or a distinguished foreign visitor who has been so designated by the Department of State. This exception only applies if the firearm or ammunition is shipped, transported, possessed, or received in the official's or visitor's official capacity, except if the visitor is a private individual who does not have an official capacity; or

(D) A foreign law enforcement officer of a friendly foreign government entering the United States on official law enforcement business,

(6) Has been discharged from the Armed Forces under dishonorable conditions,

(7) Having been a citizen of the United States, has renounced citizenship,

(8) Is subject to a court order that—

(i) Was issued after a hearing of which such person received actual notice, and at which such person had an opportunity to participate;

(ii) Restrains such person from harassing, stalking, or threatening an intimate partner of such person or child of such intimate partner or person, or engaging in other conduct that would place an intimate partner in reasonable fear of bodily injury to the partner or child; and

(iii)(A) Includes a finding that such person represents a credible threat to the physical safety of such intimate partner or child; or

(B) By its terms explicitly prohibits the use, attempted use, or threatened use of physical force against such intimate partner or child that would reasonably be expected to cause bodily injury, or

(9) Has been convicted of a misdemeanor crime of domestic violence.

(b) No person who is under indictment for a crime punishable by imprisonment for a term exceeding one year may ship or transport any firearm or ammunition in interstate or foreign commerce or receive any firearm or ammunition which has been shipped or transported in interstate or foreign commerce.

(c) Any individual, who to that individual's knowledge and while being employed by any person described in paragraph (a) of this section, may not in the course of such employment receive, possess, or transport any firearm or ammunition in commerce or affecting commerce or receive any firearm or ammunition which has been shipped or transported in interstate or foreign commerce.

(d) No person may sell or otherwise dispose of any firearm or ammunition to any person knowing or having reasonable cause to believe that such person:

(1) Is under indictment for, or has been convicted in any court of, a crime punishable by imprisonment for a term exceeding 1 year,

(2) Is a fugitive from justice,

(3) Is an unlawful user of or addicted to any controlled substance (as defined in section 102 of the Controlled Substances Act, 21 U.S.C. 802),

(4) Has been adjudicated as a mental defective or has been committed to a mental institution,

(5) Being an alien—

(i) Is illegally or unlawfully in the United States; or

(ii) Except as provided in paragraph (f) of this section, has been admitted to the United States under a nonimmigrant visa: Provided, That the provisions of this paragraph

(d)(5)(ii) do not apply to any alien who has been lawfully admitted to the United States under a nonimmigrant visa, if that alien is—

(A) Admitted to the United States for lawful hunting or sporting purposes or is in possession of a hunting license or permit lawfully issued in the United States;

(B) An official representative of a foreign government who is either accredited to the United States Government or the

Government's mission to an international organization having its headquarters in the United States or en route to or from another country to which that alien is accredited. This exception only applies if the firearm or ammunition is shipped, transported, possessed, or received in the representative's official capacity;

(C) An official of a foreign government or a distinguished foreign visitor who has been so designated by the Department of State. This exception only applies if the firearm or ammunition is shipped, transported, possessed, or received in the official's or visitor's official capacity, except if the visitor is a private individual who does not have an official capacity; or

(D) A foreign law enforcement officer of a friendly foreign government entering the United States on official law enforcement business,

(6) Has been discharged from the Armed Forces under dishonorable conditions,

(7) Having been a citizen of the United States, has renounced citizenship,

(8) Is subject to a court order that restrains such person from harassing, stalking, or threatening an intimate partner of such person or child of such intimate partner or person, or engaging in other conduct that would place an intimate partner in reasonable fear of bodily injury to the partner or child: Provided, That the provisions of this paragraph shall only apply to a court order that—

(i) Was issued after a hearing of which such person received actual notice, and at which such person had the opportunity to participate; and

(ii)(A) Includes a finding that such person represents a credible threat to the physical safety of such intimate partner or child; or

(B) By its terms explicitly prohibits the use, attempted use, or threatened use of physical force against such intimate partner or child that would reasonably be expected to cause bodily injury, or

(9) Has been convicted of a misdemeanor crime of domestic violence.

(e) The actual notice required by paragraphs (a)(8)(i) and (d)(8)(i) of this section is notice expressly and actually given, and brought home to the party directly, including service of process personally served on the party and service by mail. Actual notice also includes proof of facts and circumstances that raise the inference that the party received notice including, but not limited to, proof that notice was left at the party's dwelling house or usual place of abode with some person of suitable age and discretion residing therein; or proof that the party signed a return receipt for a hearing notice which had been mailed to the party. It does not include notice published in a newspaper.

(f) Pursuant to 18 U.S.C. 922(y)(3), any individual who has been admitted to the United States under a nonimmigrant visa may receive a waiver from the prohibition contained in paragraph (a)(5)(ii) of this section if the Attorney General approves a petition for the waiver.

§ 478.33 Stolen firearms and ammunition.

No person shall transport or ship in interstate or foreign commerce any stolen firearm or stolen ammunition knowing or having reasonable cause to believe that the firearm or ammunition was stolen, and no person shall receive, possess, conceal, store, barter, sell, or dispose of any stolen firearm or stolen ammunition, or pledge or accept as security for a loan any stolen firearm or stolen ammunition, which is moving as, which is a part of, which constitutes, or which has been shipped or transported in, interstate or foreign commerce, either before or after it was stolen, knowing or having reasonable cause to believe that the firearm or ammunition was stolen.

§ 478.33a Theft of firearms.

No person shall steal or unlawfully take or carry away from the person or the premises of a person who is licensed to engage in the business of importing, manufacturing, or dealing in firearms, any firearm in the licensee's business inventory that has been shipped or transported in interstate or foreign commerce.

§ 478.34 Removed, obliterated, or altered serial number.

No person shall knowingly transport, ship, or receive in interstate or foreign commerce any firearm which has had the importer's or manufacturer's serial number removed, obliterated, or altered, or possess or receive any firearm which has had the importer's or manufacturer's serial number removed, obliterated, or altered and has, at any time, been shipped or transported in interstate or foreign commerce.

§ 478.35 Skeet, trap, target, and similar shooting activities.

Licensing and recordkeeping requirements, including permissible alternate records, for skeet, trap, target, and similar organized activities shall be determined by the Director of Industry Operations on a case by case basis.

§ 478.36 Transfer or possession of machine guns.

No person shall transfer or possess a machine gun except:

(a) A transfer to or by, or possession by or under the authority of, the United States, or any department or agency thereof, or a State, or a department, agency, or political subdivision thereof. (See Part 479 of this chapter); or

(b) Any lawful transfer or lawful possession of a machine gun that was lawfully possessed before May 19, 1986 (See Part 479 of this chapter).

§ 478.37 Manufacture, importation and sale of armor piercing ammunition.

No person shall manufacture or import, and no manufacturer or importer shall sell or deliver, armor piercing ammunition, except:

(a) The manufacture or importation, or the sale or delivery by any manufacturer or importer, of armor piercing ammunition for the use of the United States or any department or agency thereof or any State or any department, agency or political subdivision thereof;

(b) The manufacture, or the sale or delivery by a manufacturer or importer, of armor piercing ammunition for the purpose of exportation; or

(c) The sale or delivery by a manu-

facturer or importer of armor piercing ammunition for the purposes of testing or experimentation as authorized by the Director under the provisions of §478.149.

§ 478.38 Transportation of firearms.

Notwithstanding any other provision of any law or any rule or regulation of a State or any political subdivision thereof, any person who is not otherwise prohibited by this chapter from transporting, shipping, or receiving a firearm shall be entitled to transport a firearm for any lawful purpose from any place where such person may lawfully possess and carry such firearm to any other place where such person may lawfully possess and carry such firearm if, during such transportation the firearm is unloaded, and neither the firearm nor any ammunition being transported is readily accessible or is directly accessible from the passenger compartment of such transporting vehicle: Provided, That in the case of a vehicle without a compartment separate from the driver's compartment the firearm or ammunition shall be contained in a locked container other than the glove compartment or console.

§ 478.39 Assembly of semiautomatic rifles or shotguns.

(a) No person shall assemble a semiautomatic rifle or any shotgun using more than 10 of the imported parts listed in paragraph (c) of this section if the assembled firearm is prohibited from importation under section 925(d)(3) as not being particularly suitable for or readily adaptable to sporting purposes.

(b) The provisions of this section shall not apply to:

(1) The assembly of such rifle or shotgun for sale or distribution by a licensed manufacturer to the United States or any department or agency thereof or to any State or any department, agency, or political subdivision thereof; or

(2) The assembly of such rifle or shotgun for the purposes of testing or experimentation authorized by the Director under the provisions of §478.151; or

(3) The repair of any rifle or shotgun which had been imported into or assembled in the United States prior to November 30, 1990, or the replacement of any part of such firearm.

(c) For purposes of this section, the term imported parts are:

(1) Frames, receivers, receiver castings, forgings or stampings

(2) Barrels

(3) Barrel extensions

(4) Mounting blocks (trunions)

(5) Muzzle attachments

(6) Bolts

(7) Bolt carriers

(8) Operating rods

(9) Gas pistons

(10) Trigger housings

(11) Triggers

(12) Hammers

(13) Sears

(14) Disconnectors

(15) Buttstocks

(16) Pistol grips

(17) Forearms, handguards

(18) Magazine bodies

(19) Followers

(20) Floorplates

§478.39a Reporting theft or loss of firearms.

Each licensee shall report the theft or loss of a firearm from the licensee's inventory (including any firearm which has been transferred from the licensee's inventory to a personal collection and held as a personal firearm for at least 1 year), or from the collection of a licensed collector, within 48 hours after the theft or loss is discovered. Licensees shall report thefts or losses by telephoning 1–888–930–9275 (nationwide toll free number) and by preparing ATF Form 3310.11, Federal Firearms Licensee Theft/Loss Report, in accordance with the instructions on the form. The original of the report shall be forwarded to the office specified thereon, and Copy 1 shall be retained by the licensee as part of the licensee's permanent records. Theft or loss of any firearm shall also be reported to the appropriate local authorities.

(Approved by the Office of Management and Budget under control number 1140–0039)

§ 478.40 Manufacture, transfer, and possession of semiautomatic assault weapons.

Editor's Note:

Section 478.40 is not applicable on or after September 13, 2004.

§ 478.40a Transfer and possession of large capacity ammunition feeding devices.

Editor's Note:

Section 478.40a is not applicable on or after September 13, 2004.

Subpart D—Licenses

§ 478.41 General.

(a) Each person intending to engage in business as an importer or manufacturer of firearms or ammunition, or a dealer in firearms shall, before commencing such business, obtain the license required by this subpart for the business to be operated. Each person who desires to obtain a license as a collector of curios or relics may obtain such a license under the provisions of this subpart.

(b) Each person intending to engage in business as a firearms or ammunition importer or manufacturer, or dealer in firearms shall file an application, with the required fee (see §478.42), with ATF in accordance with the instructions on the form (see §478.44), and, pursuant to §478.47, receive the license required for such business from the Chief, Federal Firearms Licensing Center. Except as provided in §478.50, a license must be obtained for each business and each place at which the applicant is to do business. A license as an importer or manufacturer of firearms or ammunition, or a dealer in firearms shall, subject to the provisions of the Act and other applicable provisions of law, entitle the licensee to transport, ship, and receive firearms and ammunition covered by such license in interstate or foreign commerce and to engage in the business specified by the license, at the location described on the license, and for the period stated on the license. However, it shall not be necessary for a licensed importer or a licensed manufacturer to also obtain a dealer's license in order to engage in business on the licensed premises as a dealer in the same type of firearms authorized by the license to be imported or manufactured. Payment of the license fee as an importer or manufacturer of destructive devices, ammunition for destructive devices or armor piercing ammunition or as a dealer in destructive devices includes the privilege of importing or manufacturing firearms other than destructive devices and ammunition for other than destructive devices or ammunition other than armor piercing ammunition, or dealing in firearms other than destructive devices, as the case may be, by such a licensee at the licensed premises.

(c) Each person seeking the privileges of a collector licensed under this part shall file an application, with the required fee (see §478.42), with ATF in accordance with the instructions on the form (see §478.44), and pursuant to §478.47, receive from the Chief, Federal Firearms Licensing Center, the license covering the collection of curios and relics. A separate license may be obtained for each collection premises, and such license shall, subject to the provisions of the Act and other applicable provisions of law, entitle the licensee to transport, ship, receive, and acquire curios and relics in interstate or foreign commerce, and to make disposition of curios and relics in interstate or foreign commerce, to any other person licensed under the provisions of this part, for the period stated on the license.

(d) The collector license provided by this part shall apply only to transactions related to a collector's activity in acquiring, holding or disposing of curios and relics. A collector's license does not authorize the collector to engage in a business required to be licensed under the Act or this part. Therefore, if the acquisitions and dispositions of curios and relics by a collector bring the collector within the definition of a manufacturer, importer, or dealer under this part, he shall qualify as such. (See also §478.93 of this part.)

§ 478.42 License fees.

Each applicant shall pay a fee for obtaining a firearms license or ammunition license, a separate fee being required for each business or collecting activity at each place of such business or activity, as follows:

(a) For a manufacturer:

(1) Of destructive devices, ammunition for destructive devices or armor piercing ammunition—$1,000 per year.

(2) Of firearms other than destructive devices—$50 per year.

(3) Of ammunition for firearms other than ammunition for destructive devices or armor piercing ammunition—$10 per year.

(b) For an importer:

(1) Of destructive devices, ammunition for destructive devices or armor piercing ammunition—$1,000 per year.

(2) Of firearms other than destructive devices or ammunition for firearms other than destructive devices or ammunition other than armor piercing ammunition—$50 per year.

(c) For a dealer:

(1) In destructive devices—$1,000 per year.

(2) Who is not a dealer in destructive devices—$200 for 3 years, except that the fee for renewal of a valid license shall be $90 for 3 years.

(d) For a collector of curios and relics—$10 per year.

§ 478.43 License fee not refundable.

No refund of any part of the amount paid as a license fee shall be made where the operations of the licensee are, for any reason, discontinued during the period of an issued license. However, the license fee submitted with an application for a license shall be refunded if that application is denied or withdrawn by the applicant prior to being acted upon.

§ 478.44 Original license.

(a) (1) Any person who intends to engage in business as a firearms or ammunition importer or manufacturer, or firearms dealer, or who has not previously been licensed under the provisions of this part to so engage in business, or who has not timely submitted an application for renewal of the previous license issued under this part, must file an application for license, ATF Form 7 (Firearms), in duplicate, with ATF in accordance with the instructions on the form. The application must:

(i) Be executed under the penalties of perjury and the penalties imposed by 18 U.S.C. 924;

(ii) Include a photograph and fingerprints as required in the instructions on the form;

(iii) If the applicant (including, in the case of a corporation, partnership, or association, any individual possessing, directly or indirectly, the power to direct or cause the direction of the management and policies of the corporation, partnership, or association) is an alien who has been admitted to the United States under a nonimmigrant visa, applicable documentation demonstrating that the alien falls within an exception specified in 18 U.S.C. 922(y)(2) (e.g., a hunting license or permit lawfully issued in the United States) or has obtained a waiver as specified in 18 U.S.C. 922(y)(3); and

(iv) Include the appropriate fee in the form of money order or check made payable to the **"Bureau of Alcohol, Tobacco, Firearms, and Explosives"**.

(2) ATF Form 7 may be obtained by contacting the ATF Distribution Center (See §478.21).

(b) Any person who desires to obtain a license as a collector under the Act and this part, or who has not timely submitted an application for renewal of the previous license issued under this part, shall file an application, ATF Form 7CR (Curios and Relics), with ATF in accordance with the instructions on the form. If the applicant (including, in the case of a corporation, partnership, or association, any individual possessing, directly or indirectly, the power to direct or cause the direction of the management and policies of the corporation, partnership, or association) is an alien who has been admitted to the United States under a nonimmigrant visa, the application must include applicable documentation demonstrating that the alien falls within an exception specified in 18 U.S.C. 922(y)(2)(e.g., a hunting license or permit lawfully issued in the United States) or has obtained a waiver as specified in 18 U.S.C. 922(y)

(3). The application must be executed under the penalties of perjury and the penalties imposed by 18 U.S.C. 924. The application shall include the appropriate fee in the form of a money order or check made payable to the Bureau of Alcohol, Tobacco, Firearms, and Explosives. ATF Form 7CR (Curios and Relics) may be obtained by contacting the ATF Distribution Center (See §478.21).

(Paragraphs (a) and (b) approved by the Office of Management and Budget under control number 1140–0060)

§ 478.45 Renewal of license.

If a licensee intends to continue the business or activity described on a license issued under this part during any portion of the ensuing year, the licensee shall, unless otherwise notified in writing by the Chief, Federal Firearms Licensing Center, execute and file with ATF prior to the expiration of the license an application for a license renewal, ATF Form 8 Part II, in accordance with the instructions on the form, and the required fee. If the applicant is an alien who has been admitted to the United States under a nonimmigrant visa, the application must include applicable documentation demonstrating that the alien falls within an exception specified in 18 U.S.C. 922(y)(2) (e.g., a hunting license or permit lawfully issued in the United States) or has obtained a waiver as specified in 18 U.S.C. 922(y)(3). The Chief, Federal Firearms Licensing Center may, in writing, require the applicant for license renewal to also file completed ATF Form 7 or ATF Form 7CR in the manner required by §478.44. In the event the licensee does not timely file an ATF Form 8 Part II, the licensee must file an ATF Form 7 or ATF Form 7CR as required by §478.44, and obtain the required license before continuing business or collecting activity. If an ATF Form 8 Part II is not timely received through the mails, the licensee should so notify the Chief, Federal Firearms Licensing Center.

(Approved by the Office of Management and Budget under control number 1140–0060)

§ 478.46 Insufficient fee.

If an application is filed with an insufficient fee, the application and any fee submitted will be returned to the applicant.

§ 478.47 Issuance of license.

(a) Upon receipt of a properly executed application for a license on ATF Form 7, ATF Form 7CR, or ATF Form 8 Part II, the Chief, Federal Firearms Licensing Center, shall, upon finding through further inquiry or investigation, or otherwise, that the applicant is qualified, issue the appropriate license. Each license shall bear a serial number and such number may be assigned to the licensee to whom issued for so long as the licensee maintains continuity of renewal in the same location (State).

(b) The Chief, Federal Firearms Licensing Center, shall approve a properly executed application for license on ATF Form 7, ATF Form 7CR, or ATF Form 8 Part II, if:

(1) The applicant is 21 years of age or over;

(2) The applicant (including, in the case of a corporation, partnership, or association, any individual possessing, directly or indirectly, the power to direct or cause the direction of the management and policies of the corporation, partnership, or association) is not prohibited under the provisions of the Act from shipping or transporting in interstate or foreign commerce, or possessing in or affecting commerce, any firearm or ammunition, or from receiving any firearm or ammunition which has been shipped or transported in interstate or foreign commerce;

(3) The applicant has not willfully violated any of the provisions of the Act or this part;

(4) The applicant has not willfully failed to disclose any material information required, or has not made any false statement as to any material fact, in connection with his application; and

(5) The applicant has in a State (i) premises from which he conducts business subject to license under the Act or from which he intends to conduct such business within a reasonable period of time, or (ii) in the case of a collector, premises from which he conducts his collecting subject to license under the Act or from which he intends to conduct such collecting within a reasonable period of time.

(c) The Chief, Federal Firearms Licensing Center, shall approve or the Director of Industry Operations shall deny an application for a license within the 60–day period beginning on the date the properly executed application was received: Provided, That when an applicant for license renewal is a person who is, pursuant to the provisions of §478.78, §478.143, or §478.144, conducting business or collecting activity under a previously issued license, action regarding the application will be held in abeyance pending the completion of the proceedings against the applicant's existing license or license application, final determination of the applicant's criminal case, or final action by the Director on an application for relief submitted pursuant to §478.144, as the case may be.

(d) When the Director of Industry Operations or the Chief, Federal Firearms Licensing Center fails to act on an application for a license within the 60–day period prescribed by paragraph (c) of this section, the applicant may file an action under section 1361 of title 28, United States Code, to compel ATF to act upon the application.

§ 478.48 Correction of error on license.

(a) Upon receipt of a license issued under the provisions of this part, each licensee shall examine same to ensure that the information contained thereon is accurate. If the license is incorrect, the licensee shall return the license to the Chief, Federal Firearms Licensing Center, with a statement showing the nature of the error. The Chief, Federal Firearms Licensing Center, shall correct the error, if the error was made in his office, and return the license. However, if the error resulted from information contained in the licensee's application for the license, the Chief, Federal Firearms Licensing Center, shall require the licensee to file an amended application setting forth the correct information and a statement explaining the error contained in the application. Upon receipt of the amended application and a satisfactory explanation of the error, the Chief, Federal Firearms Licensing Center, shall make the correction on the license and return same to the licensee.

(b) When the Chief, Federal Firearms Licensing Center, finds through any means other than notice from the licensee that an incorrect license has been issued, the Chief, Federal Firearms Licensing Center, may require the holder of the incorrect license to (1) return the license for correction, and (2) if the error resulted from information contained in the licensee's application for the license, the Chief, Federal Firearms Licensing Center, shall require the licensee to file an amended application setting forth the correct information, and a statement explaining the error contained in the application. The Chief, Federal Firearms Licensing Center, then shall make the correction on the license and return same to the licensee.

§ 478.49 Duration of license.

The license entitles the person to whom issued to engage in the business or activity specified on the license, within the limitations of the Act and the regulations contained in this part, for a three year period, unless terminated sooner.

§ 478.50 Locations covered by license.

The license covers the class of business or the activity specified in the license at the address specified therein. A separate license must be obtained for each location at which a firearms or ammunition business or activity requiring a license under this part is conducted except:

(a) No license is required to cover a separate warehouse used by the licensee solely for storage of firearms or ammunition if the records required by this part are maintained at the licensed premises served by such warehouse;

(b) A licensed collector may acquire curios and relics at any location, and dispose of curios or relics to any licensee or to other persons who are residents of the State where the collector's license is held and the disposition is made;

(c) A licensee may conduct business at a gun show pursuant to the provision of §478.100; or

(d) A licensed importer, manufacturer, or dealer may engage in the business of dealing in curio or relic firearms with another licensee at any location pursuant to the provisions of §478.100.

§ 478.51 License not transferable.

Licenses issued under this part are not transferable. In the event of the lease, sale, or other transfer of the operations authorized by the license, the successor must obtain the license required by this part prior to commencing such operations. However, for rules on right of succession, see §478.56.

§ 478.52 Change of address.

(a) Licensees may during the term of their current license remove their business or activity to a new location at which they intend regularly to carry on such business or activity by filing an Application for an Amended Federal Firearms License, ATF Form 5300.38, in duplicate, not less than 30 days prior to such removal with the Chief, Federal Firearms Licensing Center. The ATF Form 5300.38 shall be completed in accordance with the instructions on the form. The application must be executed under the penalties of perjury and penalties imposed by 18 U.S.C. 924. The application shall be accompanied by the licensee's original license. The Chief, Federal Firearms Licensing Center, may, in writing, require the applicant for an amended license to also file completed ATF Form 7 or ATF Form 7CR, or portions thereof, in the manner required by §478.44.

(b) Upon receipt of a properly executed application for an amended license, the Chief, Federal Firearms Licensing Center, shall, upon finding through further inquiry or investigation, or otherwise, that the applicant is qualified at the new location, issue the amended license, and return it to the applicant. The license shall be valid for the remainder of the term of the original license. The Chief, Federal Firearms Licensing Center, shall, if the applicant is not qualified, refer the application for amended license to the Director of Industry Operations for denial in accordance with §478.71.

(Approved by the Office of Management and Budget under control number 1140–0040)

§ 478.53 Change in trade name.

A licensee continuing to conduct business at the location shown on his license is not required to obtain a new license by reason of a mere change in trade name under which he conducts his business: Provided, That such licensee furnishes his license for endorsement of such change to the Chief, Federal Firearms Licensing Center within 30 days from the date the licensee begins his business under the new trade name.

§ 478.54 Change of control.

In the case of a corporation or association holding a license under this part, if actual or legal control of the corporation or association changes, directly or indirectly, whether by reason of change in stock ownership or control (in the li-

censed corporation or in any other corporation), by operations of law, or in any other manner, the licensee shall, within 30 days of such change, give written notification thereof, executed under the penalties of perjury, to the Chief, Federal Firearms Licensing Center. Upon expiration of the license, the corporation or association must file a Form 7 (Firearms) as required by §478.44.

§ 478.55 Continuing partnerships.

Where, under the laws of the particular State, the partnership is not terminated on death or insolvency of a partner, but continues until the winding up of the partnership affairs is completed, and the surviving partner has the exclusive right to the control and possession of the partnership assets for the purpose of liquidation and settlement, such surviving partner may continue to operate the business under the license of the partnership. If such surviving partner acquires the business on completion of the settlement of the partnership, he shall obtain a license in his own name from the date of acquisition, as provided in §478.44. The rule set forth in this section shall also apply where there is more than one surviving partner.

§ 478.56 Right of succession by certain persons.

(a) Certain persons other than the licensee may secure the right to carry on the same firearms or ammunition business at the same address shown on, and for the remainder of the term of, a current license. Such persons are:

> **(1)** The surviving spouse or child, or executor, administrator, or other legal representative of a deceased licensee; and

> **(2)** A receiver or trustee in bankruptcy, or an assignee for benefit of creditors.

(b) In order to secure the right provided by this section, the person or persons continuing the business shall furnish the license for that business for endorsement of such succession to the Chief, Federal Firearms Licensing Center, within 30 days from the date on which the successor begins to carry on the business.

§ 478.57 Discontinuance of business.

(a) Where a firearm or ammunition

business is either discontinued or succeeded by a new owner, the owner of the business discontinued or succeeded shall within 30 days thereof furnish to the Chief, Federal Firearms Licensing Center notification of the discontinuance or succession. (See also §478.127.)

Editor's Note:

Paragraphs (b) and (c) of section 478.57 are not applicable on or after September 13, 2004.

§ 478.58 State or other law.

A license issued under this part confers no right or privilege to conduct business or activity contrary to State or other law. The holder of such a license is not by reason of the rights and privileges granted by that license immune from punishment for operating a firearm or ammunition business or activity in violation of the provisions of any State or other law. Similarly, compliance with the provisions of any State or other law affords no immunity under Federal law or regulations.

§ 478.59 Abandoned application.

Upon receipt of an incomplete or improperly executed application on ATF form 7 (5310.12), or ATF Form 8 (5310.11) Part II, the applicant shall be notified of the deficiency in the application. If the application is not corrected and returned within 30 days following the date of notification, the application shall be considered as having been abandoned and the license fee returned.

§ 478.60 Certain continuances of business.

A licensee who furnishes his license to the Chief, Federal Firearms Licensing Center for correction or endorsement in compliance with the provisions contained in this subpart may continue his operations while awaiting its return.

Subpart E—License Proceedings

§ 478.71 Denial of an application for license.

Whenever the Director has reason to believe that an applicant is not qualified to receive a license under the provisions of §478.47, he may issue a notice of denial, on Form 4498, to the applicant. The notice shall set forth the matters of fact and law relied upon in determining that the application should be denied, and shall afford the applicant 15 days from the date of receipt of the notice in which to request a hearing to review the denial. If no request for a hearing is filed within such time, the application shall be disapproved and a copy, so marked, shall be returned to the applicant.

§ 478.72 Hearing after application denial.

If the applicant for an original or renewal license desires a hearing to review the denial of his application, he shall file a request therefor, in duplicate, with the Director of Industry Operations within 15 days after receipt of the notice of denial. The request should include a statement of the reasons therefor. On receipt of the request, the Director of Industry Operations shall, as expeditiously as possible, make the necessary arrangements for the hearing and advise the applicant of the date, time, location, and the name of the officer before whom the hearing will be held. Such notification shall be made not less than 10 days in advance of the date set for the hearing. On conclusion of the hearing and consideration of all relevant facts and circumstances presented by the applicant or his representative, the Director shall render his decision confirming or reversing the denial of the application. If the decision is that the denial should stand, a certified copy of the Director's findings and conclusions shall be furnished to the applicant with a final notice of denial, Form 5300.13. A copy of the application, marked **"Disapproved,"** will be returned to the applicant. If the decision is that the license applied for should be issued, the applicant shall be so notified, in writing, and the license shall be issued as provided by §478.47.

§ 478.73 Notice of revocation, suspension, or imposition of civil fine.

(a) Basis for action. Whenever the Director has reason to believe that a licensee has willfully violated any provision of the Act or this part, a notice of revocation of the license, ATF Form 4500, may be issued. In addition, a notice of revocation, suspension, or imposition of a civil fine may be issued on ATF Form 4500 whenever the Director has reason to believe that a licensee has knowingly transferred a firearm to an unlicensed person and knowingly failed to comply with the requirements of 18 U.S.C. 922(t)(1) with respect to the transfer and, at the time that the transferee most recently proposed the transfer, the national instant criminal background check system was operating and information was available to the system demonstrating that the transferee's receipt of a firearm would violate 18 U.S.C. 922(g) or 922(n) or State law.

(b) Issuance of notice. The notice shall set forth the matters of fact constituting the violations specified, dates, places, and the sections of law and regulations violated. The Director shall afford the licensee 15 days from the date of receipt of the notice in which to request a hearing prior to suspension or revocation of the license, or imposition of a civil fine. If the licensee does not file a timely request for a hearing, the Director shall issue a final notice of suspension or revocation and/or imposition of a civil fine on ATF Form 5300.13, as provided in §478.74.

§ 478.74 Request for hearing after notice of suspension, revocation, or imposition of civil fine.

If a licensee desires a hearing after receipt of a notice of suspension or revocation of a license, or imposition of a civil fine, the licensee shall file a request, in duplicate, with the Director of Industry Operations within 15 days after receipt of the notice of suspension or revocation of a license, or imposition of a civil fine. On receipt of such request, the Director of Industry Operations shall, as expeditiously as possible, make necessary arrangements for the hearing and advise the licensee of the date, time, location and the name of the officer before whom the hearing will be held. Such notification shall be made no less than 10 days in advance of the date set for the hearing. On conclusion of the hearing and consideration of all the relevant presentations made by the licensee or the licensee's representative, the Director shall render a decision and shall prepare a brief summary of the findings and conclusions on which the decision is based. If the decision is that the license should be revoked, or, in actions under 18 U.S.C. 922(t)(5), that the license should be revoked or suspended, and/or that a civil fine should be imposed, a certified copy of the summary shall be furnished to the licensee with the final notice of revocation, suspension, or imposition of a civil fine on ATF Form 5300.13. If the decision is that the license should not be revoked, or in actions under 18 U.S.C. 922(t)(5), that the license should not be revoked or suspended, and a civil fine should not be imposed, the licensee shall be notified in writing.

§ 478.75 Service on applicant or licensee.

All notices and other documents required to be served on an applicant or licensee under this subpart shall be served by certified mail or by personal delivery. Where service is by certified mail, a signed duplicate original copy of the formal document shall be mailed, with return receipt requested, to the applicant or licensee at the address stated in his application or license, or at his last known address. Where service is by personal delivery, a signed duplicate original copy of the formal document shall be delivered to the applicant or licensee, or, in the case of a corporation, partnership, or association, by delivering it to an officer, manager, or general agent thereof, or to its attorney of record.

§ 478.76 Representation at a hearing.

An applicant or licensee may be represented by an attorney, certified public accountant, or other person recognized to practice before the Bureau of Alcohol, Tobacco, Firearms, Explosives as provided in 31 CFR Part 8 (Practice Before the Bureau of Alcohol, Tobacco and Firearms), if he has otherwise complied with the applicable requirements of 26 CFR 601.521 through 601.527 (conference and practice requirements for alcohol, tobacco, and firearms activities) of this chapter. The Director may be represented in proceedings by an attorney in the office of the Chief Counsel who is authorized to execute and file motions, briefs and other papers in the proceeding, on behalf of the Director, in his own name as **"Attorney for the Government."**

§ 478.77 Designated place of hearing.

The designated place of the hearing shall be a location convenient to the aggrieved party.

§ 478.78 Operations by licensee after notice.

In any case where denial, suspension, or revocation proceedings are pending before the Bureau of Alcohol, Tobacco, Firearms and, Explosives, or notice of denial, suspension, or revocation has been served on the licensee and he has filed timely request for a hearing, the license in the possession of the licensee shall remain in effect even though such license has expired, or the suspension or revocation date specified in the notice of revocation on Form 4500 served on the licensee has passed: **Provided**, That with respect to a license that has expired, the licensee has timely filed an application for the renewal of his license. If a licensee is dissatisfied with a post hearing decision revoking or suspending the license or denying the application or imposing a civil fine, as the case may be, he may, pursuant to 18 U.S.C. 923(f)(3), within 60 days after receipt of the final notice denying the application or revoking or suspending the license or imposing a civil fine, file a petition for judicial review of such action. Such petition should be filed with the U.S. district court for the district in which the applicant or licensee resides or has his principal place of business. In such case, when the Director finds that justice so requires, he may postpone the effective date of suspension or revocation of a license or authorize continued operations under the expired license, as applicable, pending judicial review.

Subpart F—Conduct of Business

§ 478.91 Posting of license.

Any license issued under this part shall be kept posted and kept available for inspection on the premises covered by the license.

§ 478.92 How must licensed manufacturers and licensed importers identify firearms, armor piercing ammunition, and large capacity ammunition feeding devices?

(a) (1) **Firearms.** You, as a licensed manufacturer or licensed importer of firearms, must legibly identify each firearm manufactured or imported as follows:

(i) By engraving, casting, stamping (impressing), or otherwise conspicuously placing or causing to be engraved, cast, stamped (impressed) or placed on the frame or receiver thereof an individual serial number. The serial number must be placed in a manner not susceptible of being readily obliterated, altered, or removed, and must not duplicate any serial number placed by you on any other firearm. For firearms manufactured or imported on and after January 30, 2002, the engraving, casting, or stamping (impressing) of the serial number must be to a minimum depth of .003 inch and in a print size no smaller than 1/16 inch; and

(ii) By engraving, casting, stamping (impressing), or otherwise conspicuously placing or causing to be engraved, cast, stamped (impressed) or placed on the frame, receiver, or barrel thereof certain additional information. This information must be placed in a manner not susceptible of being readily obliterated, altered, or removed. For firearms manufactured or imported on and after January 30, 2002, the engraving, casting, or stamping (impressing) of this information must be to a minimum depth of .003 inch. The additional information includes:

(A) The model, if such designation has been made;

(B) The caliber or gauge;

(C) Your name (or recognized abbreviation) and also, when applicable, the name of the foreign manufacturer;

(D) In the case of a domestically made firearm, the city and State (or recognized abbreviation thereof) where you as the manufacturer maintain your place of business; and

(E) In the case of an imported firearm, the name of the country in which it was manufactured and the city and State (or recognized abbreviation thereof) where you as the importer maintain your place of business. For additional requirements relating to imported firearms, see Customs regulations at 19 CFR part 134.

(2) **Firearm frames or receivers.** A firearm frame or receiver that is not a component part of a complete weapon at the time it is sold, shipped, or otherwise disposed of by you must be identified as required by this section.

(4) **Exceptions.**

(i) **Alternate means of identification.** The Director may authorize other means of identification upon receipt of a letter application from you, submitted in duplicate, showing that such other identification is reasonable and will not hinder the effective administration of this part.

(ii) **Destructive devices.** In the case of a destructive device, the Director may authorize other means of identifying that weapon upon receipt of a letter application from you, submitted in duplicate, showing that engraving, casting, or stamping (impressing) such a weapon would be dangerous or impracticable.

(iii) **Machine guns, silencers, and parts.** Any part defined as a machine gun, firearm muffler, or firearm silencer in §478.11, that is not a component part of a complete weapon at the time it is sold, shipped, or otherwise disposed of by you, must be identified as required by this section. The Director may authorize other means of identification of parts defined as machine guns other than frames or receivers and parts defined as mufflers or silencers upon receipt of a letter application from you, submitted in duplicate, showing that such other identification is reasonable and will not hinder the effective administration of this part.

(5) **Measurement of height and depth of markings.** The depth of all markings required by this section will be measured from the flat surface of the metal and not the peaks or ridges. The height of serial numbers required by paragraph (a)(1)(i) of this section will be measured as the distance between the latitudinal ends of the character impression bottoms (bases).

(b) **Armor piercing ammunition**

(1) **Marking of ammunition.** Each licensed manufacturer or licensed importer of armor piercing ammunition shall identify such ammunition by means of painting, staining or dying the exterior of the projectile with an opaque black coloring. This coloring must completely cover the point of the projectile and at least 50 percent of that portion of the projectile which is visible when the projectile is loaded into a cartridge case.

(2) **Labeling of packages.** Each licensed manufacturer or licensed importer of armor piercing ammunition shall clearly and conspicuously label each package in which armor piercing ammunition is contained, e.g., each box, carton, case, or other container. The label shall include the words **"ARMOR PIERCING"** in block letters at least 1/4 inch in height. The lettering shall be located on the exterior surface of the package which contains information concerning

the caliber or gauge of the ammunition. There shall also be placed on the same surface of the package in block lettering at least ⅛ inch in height the words **"FOR GOVERNMENTAL ENTITIES OR EXPORTATION ONLY."** The statements required by this subparagraph shall be on a contrasting background.

(Approved by the Office of Management and Budget under control number 1140–0050)

Editor's Note:

Paragraphs (a)(3) and (c) of section 478.92 are not applicable on or after September 13, 2004

§ 478.93 Authorized operations by a licensed collector.

The license issued to a collector of curios or relics under the provisions of this part shall cover only transactions by the licensed collector in curios and relics. The collector's license is of no force or effect and a licensed collector is of the same status under the Act and this part as a nonlicensee with respect to (a) any acquisition or disposition of firearms other than curios or relics, or any transportation, shipment, or receipt of firearms other than curios or relics in interstate or foreign commerce, and (b) any transaction with a nonlicensee involving any firearm other than a curio or relic. (See also §478.50.) A collectors license is not necessary to receive or dispose of ammunition, and a licensed collector is not precluded by law from receiving or disposing of armor piercing ammunition. However, a licensed collector may not dispose of any ammunition to a person prohibited from receiving or possessing ammunition (see §478.99(c)). Any licensed collector who disposes of armor piercing ammunition must record the disposition as required by §478.125 (a) and (b).

§ 478.94 Sales or deliveries between licensees.

A licensed importer, licensed manufacturer, or licensed dealer selling or otherwise disposing of firearms, and a licensed collector selling or otherwise disposing of curios or relics, to another licensee shall verify the identity and licensed status of the transferee prior to making the transaction. Verification shall be established by the transferee furnishing to the transferor a certified copy of the transferee's license and by such other means as the transferor deems necessary: **Provided**, That it shall not be required (a) for a transferee who

has furnished a certified copy of its license to a transferor to again furnish such certified copy to that transferor during the term of the transferee's current license, (b) for a licensee to furnish a certified copy of its license to another licensee if a firearm is being returned either directly or through another licensee to such licensee and (c) for licensees of multilicensed business organizations to furnish certified copies of their licenses to other licensed locations operated by such organization: Provided further, That a multilicensed business organization may furnish to a transferor, in lieu of a certified copy of each license, a list, certified to be true, correct and complete, containing the name, address, license number, and the date of license expiration of each licensed location operated by such organization, and the transferor may sell or otherwise dispose of firearms as provided by this section to any licensee appearing on such list without requiring a certified copy of a license therefrom. A transferor licensee who has the certified information required by this section may sell or dispose of firearms to a licensee for not more than 45 days following the expiration date of the transferee's license.

(Approved by the Office of Management and Budget under control number 1140–0032)

§ 478.95 Certified copy of license.

The license furnished to each person licensed under the provisions of this part contains a purchasing certification statement. This original license may be reproduced and the reproduction then certified by the licensee for use pursuant to §178.94. If the licensee desires an additional copy of the license for certification (instead of making a reproduction of the original license), the licensee may submit a request, in writing, for a certified copy or copies of the license to the Chief, Federal Firearms Licensing Center. The request must set forth the name, trade name (if any) and address of the licensee, and the number of license copies desired. There is a charge of $1 for each copy. The fee paid for copies of the license must accompany the request for copies. The fee may be paid by (a) cash, or (b) money order or check made payable to the Bureau of Alcohol, Tobacco, Firearms, and Explosives.

(Approved by the Office of Management and Budget under control number 1140–0032)

§ 478.96 Out–of–State and mail order sales.

(a) The provisions of this section shall

apply when a firearm is purchased by or delivered to a person not otherwise prohibited by the Act from purchasing or receiving it.

(b) A licensed importer, licensed manufacturer, or licensed dealer may sell a firearm that is not subject to the provisions of §478.102(a) to a nonlicensee who does not appear in person at the licensee's business premises if the nonlicensee is a resident of the same State in which the licensee's business premises are located, and the nonlicensee furnishes to the licensee the firearms transaction record, Form 4473, required by §478.124. The nonlicensee shall attach to such record a true copy of any permit or other information required pursuant to any statute of the State and published ordinance applicable to the locality in which he resides. The licensee shall prior to shipment or delivery of the firearm, forward by registered or certified mail (return receipt requested) a copy of the record, Form 4473, to the chief law enforcement officer named on such record, and delay shipment or delivery of the firearm for a period of at least 7 days following receipt by the licensee of the return receipt evidencing delivery of the copy of the record to such chief law enforcement officer, or the return of the copy of the record to him due to the refusal of such chief law enforcement officer to accept same in accordance with U.S. Postal Service regulations. The original Form 4473, and evidence of receipt or rejection of delivery of the copy of the Form 4473 sent to the chief law enforcement officer shall be retained by the licensee as a part of the records required of him to be kept under the provisions of subpart H of this part.

(c) (1) A licensed importer, licensed manufacturer, or licensed dealer may sell or deliver a rifle or shotgun, and a licensed collector may sell or deliver a rifle or shotgun that is a curio or relic to a nonlicensed resident of a State other than the State in which the licensee's place of business is located if—

(i) The purchaser meets with the licensee in person at the licensee's premises to accomplish the transfer, sale, and delivery of the rifle or shotgun;

(ii) The licensed importer, licensed manufacturer, or licensed dealer complies with the provisions of §478.102;

(iii) The purchaser furnishes to the licensed importer, licensed

51

manufacturer, or licensed dealer the firearms transaction record, Form 4473, required by §478.124; and

(iv) The sale, delivery, and receipt of the rifle or shotgun fully comply with the legal conditions of sale in both such States.

(2) For purposes of paragraph (c) of this section, any licensed manufacturer, licensed importer, or licensed dealer is presumed, in the absence of evidence to the contrary, to have had actual knowledge of the State laws and published ordinances of both such States.

(Approved by the Office of Management and Budget under control number 1140–0021)

§ 478.97 Loan or rental of firearms.

(a) A licensee may lend or rent a firearm to any person for temporary use off the premises of the licensee for lawful sporting purposes: **Provided**, That the delivery of the firearm to such person is not prohibited by §478.99(b) or §478.99(c), the licensee complies with the requirements of §478.102, and the licensee records such loan or rental in the records required to be kept by him under Subpart H of this part.

(b) A club, association, or similar organization temporarily furnishing firearms (whether by loan, rental, or otherwise) to participants in a skeet, trap, target, or similar shooting activity for use at the time and place such activity is held does not, unattended by other circumstances, cause such club, association, or similar organization to be engaged in the business of a dealer in firearms or as engaging in firearms transactions. Therefore, licensing and recordkeeping requirements contained in this part pertaining to firearms transactions would not apply to this temporary furnishing of firearms for use on premises on which such an activity is conducted.

§ 478.98 Sales or deliveries of destructive devices and certain firearms.

The sale or delivery by a licensee of any destructive device, machine gun, short–barreled shotgun, or short–barreled rifle, to any person other than another licensee who is licensed under this part to deal in such device or firearm, is prohibited unless the person to receive such device or firearm furnishes to the licensee a sworn statement setting forth

(a) The reasons why there is a reasonable necessity for such person to purchase or otherwise acquire the device or weapon; and

(b) That such person's receipt or possession of the device or weapon would be consistent with public safety. Such sworn statement shall be made on the application to transfer and register the firearm required by Part 479 of this chapter. The sale or delivery of the device or weapon shall not be made until the application for transfer is approved by the Director and returned to the licensee (transferor) as provided in Part 479 of this chapter.

§ 478.99 Certain prohibited sales or deliveries.

(a) **Interstate sales or deliveries.** A licensed importer, licensed manufacturer, licensed dealer, or licensed collector shall not sell or deliver any firearm to any person not licensed under this part and who the licensee knows or has reasonable cause to believe does not reside in (or if a corporation or other business entity, does not maintain a place of business in) the State in which the licensee's place of business or activity is located: Provided, That the foregoing provisions of this paragraph (1) shall not apply to the sale or delivery of a rifle or shotgun (curio or relic, in the case of a licensed collector) to a resident of a State other than the State in which the licensee's place of business or collection premises is located if the requirements of §478.96(c) are fully met, and (2) shall not apply to the loan or rental of a firearm to any person for temporary use for lawful sporting purposes (see §478.97).

(b) **Sales or deliveries to under-aged persons.** A licensed importer, licensed manufacturer, licensed dealer, or licensed collector shall not sell or deliver (1) any firearm or ammunition to any individual who the importer, manufacturer, dealer, or collector knows or has reasonable cause to believe is less than 18 years of age, and, if the firearm, or ammunition, is other than a shotgun or rifle, or ammunition for a shotgun or rifle, to any individual who the importer, manufacturer, dealer, or collector knows or has reasonable cause to believe is less than 21 years of age, or (2) any firearm to any person in any State where the purchase or possession by such person of such firearm would be in violation of any State law or any published ordi-

nance applicable at the place of sale, delivery, or other disposition, unless the importer, manufacturer, dealer, or collector knows or has reasonable cause to believe that the purchase or possession would not be in violation of such State law or such published ordinance.

(c) **Sales or deliveries to prohibited categories of persons.** A licensed manufacturer, licensed importer, licensed dealer, or licensed collector shall not sell or otherwise dispose of any firearm or ammunition to any person knowing or having reasonable cause to believe that such person:

(1) Is, except as provided by §478.143, under indictment for, or, except as provided by §478.144, has been convicted in any court of a crime punishable by imprisonment for a term exceeding 1 year;

(2) Is a fugitive from justice;

(3) Is an unlawful user of or addicted to any controlled substance (as defined in section 102 of the Controlled Substance Act, 21 U.S.C. 802);

(4) Has been adjudicated as a mental defective or has been committed to any mental institution;

(5) Is an alien illegally or unlawfully in the United States or, except as provided in §478.32(f), is an alien who has been admitted to the United States under a nonimmigrant visa: Provided, That the provisions of this paragraph (c)(5) do not apply to any alien who has been lawfully admitted to the United States under a nonimmigrant visa if that alien is—

(i) Admitted to the United States for lawful hunting or sporting purposes or is in possession of a hunting license or permit lawfully issued in the United States;

(ii) An official representative of a foreign government who is either accredited to the United States Government or the Government's mission to an international organization having its headquarters in the United States or en route to or from another country to which that alien is accredited. This exception only applies if the firearm or ammunition is shipped, transported, possessed, or received in the representative's official capacity;

(iii) An official of a foreign government or a distinguished foreign visitor who has been so designated by the Department of State. This exception only applies if the firearm or ammunition is shipped, transported, possessed, or received in the official's or visitor's official capacity, except if the visitor is a private individual who does not have an official capacity; or

(iv) A foreign law enforcement officer of a friendly foreign government entering the United States on official law enforcement business;

(6) Has been discharged from the Armed Forces under dishonorable conditions;

(7) Who, having been a citizen of the United States, has renounced citizenship;

(8) Is subject to a court order that restrains such person from harassing, stalking, or threatening an intimate partner of such person or child of such intimate partner or person, or engaging in other conduct that would place an intimate partner in reasonable fear of bodily injury to the partner or child, except that this paragraph shall only apply to a court order that—

(i) Was issued after a hearing of which such person received actual notice, and at which such person had the opportunity to participate; and

(ii)(A) Includes a finding that such person represents a credible threat to the physical safety of such intimate partner or child; or

(B) By its terms explicitly prohibits the use, attempted use, or threatened use of physical force against such intimate partner or child that would reasonably be expected to cause bodily injury, or

(9) Has been convicted of a misdemeanor crime of domestic violence.

(d) Manufacture, importation, and sale of armor piercing ammunition by licensed importers and licensed manufacturers. A licensed importer or licensed manufacturer shall not import or manufacture armor piercing ammunition or sell or deliver such ammunition, except:

(1) For use of the United States or any department or agency thereof or any State or any department, agency, or political subdivision thereof;

(2) For the purpose of exportation; or

(3) For the purpose of testing or experimentation authorized by the Director under the provisions of §478.149.

(e) Transfer of armor piercing ammunition by licensed dealers. A licensed dealer shall not willfully transfer armor piercing ammunition: Provided, That armor piercing ammunition received and maintained by the licensed dealer as business inventory prior to August 28, 1986, may be transferred to any department or agency of the United States or any State or political subdivision thereof if a record of such ammunition is maintained in the form and manner prescribed by §478.125(c). Any licensed dealer who violates this paragraph is subject to license revocation. See subpart E of this part. For purposes of this paragraph, the Director shall furnish each licensed dealer information defining which projectiles are considered armor piercing. Such information may not be all–inclusive for purposes of the prohibition on manufacture, importation, or sale or delivery by a manufacturer or importer of such ammunition or 18 U.S.C. 929 relating to criminal misuse of such ammunition.

§ 478.100 Conduct of business away from licensed premises.

(a)(1) A licensee may conduct business temporarily at a gun show or event as defined in paragraph (b) if the gun show or event is located in the same State specified on the license: Provided, That such business shall not be conducted from any motorized or towed vehicle. The premises of the gun show or event at which the licensee conducts business shall be considered part of the licensed premises. Accordingly, no separate fee or license is required for the gun show or event locations. However, licensees shall comply with the provisions of §478.91 relating to posting of licenses (or a copy thereof) while conducting business at the gun show or event.

(2) A licensed importer, manufacturer, or dealer may engage in the business of dealing in curio or relic firearms with another licensee at any location.

(b) A gun show or an event is a function sponsored by any national, State, or local organization, devoted to the collection, competitive use, or other sporting use of firearms, or an organization or association that sponsors functions devoted to the collection, competitive use, or other sporting use of firearms in the community.

(c) Licensees conducting business at locations other than the premises specified on their license under the provisions of paragraph (a) of this section shall maintain firearms records in the form and manner prescribed by subpart H of this part. In addition, records of firearms transactions conducted at such locations shall include the location of the sale or other disposition, be entered in the acquisition and disposition records of the licensee, and retained on the premises specified on the license.

§ 478.101 Record of transactions.

Every licensee shall maintain firearms and armor piercing ammunition records in such form and manner as is prescribed by subpart H of this part.

§ 478.102 Sales or deliveries of firearms on and after November 30, 1998.

(a) Background check. Except as provided in paragraph (d) of this section, a licensed importer, licensed manufacturer, or licensed dealer (the licensee) shall not sell, deliver, or transfer a firearm to any other person who is not licensed under this part unless the licensee meets the following requirements:

(1) Before the completion of the transfer, the licensee has contacted NICS;

(2)(i) NICS informs the licensee that it has no information that receipt of the firearm by the transferee would be in violation of Federal or State law and provides the licensee with a unique identification number; or

(ii) Three business days (meaning days on which State offices are open) have elapsed from the date the licensee contacted NICS and NICS has not notified the licensee that receipt of the firearm by the transferee would be in violation of law; and

(3) The licensee verifies the identity of the transferee by examining the identification document presented in accordance with the provisions of §478.124(c).

Example for paragraph *(a)*. A licensee contacts NICS on Thursday, and gets a **"delayed"** response. The licensee does not get a further response from NICS. If State offices are not open on Saturday and Sunday, 3 business days would have elapsed on the following Tuesday. The licensee may transfer the firearm on the next day, Wednesday.

(b) Transaction number. In any transaction for which a licensee receives a transaction number from NICS (which shall include either a NICS transaction number or, in States where the State is recognized as a point of contact for NICS checks, a State transaction number), such number shall be recorded on a firearms transaction record, Form 4473, which shall be retained in the records of the licensee in accordance with the provisions of §478.129. This applies regardless of whether the transaction is approved or denied by NICS, and regardless of whether the firearm is actually transferred.

(c) Time limitation on NICS checks. A NICS check conducted in accordance with paragraph (a) of this section may be relied upon by the licensee only for use in a single transaction, and for a period not to exceed 30 calendar days from the date that NICS was initially contacted. If the transaction is not completed within the 30–day period, the licensee shall initiate a new NICS check prior to completion of the transfer.

Example 1 for paragraph (c). A purchaser completes the Form 4473 on December 15, 1998, and a NICS check is initiated by the licensee on that date. The licensee is informed by NICS that the information available to the system does not indicate that receipt of the firearm by the transferee would be in violation of law, and a unique identification number is provided. However, the State imposes a 7–day waiting period on all firearms transactions, and the purchaser does not return to pick up the firearm until January 22, 1999. The licensee must conduct another NICS check before transferring the firearm to the purchaser.

Example 2 for paragraph (c). A purchaser completes the Form 4473 on January 25, 1999, and arranges for the purchase of a single firearm. A NICS check is initiated by the licensee on that date. The licensee is informed by NICS that the information available to the system does not indicate that receipt of the firearm by the transferee would be in violation of law, and a unique identification

number is provided. The State imposes a 7–day waiting period on all firearms transactions, and the purchaser returns to pick up the firearm on February 15, 1999. Before the licensee executes the Form 4473, and the firearm is transferred, the purchaser decides to purchase an additional firearm. The transfer of these two firearms is considered a single transaction; accordingly, the licensee may add the second firearm to the Form 4473, and transfer that firearm without conducting another NICS check.

Example 3 for paragraph (c). A purchaser completes a Form 4473 on February 15, 1999. The licensee receives a unique identification number from NICS on that date, the Form 4473 is executed by the licensee, and the firearm is transferred. On February 20, 1999, the purchaser returns to the licensee's premises and wishes to purchase a second firearm. The purchase of the second firearm is a separate transaction; thus, a new NICS check must be initiated by the licensee.

(d) Exceptions to NICS check. The provisions of paragraph (a) of this section shall not apply if—

(1) The transferee has presented to the licensee a valid permit or license that—

(i) Allows the transferee to possess, acquire, or carry a firearm;

(ii) Was issued not more than 5 years earlier by the State in which the transfer is to take place; and

(iii) The law of the State provides that such a permit or license is to be issued only after an authorized government official has verified that the information available to such official does not indicate that possession of a firearm by the transferee would be in violation of Federal, State, or local law: Provided, That on and after November 30, 1998, the information available to such official includes the NICS;

(2) The firearm is subject to the provisions of the National Firearms Act and has been approved for transfer under 27 CFR part 479; or

(3) On application of the licensee, in accordance with the provisions of §478.150, the Director has certified that compliance with paragraph (a)(1) of this section is impracticable.

(e) The document referred to in paragraph (d)(1) of this section (or a copy thereof) shall be retained or the required information from the document shall be recorded on the firearms transaction record in accordance with the provisions of §478.131.

(Approved by the Office of Management and Budget under control number 1140–0045)

§ 478.103 Posting of signs and written notification to purchasers of handguns.

(a) Each licensed importer, manufacturer, dealer, or collector who delivers a handgun to a nonlicensee shall provide such nonlicensee with written notification as described in paragraph (b) of this section.

(b) The written notification (ATF I 5300.2) required by paragraph (a) of this section shall state as follows:

(1) The misuse of handguns is a leading contributor to juvenile violence and fatalities.

(2) Safely storing and securing firearms away from children will help prevent the unlawful possession of handguns by juveniles, stop accidents, and save lives.

(3) Federal law prohibits, except in certain limited circumstances, anyone under 18 years of age from knowingly possessing a handgun, or any person from transferring a handgun to a person under 18.

(4) A knowing violation of the prohibition against selling, delivering, or otherwise transferring a handgun to a person under the age of 18 is, under certain circumstances, punishable by up to 10 years in prison.

FEDERAL LAW

The Gun Control Act of 1968, 18 U.S.C. Chapter 44, provides in pertinent part as follows:

18 U.S.C. 922(x)

(x)(1) It shall be unlawful for a person to sell, deliver, or otherwise transfer to a person who the transferor knows or has reasonable cause to believe is a juvenile—

(A) a handgun; or

(B) ammunition that is suitable for use only in a handgun.

(2) It shall be unlawful for any person who is a juvenile to knowingly possess—

(A) a handgun; or

(B) ammunition that is suitable for use only in a handgun.

(3) This subsection does not apply to—

(A) a temporary transfer of a handgun or ammunition to a juvenile or to the possession or use of a handgun or ammunition by a juvenile if the handgun and ammunition are possessed and used by the juvenile—

(i) in the course of employment, in the course of ranching or farming related to activities at the residence of the juvenile (or on property used for ranching or farming at which the juvenile, with the permission of the property owner or lessee, is performing activities related to the operation of the farm or ranch), target practice, hunting, or a course of instruction in the safe and lawful use of a handgun;

(ii) with the prior written consent of the juvenile's parent or guardian who is not prohibited by Federal, State, or local law from possessing a firearm, except—

(I) during transportation by the juvenile of an unloaded handgun in a locked container directly from the place of transfer to a place at which an activity described in clause (i) is to take place and transportation by the juvenile of that handgun, unloaded and in a locked container, directly from the place at which such an activity took place to the transferor; or

(II) with respect to ranching or farming activities as described in clause (i) a juvenile may possess and use a handgun or ammunition with the prior written approval of the juvenile's parent or legal guardian and at the direction of an adult who is not prohib-

ited by Federal, State, or local law from possessing a firearm;

(iii) the juvenile has the prior written consent in the juvenile's possession at all times when a handgun is in the possession of the juvenile; and

(iv) in accordance with State and local law;

(B) a juvenile who is a member of the Armed Forces of the United States or the National Guard who possesses or is armed with a handgun in the line of duty;

(C) a transfer by inheritance of title (but not possession) of a handgun or ammunition to a juvenile; or

(D) the possession of a handgun or ammunition by a juvenile taken in defense of the juvenile or other persons against an intruder into the residence of the juvenile or a residence in which the juvenile is an invited guest.

(4) A handgun or ammunition, the possession of which is transferred to a juvenile in circumstances in which the transferor is not in violation of this subsection shall not be subject to permanent confiscation by the Government if its possession by the juvenile subsequently becomes unlawful because of the conduct of the juvenile, but shall be returned to the lawful owner when such handgun or ammunition is no longer required by the Government for the purposes of investigation or prosecution.

(5) For purposes of this subsection, the term **"juvenile"** means a person who is less than 18 years of age.

(6)(A) In a prosecution of a violation of this subsection, the court shall require the presence of a juvenile defendant's parent or legal guardian at all proceedings.

(B) The court may use the contempt power to enforce subparagraph (A).

(C) The court may excuse attendance of a parent or legal guardian of a juvenile defendant at a proceeding in a prosecution of a violation of this subsection for good cause shown.

18 U.S.C. 924(a)(6)

(6)(A)(i) A juvenile who violates section 922(x) shall be fined under this title, imprisoned not more than 1 year, or both, except that a juvenile described in clause (ii) shall be sentenced to probation on appropriate conditions and shall not be incarcerated unless the juvenile fails to comply with a condition of probation.

(ii) A juvenile is described in this clause if—

(I) the offense of which the juvenile is charged is possession of a handgun or ammunition in violation of section 922(x)(2); and

(II) the juvenile has not been convicted in any court of an offense (including an offense under section 922(x) or a similar State law, but not including any other offense consisting of conduct that if engaged in by an adult would not constitute an offense) or adjudicated as a juvenile delinquent for conduct that if engaged in by an adult would constitute an offense.

(B) A person other than a juvenile who knowingly violates section 922(x)—

(i) shall be fined under this title, imprisoned not more than 1 year, or both; and

(ii) if the person sold, delivered, or otherwise transferred a handgun or ammunition to a juvenile knowing or having reasonable cause to know that the juvenile intended to carry or otherwise possess or discharge or otherwise use the handgun or ammunition in the commission of a crime of violence, shall be fined under this title, imprisoned not more than 10 years, or both.

(c) This written notification shall be delivered to the nonlicensee on ATF I 5300.2, or in the alternative, the same written notification may be delivered to the nonlicensee on another type of written notification, such as a manufacturer's or importer's brochure accompanying the handgun; a manufacturer's or importer's operational manual accompanying the handgun; or a sales receipt or invoice applied to the handgun package or container delivered to a nonlicensee. Any written notification delivered to a nonlicensee other than on ATF I 5300.2 shall include the language set forth in paragraph (b) of this section in its entirety. Any written notification other than ATF I 5300.2 shall be legible, clear, and conspicuous, and the required language shall appear in type size no smaller than 10–point type.

(d) Except as provided in paragraph (f) of this section, each licensed importer, manufacturer, or dealer who delivers a handgun to a nonlicensee shall display at its licensed premises (including temporary business locations at gun shows) a sign as described in paragraph (e) of this section. The sign shall be displayed where customers can readily see it. Licensed importers, manufacturers, and dealers will be provided with such signs by ATF. Replacement signs may be requested from the ATF Distribution Center.

(e) The sign (ATF I 5300.1) required by paragraph (d) of this section shall state as follows:

(1) The misuse of handguns is a leading contributor to juvenile violence and fatalities.

(2) Safely storing and securing firearms away from children will help prevent the unlawful possession of handguns by juveniles, stop accidents, and save lives.

(3) Federal law prohibits, except in certain limited circumstances, anyone under 18 years of age from knowingly possessing a handgun, or any person from transferring a handgun to a person under 18.

(4) A knowing violation of the prohibition against selling, delivering, or otherwise transferring a handgun to a person under the age of 18 is, under certain circumstances, punishable by up to 10 years in prison.

Editor's Note:

ATF I 5300.2 provides the complete language of the statutory prohibitions and exceptions provided in 18 U.S.C. 922(x) and the penalty provisions of 18 U.S.C. 924(a) (6). The Federal firearms licensee posting this sign will provide you with a copy of this publication upon request. Requests for additional copies of ATF I 5300.2 should be submitted to the ATF Distribution Center (http://www.atf.gov) or by calling (703) 870-7526 or (703) 870-7528.

(f) The sign required by paragraph (d) of this section need not be posted on the premises of any licensed importer, manufacturer, or dealer whose only dispositions of handguns to nonlicensees are to nonlicensees who do not appear at the licensed premises and the dispositions otherwise comply with the provisions of this part.

Subpart G—Importation

§ 478.111 General.

(a) Section 922(a)(3) of the Act makes it unlawful, with certain exceptions not pertinent here, for any person other than a licensee to transport into or receive in the State where the person resides any firearm purchased or otherwise obtained by the person outside of that State. However, section 925(a) (4) provides a limited exception for the transportation, shipment, receipt or importation of certain firearms and ammunition by certain members of the United States Armed Forces. Section 922(1) of the Act makes it unlawful for any person knowingly to import or bring into the United States or any possession thereof any firearm or ammunition except as provided by section 925(d) of the Act, which section provides standards for importing or bringing firearms or ammunition into the United States. Section 925(d) also provides standards for importing or bringing firearm barrels into the United States. Accordingly, no firearm, firearm barrel, or ammunition may be imported or brought into the United States except as provided by this part.

(b) Where a firearm, firearm barrel, or ammunition is imported and the authorization for importation required by this subpart has not been obtained by the person importing same, such person shall:

(1) Store, at the person's expense, such firearm, firearm barrel, or ammunition at a facility designated by U.S Customs or the Director of Industry Operations to await the issuance of the required authorization or other disposition; or

(2) Abandon such firearm, firearm barrel, or ammunition to the U.S. Government; or

(3) Export such firearm, firearm barrel, or ammunition.

(c) Any inquiry relative to the provisions or procedures under this subpart, other than that pertaining to the pay-

ment of customs duties or the release from Customs custody of firearms, firearm barrels, or ammunition authorized by the Director to be imported, shall be directed to the Director of Industry Operations for reply.

§ 478.112 Importation by a licensed importer.

(a) No firearm, firearm barrel, or ammunition shall be imported or brought into the United States by a licensed importer (as defined in §478.11) unless the Director has authorized the importation of the firearm, firearm barrel, or ammunition.

(b)(1) An application for a permit, ATF Form 6–Part I, to import or bring a firearm, firearm barrel, or ammunition into the United States or a possession thereof under this section must be filed, in triplicate, with the Director. The application must be signed and dated and must contain the information requested on the form, including:

(i) The name, address, telephone number, and license number (including expiration date) of the importer;

(ii) The country from which the firearm, firearm barrel, or ammunition is to be imported;

(iii) The name and address of the foreign seller and foreign shipper;

(iv) A description of the firearm, firearm barrel, or ammunition to be imported, including:

(A) The name and address of the manufacturer;

(B) The type (e.g., rifle, shotgun, pistol, revolver and, in the case of ammunition only, ball, wadcutter, shot, etc.);

(C) The caliber, gauge, or size;

(D) The model;

(E) The barrel length, if a firearm or firearm barrel (in inches);

(F) The overall length, if a firearm (in inches);

(G) The serial number, if known;

(H) Whether the firearm is new or used;

(I) The quantity;

(J) The unit cost of the firearm, firearm barrel, or ammunition to be imported;

(v) The specific purpose of importation, including final recipient information if different from the importer;

(vi) Verification that if a firearm, it will be identified as required by this part; and

(vii)(A) If a firearm or ammunition imported or brought in for scientific or research purposes, a statement describing such purpose; or

(B) If a firearm or ammunition for use in connection with competition or training pursuant to Chapter 401 of Title 10, U.S.C., a statement describing such intended use; or

(C) If an unserviceable firearm (other than a machine gun) being imported as a curio or museum piece, a description of how it was rendered unserviceable and an explanation of why it is a curio or museum piece; or

(D) If a firearm other than a surplus military firearm, of a type that does not fall within the definition of a firearm under section 5845(a) of the Internal Revenue Code of 1986, and is for sporting purposes, an explanation of why the firearm is generally recognized as particularly suitable for or readily adaptable to sporting purposes; or

(E) If ammunition being imported for sporting purposes, a statement why the ammunition is particularly suitable for or readily adaptable to sporting purposes; or

(F) If a firearm barrel for a handgun, an explanation why the handgun is generally recognized as particularly suitable for or readily adaptable to sporting purposes.

(2)(i) If the Director approves the application, such approved application

will serve as the permit to import the firearm, firearm barrel, or ammunition described therein, and importation of such firearms, firearm barrels, or ammunition may continue to be made by the licensed importer under the approved application (permit) during the period specified thereon. The Director will furnish the approved application (permit) to the applicant and retain two copies thereof for administrative use.

(ii) If the Director disapproves the application, the licensed importer will be notified of the basis for the disapproval.

(c) A firearm, firearm barrel, or ammunition imported or brought into the United States or a possession thereof under the provisions of this section by a licensed importer may be released from Customs custody to the licensed importer upon showing that the importer has obtained a permit from the Director for the importation of the firearm, firearm barrel, or ammunition to be released. The importer will also submit to Customs a copy of the export license authorizing the export of the firearm, firearm barrel, or ammunition from the exporting country. If the exporting country does not require issuance of an export license, the importer must submit a certification, under penalty of perjury, to that effect.

(1) In obtaining the release from Customs custody of a firearm, firearm barrel, or ammunition authorized by this section to be imported through the use of a permit, the licensed importer will prepare ATF Form 6A, in duplicate, and furnish the original ATF Form 6A to the Customs officer releasing the firearm, firearm barrel, or ammunition. The Customs officer will, after certification, forward the ATF Form 6A to the address specified on the form.

(2) The ATF Form 6A must contain the information requested on the form, including:

(i) The name, address, and license number of the importer;

(ii) The name of the manufacturer of the firearm, firearm barrel, or ammunition;

(iii) The country of manufacture;

(iv) The type;

(v) The model;

(vi) The caliber, gauge, or size;

(vii) The serial number in the case of firearms, if known; and

(viii) The number of firearms, firearm barrels, or rounds of ammunition released.

(d) Within 15 days of the date of release from Customs custody, the licensed importer must:

(1) Forward to the address specified on the form a copy of ATF Form 6A on which must be reported any error or discrepancy appearing on the ATF Form 6A certified by Customs and serial numbers if not previously provided on ATF Form 6A;

(2) Pursuant to §478.92, place all required identification data on each imported firearm if same did not bear such identification data at the time of its release from Customs custody; and

(3) Post in the records required to be maintained by the importer under subpart H of this part all required information regarding the importation.

(Paragraph (b) approved by the Office of Management and Budget under control number 1140–0005; paragraphs (c) and (d) approved by the Office of Management and Budget under control number 1140–0007)

§ 478.113 Importation by other licensees.

(a) No person other than a licensed importer (as defined in §478.11) shall engage in the business of importing firearms or ammunition. Therefore, no firearm or ammunition shall be imported or brought into the United States or a possession thereof by any licensee other than a licensed importer unless the Director issues a permit authorizing the importation of the firearm or ammunition. No barrel for a handgun not generally recognized as particularly suitable for or readily adaptable to sporting purposes shall be imported or brought into the United States or a possession thereof by any person. Therefore, no firearm barrel shall be imported or brought into the United States or possession thereof by any licensee other than a licensed importer unless the Director issues a permit authorizing the importation of the firearm barrel.

(b)(1) An application for a permit, ATF Form 6–Part I, to import or bring a firearm, firearm barrel, or ammunition into the United States or a possession thereof by a licensee, other than a licensed importer, must be filed, in triplicate, with the Director. The application must be signed and dated and must contain the information requested on the form, including:

(i) The name, address, telephone number, and license number (including expiration date) of the applicant;

(ii) The country from which the firearm, firearm barrel, or ammunition is to be imported;

(iii) The name and address of the foreign seller and foreign shipper;

(iv) A description of the firearm, firearm barrel, or ammunition to be imported, including:

(A) The name and address of the manufacturer;

(B) The type (e.g., rifle, shotgun, pistol, revolver and, in the case of ammunition only, ball, wadcutter, shot, etc.);

(C) The caliber, gauge, or size;

(D) The model;

(E) The barrel length, if a firearm or firearm barrel (in inches);

(F) The overall length, if a firearm (in inches);

(G) The serial number, if known;

(H) Whether the firearm is new or used;

(I) The quantity;

(J) The unit cost of the firearm, firearm barrel, or ammunition to be imported;

(v) The specific purpose of importation, including final recipient information if different from the applicant; and

(vi)(A) If a firearm or ammunition imported or brought in for scientific or research purposes, a statement describing such purpose; or

(B) If a firearm or ammunition for use in connection with competition or training pursuant to Chapter 401 of Title 10, U.S.C., a statement describing such intended use; or

(C) If an unserviceable firearm (other than a machine gun) being imported as a curio or museum piece, a description of how it was rendered unserviceable and an explanation of why it is a curio or museum piece; or

(D) If a firearm other than a surplus military firearm, of a type that does not fall within the definition of a firearm under section 5845(a) of the Internal Revenue Code of 1986, and is for sporting purposes, an explanation of why the firearm is generally recognized as particularly suitable for or readily adaptable to sporting purposes; or

(E) If ammunition being imported for sporting purposes, a statement why the ammunition is particularly suitable for or readily adaptable to sporting purposes; or

(F) If a firearm barrel for a handgun, an explanation why the handgun is generally recognized as particularly suitable for or readily adaptable to sporting purposes.

(2)(i) If the Director approves the application, such approved application will serve as the permit to import the firearm, firearm barrel, or ammunition described therein, and importation of such firearms, firearm barrels, or ammunition may continue to be made by the applicant under the approved application (permit) during the period specified thereon. The Director will furnish the approved application (permit) to the applicant and retain two copies thereof for administrative use.

(ii) If the Director disapproves the application, the applicant will be notified of the basis for the disapproval.

(c) A firearm, firearm barrel, or ammunition imported or brought into the United States or a possession thereof under the provisions of this section may be released from Customs custody to the licensee upon showing that the licensee has obtained a permit from the Director for the importation of the firearm, firearm barrel, or ammunition to be released.

(1) In obtaining the release from Customs custody of a firearm, firearm barrel, or ammunition authorized by this section to be imported through the use of a permit, the licensee will prepare ATF Form 6A, in duplicate, and furnish the original ATF Form 6A to the Customs officer releasing the firearm, firearm barrel, or ammunition. The Customs officer will, after certification, forward the ATF Form 6A to the address specified on the form.

(2) The ATF Form 6A must contain the information requested on the form, including:

(i) The name, address, and license number of the licensee;

(ii) The name of the manufacturer of the firearm, firearm barrel, or ammunition;

(iii) The country of manufacture;

(iv) The type;

(v) The model;

(vi) The caliber, gauge, or size;

(vii) The serial number in the case of firearms; and

(viii) The number of firearms, firearm barrels, or rounds of ammunition released.

(Paragraph (b) approved by the Office of Management and Budget under control number 1140–0005; paragraph (c) approved by the Office of Management and Budget under control number 1140–0007)

§ 478.113a Importation of firearm barrels by nonlicensees.

(a) A permit will not be issued for a firearm barrel for a handgun not generally recognized as particularly suitable for or readily adaptable to sporting purposes. No firearm barrel shall be imported or brought into the United States or possession thereof by any nonlicensee unless the Director issues a permit authorizing the importation of the firearm barrel.

(b)(1) An application for a permit, ATF Form 6–Part I, to import or bring a firearm barrel into the United States or a possession thereof under this section must be filed, in triplicate, with the Director. The application must be signed and dated and must contain the information requested on the form, including:

(i) The name, address, and telephone number of the applicant;

(ii) The country from which the firearm barrel is to be imported;

(iii) The name and address of the foreign seller and foreign shipper;

(iv) A description of the firearm barrel to be imported, including:

(A) The name and address of the manufacturer;

(B) The type (e.g., rifle, shotgun, pistol, revolver);

(C) The caliber, gauge, or size;

(D) The model;

(E) The barrel length (in inches);

(F) The quantity;

(G) The unit cost of the firearm barrel;

(v) The specific purpose of importation, including final recipient information if different from the importer; and

(vi) If a handgun barrel, an explanation of why the barrel is for a handgun that is generally recognized as particularly suitable for or readily adaptable to sporting purposes.

(2)(i) If the Director approves the application, such approved application will serve as the permit to import the firearm barrel, and importation of such firearm barrels may continue to be made by the applicant under the approved application (permit) during the period specified thereon. The Director will furnish the approved application (permit) to the applicant and retain two copies thereof for administrative use.

(ii) If the Director disapproves the application, the applicant will be notified of the basis for the disapproval.

(c) A firearm barrel imported or brought into the United States or a pos-

session thereof under the provisions of this section may be released from Customs custody to the person importing the firearm barrel upon showing that the person has obtained a permit from the Director for the importation of the firearm barrel to be released.

(1) In obtaining the release from Customs custody of a firearm barrel authorized by this section to be imported through the use of a permit, the person importing the firearm barrel will prepare ATF Form 6A, in duplicate, and furnish the original ATF Form 6A to the Customs officer releasing the firearm barrel. The Customs officer will, after certification, forward the ATF Form 6A to the address specified on the form.

(2) The ATF Form 6A must contain the information requested on the form, including:

(i) The name and address of the person importing the firearm barrel;

(ii) The name of the manufacturer of the firearm barrel;

(iii) The country of manufacture;

(iv) The type;

(v) The model;

(vi) The caliber or gauge of the firearm barrel so released; and

(vii) The number of firearm barrels released.

(Paragraph (b) approved by the Office of Management and Budget under control number 1140–0005; paragraph (c) approved by the Office of Management and Budget under control number 1140–0007)

§ 478.114 Importation by members of the U.S. Armed Forces.

(a) The Director may issue a permit authorizing the importation of a firearm or ammunition into the United States to the place of residence of any military member of the U.S. Armed Forces who is on active duty outside the United States, or who has been on active duty outside the United States within the 60–day period immediately preceding the intended importation: **Provided**, That such firearm or ammunition is generally recognized as particularly suitable for or readily adaptable to sporting purposes

and is intended for the personal use of such member.

(1) An application for a permit, ATF Form 6–Part II, to import a firearm or ammunition into the United States under this section must be filed, in triplicate, with the Director. The application must be signed and dated and must contain the information requested on the form, including:

(i) The name, current address, and telephone number of the applicant;

(ii) Certification that the transportation, receipt, or possession of the firearm or ammunition to be imported would not constitute a violation of any provision of the Act or of any State law or local ordinance at the place of the applicant's residence;

(iii) The country from which the firearm or ammunition is to be imported;

(iv) The name and address of the foreign seller and foreign shipper;

(v) A description of the firearm or ammunition to be imported, including:

(A) The name and address of the manufacturer;

(B) The type (e.g., rifle, shotgun, pistol, revolver and, in the case of ammunition only, ball, wadcutter, shot, etc.);

(C) The caliber, gauge, or size;

(D) The model;

(E) The barrel length, if a firearm (in inches);

(F) The overall length, if a firearm (in inches);

(G) The serial number;

(H) Whether the firearm is new or used;

(I) The quantity;

(J) The unit cost of the firearm or ammunition to imported;

(vi) The specific purpose of importation, that is—

(A) That the firearm or ammunition being imported is for the personal use of the applicant; and

(B) If a firearm, a statement that it is not a surplus military firearm, that it does not fall within the definition of a firearm under section 5845(a) of the Internal Revenue Code of 1986, and an explanation of why the firearm is generally recognized as particularly suitable for or readily adaptable to sporting purposes; or

(C) If ammunition, a statement why it is generally recognized as particularly suitable for or readily adaptable to sporting purposes; and

(vii) The applicant's date of birth;

(viii) The applicant's rank or grade;

(ix) The applicant's place of residence;

(x) The applicant's present foreign duty station or last foreign duty station, as the case may be;

(xi) The date of the applicant's reassignment to a duty station within the United States, if applicable; and

(xii) The military branch of which the applicant is a member.

(2)(i) If the Director approves the application, such approved application will serve as the permit to import the firearm or ammunition described therein. The Director will furnish the approved application (permit) to the applicant and retain two copies thereof for administrative use.

(ii) If the Director disapproves the application, the applicant will be notified of the basis for the disapproval.

(b) Except as provided in paragraph (b)(3) of this section, a firearm or ammunition imported into the United States under the provisions of this section by the applicant may be released from Customs custody to the applicant upon showing that the applicant has obtained a permit from the Director for the importation of the firearm or ammunition to be released.

(1) In obtaining the release from Customs custody of a firearm or ammunition authorized by this section to be imported through the use of a permit, the military member of the U.S. Armed Forces will prepare ATF Form 6A and furnish the completed form to the Customs officer releasing the firearm or ammunition. The Customs officer will, after certification, forward the ATF Form 6A to the address specified on the form.

(2) The ATF Form 6A must contain the information requested on the form, including:

(i) The name and address of the military member;

(ii) The name of the manufacturer of the firearm or ammunition;

(iii) The country of manufacture;

(iv) The type;

(v) The model;

(vi) The caliber, gauge, or size;

(vii) The serial number in the case of firearms; and

(viii) If applicable, the number of firearms or rounds of ammunition released.

(3) When such military member is on active duty outside the United States, the military member may appoint, in writing, an agent to obtain the release of the firearm or ammunition from Customs custody for such member. Such agent will present sufficient identification of the agent and the written authorization to act on behalf of such military member to the Customs officer who is to release the firearm or ammunition.

(c) Firearms determined by the Department of Defense to be war souvenirs may be imported into the United States by the military members of the U.S. Armed Forces under such provisions and procedures as the Department of Defense may issue.

(Paragraph (a) approved by the Office of Management and Budget under control number 1140–0006; paragraph (b) approved by the Office of Management and Budget under control number 1140–0007)

§ 478.115 Exempt importation.

(a) Firearms and ammunition may be brought into the United States or any possession thereof by any person who can establish to the satisfaction of Customs that such firearm or ammunition was previously taken out of the United States or any possession thereof by such person. Registration on Customs Form 4457 or on any other registration document available for this purpose may be completed before departure from the United States at any U.S. customhouse or any office of an Director of Industry Operations. A bill of sale or other commercial document showing transfer of the firearm or ammunition in the United States to such person also may be used to establish proof that the firearm or ammunition was taken out of the United States by such person. Firearms and ammunition furnished under the provisions of section 925(a)(3) of the Act to military members of the U.S. Armed Forces on active duty outside of the United States also may be imported into the United States or any possession thereof by such military members upon establishing to the satisfaction of Customs that such firearms and ammunition were so obtained.

(b) Firearms, firearm barrels, and ammunition may be imported or brought into the United States by or for the United States or any department or agency thereof, or any State or any department, agency, or political subdivision thereof. A firearm, firearm barrel or ammunition imported or brought into the United States under this paragraph may be released from Customs custody upon a showing that the firearm, firearm barrel or ammunition is being imported or brought into the United States by or for such a governmental entity.

(c) The provisions of this subpart shall not apply with respect to the importation into the United States of any antique firearm.

(d) Firearms and ammunition are not imported into the United States, and the provisions of this subpart shall not apply, when such firearms and ammunition are brought into the United States by:

(1) A nonresident of the United States for legitimate hunting or lawful sporting purposes, and such firearms and such ammunition as remains following such shooting activity are to be taken back out of the territorial limits of the United States by such person upon conclusion of the shooting activity;

(2) Foreign military personnel on official assignment to the United States who bring such firearms or ammunition into the United States for their exclusive use while on official duty in the United States, and such firearms and unexpended ammunition are taken back out of the territorial limits of the United States by such foreign military personnel when they leave the United States;

(3) Official representatives of foreign governments who are accredited to the U.S. Government or are en route to or from other countries to which accredited, and such firearms and unexpended ammunition are taken back out of the territorial limits of the United States by such official representatives of foreign governments when they leave the United States;

(4) Officials of foreign governments and distinguished foreign visitors who have been so designated by the Department of State, and such firearms and unexpended ammunition are taken back out of the territorial limits of the United States by such officials of foreign governments and distinguished foreign visitors when they leave the United States; and

(5) Foreign law enforcement officers of friendly foreign governments entering the United States on official law enforcement business, and such firearms and unexpended ammunition are taken back out of the territorial limits of the United States by such foreign law enforcement officers when they leave the United States.

(e) Notwithstanding the provisions of paragraphs (d) (1), (2), (3), (4) and (5) of this section, the Secretary of the Treasury or his delegate may in the interest of public safety and necessity require a permit for the importation or bringing into the United States of any firearms or ammunition.

§ 478.116 Conditional importation.

The Director shall permit the conditional importation or bringing into the United States or any possession thereof of any firearm, firearm barrel, ammunition, or ammunition feeding device as defined in §478.119(b) for the purpose of examining and testing the firearm, firearm barrel, ammunition, or ammunition feeding device in connection with making a determination as to whether the importation or bringing in of such firearm, firearm barrel, ammunition, or

ammunition feeding device will be authorized under this part. An application on ATF Form 6 for such conditional importation shall be filed, in duplicate, with the Director. The Director may impose conditions upon any importation under this section including a requirement that the firearm, firearm barrel, ammunition, or ammunition feeding device be shipped directly from Customs custody to the Director and that the person importing or bringing in the firearm, firearm barrel, ammunition, or ammunition feeding device must agree to either export the firearm, firearm barrel, ammunition, or ammunition feeding device or destroy same if a determination is made that the firearm, firearm barrel, ammunition, or ammunition feeding device may not be imported or brought in under this part. A firearm, firearm barrel, ammunition, or ammunition feeding device imported or brought into the United States or any possession thereof under the provisions of this section shall be released from Customs custody upon the payment of customs duties, if applicable, and in the manner prescribed in the conditional authorization issued by the Director.

Editor's Note:

The references to "ammunition feeding device" in section 478.116 are not applicable on or after September 13, 2004.

§ 478.117 Function outside a customs territory.

In the insular possessions of the United States outside customs territory, the functions performed by U.S. Customs officers under this subpart within a customs territory may be performed by the appropriate authorities of a territorial government or other officers of the United States who have been designated to perform such functions. For the purpose of this subpart, the term customs territory means the United States, the District of Columbia, and the Commonwealth of Puerto Rico.

§ 478.118 Importation of certain firearms classified as curios or relics.

Notwithstanding any other provision of this part, a licensed importer may import all rifles and shotguns classified by the Director as curios or relics, and all handguns classified by the Director as curios or relics that are determined to be generally recognized as particularly suitable for or readily adaptable to sporting purposes. The importation of such curio or relic firearms must be in accordance

with the applicable importation provisions of this part and the importation provisions of 27 CFR part 447. Curios or relics which fall within the definition of **"firearm"** under 26 U.S.C. 5845(a) must also meet the importation provisions of 27 CFR part 479 before they may be imported.

§ 478.119 Importation of ammunition feeding devices.

Editor's Note:

Section 478.119 is not applicable on or after September 13, 2004. An import permit is still needed for ammunition feeding devices pursuant to the Arms Export Control Act—see 27 CFR 447.41(a).

§ 478.120 Firearms or ammunition imported by or for a nonimmigrant alien.

(a) General. A nonimmigrant alien temporarily importing or bringing firearms or ammunition into the United States for lawful hunting or sporting purposes must first obtain an approved ATF Form 6NIA (5330.3D).

(b) Aliens admitted to the United States under a nonimmigrant visa. (1) Any alien lawfully admitted to the United States under a nonimmigrant visa who completes an ATF Form 6NIA to import firearms or ammunition into the United States, or any licensee who completes an ATF Form 6 to import firearms or ammunition for such nonimmigrant alien, must attach applicable documentation to the Form 6NIA or Form 6 establishing the nonimmigrant alien falls within an exception specified in 18 U.S.C. 922(y)(2) (e.g., a hunting license or permit lawfully issued in the United States) or has obtained a waiver as specified in 18 U.S.C. 922(y)(3). (2) Aliens admitted to the United States under a nonimmigrant visa importing or bringing firearms or ammunition into the United States must provide the United States Customs and Border Protection with applicable documentation (e.g., a hunting license or permit lawfully issued in the United States) establishing the nonimmigrant alien falls within an exception specified in 18 U.S.C. 922(y)(2) or has obtained a waiver as specified in 18 U.S.C. 922(y)(3) before the firearm or ammunition may be imported. This provision applies in all cases, whether or not a Form 6 is needed to bring the firearms or ammunition into the United States.

(Approved by the Office of Management and Budget under control number 1140–0060)

Subpart H—Records

§ 478.121 General.

(a) The records pertaining to firearms transactions prescribed by this part shall be retained on the licensed premises in the manner prescribed by this subpart and for the length of time prescribed by §478.129. The records pertaining to ammunition prescribed by this part shall be retained on the licensed premises in the manner prescribed by §478.125.

(b) ATF officers may, for the purposes and under the conditions prescribed in §478.23, enter the premises of any licensed importer, licensed manufacturer, licensed dealer, or licensed collector for the purpose of examining or inspecting any record or document required by or obtained under this part. Section 923(g) of the Act requires licensed importers, licensed manufacturers, licensed dealers, and licensed collectors to make such records available for such examination or inspection during business hours or, in the case of licensed collectors, hours of operation, as provided in §478.23.

(c) Each licensed importer, licensed manufacturer, licensed dealer, and licensed collector shall maintain such records of importation, production, shipment, receipt, sale, or other disposition, whether temporary or permanent, of firearms and such records of the disposition of ammunition as the regulations contained in this part prescribe. Section 922(m) of the Act makes it unlawful for any licensed importer, licensed manufacturer, licensed dealer, or licensed collector knowingly to make any false entry in, to fail to make appropriate entry in, or to fail to properly maintain any such record.

(d) For recordkeeping requirements for sales by licensees at gun shows see §478.100(c).

(Information collection requirements in paragraph (a) approved by the Office of Management and Budget under control number 1140–0020; information collection requirements in paragraphs (b) and (c) approved by the Office of Management and Budget under control number 1140–0032)

§ 478.122 Records maintained by importers.

(a) Each licensed importer shall, within 15 days of the date of importation or other acquisition, record the type, model, caliber or gauge, manufacturer, country of manufacture, and the serial number of each firearm imported or otherwise acquired, and the date such importation or other acquisition was made.

(b) A record of firearms disposed of by a licensed importer to another licensee and a separate record of armor piercing ammunition dispositions to governmental entities, for exportation, or for testing or experimentation authorized under the provisions of §478.149 shall be maintained by the licensed importer on the licensed premises. For firearms, the record shall show the quantity, type, manufacturer, country of manufacture, caliber or gauge, model, serial number of the firearms so transferred, the name and license number of the licensee to whom the firearms were transferred, and the date of the transaction. For armor piercing ammunition, the record shall show the date of the transaction, manufacturer, caliber or gauge, quantity of projectiles, and the name and address of the purchaser. The information required by this paragraph shall be entered in the proper record book not later than the seventh day following the date

of the transaction, and such information shall be recorded under the following formats: **(See Tables 1 & 2)**

(c) Notwithstanding the provisions of paragraph (b) of this section, the Director of Industry Operations may authorize alternate records to be maintained by a licensed importer to record the disposal of firearms and armor piercing ammunition when it is shown by the licensed importer that such alternate records will accurately and readily disclose the information required by paragraph (b) of this section. A licensed importer who proposes to use alternate records shall submit a letter application, in duplicate, to the Director of Industry Operations and shall describe the proposed alternate records and the need therefor. Such alternate records shall not be employed by the licensed importer until approval in such regard is received from the Director of Industry Operations.

(d) Each licensed importer shall maintain separate records of the sales or other dispositions made of firearms to nonlicensees. Such records shall be maintained in the form and manner as prescribed by §§478.124 and 478.125 in regard to firearms transaction records and records of acquisition and disposition of firearms.

(Approved by the Office of Management and Budget under control number 1140–0032)

§ 478.123 Records maintained by manufacturers.

(a) Each licensed manufacturer shall record the type, model, caliber or gauge, and serial number of each complete firearm manufactured or otherwise acquired, and the date such manufacture or other acquisition was made. The information required by this paragraph

TABLE 1: Importer's Firearms Disposition Record

Quantity	Type	Manufacturer	Country of manufacture	Caliber or gague	Model	Serial No.	Name and license No. of licensee to whom transferred	Date of the transaction

TABLE 2: Importer's Armor Piercing Ammunition Disposition Record

Date	Manufacturer	Caliber or gague	Quantity of projectiles	Purchaser - Name and address

shall be recorded not later than the seventh day following the date such manufacture or other acquisition was made.

(b) A record of firearms disposed of by a manufacturer to another licensee and a separate record of armor piercing ammunition dispositions to governmental entities, for exportation, or for testing or experimentation authorized under the provision of §478.149 shall be maintained by the licensed manufacturer on the licensed premises. For firearms, the record shall show the quantity, type, model, manufacturer, caliber, size or gauge, serial number of the firearms so transferred, the name and license number of the licensee to whom the firearms were transferred, and the date of the transaction. For armor piercing ammunition, the record shall show the manufacturer, caliber or gauge, quantity, the name and address of the transferee to whom the armor piercing ammunition was transferred, and the date of the transaction. The information required by this paragraph shall be entered in the proper record book not later than the seventh day following the date of the transaction, and such information shall be recorded under the format prescribed by §478.122, except that the name of the manufacturer of a firearm or armor piercing ammunition need not be recorded if the firearm or armor piercing ammunition is of the manufacturer's own manufacture.

(c) Notwithstanding the provisions of paragraph (b) of this section, the Director of Industry Operations may authorize alternate records to be maintained by a licensed manufacturer to record the disposal of firearms and armor piercing ammunition when it is shown by the licensed manufacturer that such alternate records will accurately and readily disclose the information required by paragraph (b) of this section. A licensed manufacturer who proposes to use alternate records shall submit a letter application, in duplicate, to the Director of Industry Operations and shall describe the proposed alternate record and the need therefor. Such alternate records shall not be employed by the licensed manufacturer until approval in such regard is received from the Director of Industry Operations.

(d) Each licensed manufacturer shall maintain separate records of the sales or other dispositions made of firearms to nonlicensees. Such records shall be maintained in the form and manner as prescribed by §§478.124 and 478.125 in regard to firearms transaction records and records of acquisition and disposition of firearms.

(Approved by the Office of Management and Budget under control number 1140–0067)

§ 478.124 Firearms transaction record.

(a) A licensed importer, licensed manufacturer, or licensed dealer shall not sell or otherwise dispose, temporarily or permanently, of any firearm to any person, other than another licensee, unless the licensee records the transaction on a firearms transaction record, Form 4473: Provided, That a firearms transaction record, Form 4473, shall not be required to record the disposition made of a firearm delivered to a licensee for the sole purpose of repair or customizing when such firearm or a replacement firearm is returned to the person from whom received.

(b) A licensed manufacturer, licensed importer, or licensed dealer shall retain in alphabetical (by name of purchaser), chronological (by date of disposition), or numerical (by transaction serial number) order, and as a part of the required records, each Form 4473 obtained in the course of transferring custody of the firearms.

(c)(1) Prior to making an over–the–counter transfer of a firearm to a nonlicensee who is a resident of the State in which the licensee's business premises is located, the licensed importer, licensed manufacturer, or licensed dealer so transferring the firearm shall obtain a Form 4473 from the transferee showing the transferee's name, sex, residence address (including county or similar political subdivision), date and place of birth; height, weight and race of the transferee; the transferee's country of citizenship; the transferee's INS–issued alien number or admission number; the transferee's State of residence; and certification by the transferee that the transferee is not prohibited by the Act from transporting or shipping a firearm in interstate or foreign commerce or receiving a firearm which has been shipped or transported in interstate or foreign commerce or possessing a firearm in or affecting commerce.

(2) In order to facilitate the transfer of a firearm and enable NICS to verify the identity of the person acquiring the firearm, ATF Form 4473 also requests certain optional information. This information includes the transferee's social security number. Such information may help avoid the possibility of the transferee being misidentified as a felon or other prohibited person.

(3) After the transferee has executed the Form 4473, the licensee:

(i) Shall verify the identity of the transferee by examining the identification document (as defined in §478.11) presented, and shall note on the Form 4473 the type of identification used;

(ii) [Reserved]

(iii) Must, in the case of a transferee who is an alien admitted to the United States under a nonimmigrant visa who states that he or she falls within an exception to, or has a waiver from, the prohibition in section 922(g)(5)(B) of the Act, have the transferee present applicable documentation establishing the exception or waiver, note on the Form 4473 the type of documentation provided, and attach a copy of the documentation to the Form 4473; and

(iv) Shall comply with the requirements of §478.102 and record on the form the date on which the licensee contacted the NICS, as well as any response provided by the system, including any identification number provided by the system.

(4) The licensee shall identify the firearm to be transferred by listing on the Form 4473 the name of the manufacturer, the name of the importer (if any), the type, model, caliber or gauge, and the serial number of the firearm.

(5) The licensee shall sign and date the form if the licensee does not know or have reasonable cause to believe that the transferee is disqualified by law from receiving the firearm and transfer the firearm described on the Form 4473.

(d) Prior to making an over–the–counter transfer of a shotgun or rifle under the provisions contained in §478.96(c) to a nonlicensee who is not a resident of the State in which the licensee's business premises is located, the licensee so transferring the shotgun or rifle, and such transferee, shall comply with the requirements of paragraph (c) of this section.

(e) Prior to making a transfer of a firearm to any nonlicensee who is not a resident of the State in which the licensee's business premises is located, and such nonlicensee is acquiring the firearm by loan or rental from the licensee for temporary use for lawful sporting purposes, the licensed importer, licensed manufacturer, or licensed dealer so furnishing the firearm, and such transferee, shall comply with the provisions of paragraph (c) of this section.

(f) Form 4473 shall be submitted, in duplicate, to a licensed importer, licensed manufacturer, or licensed dealer by a transferee who is purchasing or otherwise acquiring a firearm by other than an over-the-counter transaction, who is not subject to the provisions of §478.102(a), and who is a resident of the State in which the licensee's business premises are located. The Form 4473 shall show the name, address, date and place of birth, height, weight, and race of the transferee; and the title, name, and address of the principal law enforcement officer of the locality to which the firearm will be delivered. The transferee also must date and execute the sworn statement contained on the form showing, in case the firearm to be transferred is a firearm other than a shotgun or rifle, the transferee is 21 years or more of age; in case the firearm to be transferred is a shotgun or rifle, the transferee is 18 years or more of age; whether the transferee is a citizen of the United States; the transferee's State of residence; the transferee is not prohibited by the provisions of the Act from shipping or transporting a firearm in interstate or foreign commerce or receiving a firearm which has been shipped or transported in interstate or foreign commerce or possessing a firearm in or affecting commerce; and the transferee's receipt of the firearm would not be in violation of any statute of the State or published ordinance applicable to the locality in which the transferee resides. Upon receipt of such Forms 4473, the licensee shall identify the firearm to be transferred by listing in the Forms 4473 the name of the manufacturer, the name of the importer (if any), the type, model, caliber or gauge, and the serial number of the firearm to be transferred. The licensee shall prior to shipment or delivery of the firearm to such transferee, forward by registered or certified mail (return receipt requested) a copy of the Form 4473 to the principal law enforcement officer named in the Form 4473 by the transferee, and shall delay shipment or delivery of the firearm to the transferee for a period of at least 7 days following receipt by the licensee of the return receipt evidencing delivery of the copy of the Form 4473 to such principal law enforcement officer, or the return of the copy of the Form 4473 to the licensee due to the refusal of such principal law enforcement officer to accept same in accordance with U.S. Postal Service regulations. The original Form 4473, and evidence of receipt or rejection of delivery of the copy of the Form 4473 sent to the principal law enforcement officer, shall be retained by the licensee as a part of the records required to be kept under this subpart.

(g) A licensee who sells or otherwise disposes of a firearm to a nonlicensee who is other than an individual, shall obtain from the transferee the information required by this section from an individual authorized to act on behalf of the transferee. In addition, the licensee shall obtain from the individual acting on behalf of the transferee a written statement, executed under the penalties of perjury, that the firearm is being acquired for the use of and will be the property of the transferee, and showing the name and address of that transferee.

(h) The requirements of this section shall be in addition to any other recordkeeping requirement contained in this part.

(i) A licensee may obtain, upon request, an emergency supply of Forms 4473 from any Director of Industry Operations. For normal usage, a licensee should request a year's supply from the ATF Distribution Center (See §478.21).

(Paragraph (c) approved by the Office of Management and Budget under control numbers 1140–0045, 1140–0020, and 1140–0060; paragraph (f) approved by the Office of Management and Budget under control number 1140–0021; all other recordkeeping approved by the Office of Management and Budget under control number 1140–0020)

§ 478.125 Record of receipt and disposition.

(a) Armor piercing ammunition sales by licensed collectors to nonlicensees. The sale or other disposition of armor piercing ammunition by licensed collectors shall be recorded in a bound record at the time a transaction is made. The bound record shall be maintained in chronological order by date of sale or disposition of the armor piercing ammunition, and shall be retained on the licensed premises of the licensee for a period not less than two years following the date of the recorded sale or disposition of the armor piercing ammunition. The bound record entry shall show:

(1) The date of the transaction;

(2) The name of the manufacturer;

(3) The caliber or gauge;

(4) The quantity of projectiles;

(5) The name, address, and date of birth of the nonlicensee; and

(6) The method used to establish the identity of the armor piercing ammunition purchaser.

The format required for the bound record is as follows: **(See Table 3)**

However, when a commercial record is made at the time a transaction is made, a licensee may delay making an entry into the bound record if the provisions of paragraph (d) of this section are complied with.

(b) Armor piercing ammunition sales by licensed collectors to licensees. Sales or other dispositions of armor

TABLE 3: Disposition Record of Armor Piercing Ammunition

Date	Manufacturer	Caliber or gague	Qualitity of projectiles	Purchaser		Enter a (x) in the "known" column if purchaser is personally known to you. Otherwise, establish the purchaser's identification		
				Name and address	Date of birth	Known	Driver's License	Other type (specify)

piercing ammunition from a licensed collector to another licensee shall be recorded and maintained in the manner prescribed in §478.122(b) for importers: *Provided,* That the license number of the transferee may be recorded in lieu of the transferee's address.

(c) Armor piercing ammunition sales by licensed dealers to governmental entities. A record of armor piercing ammunition disposed of by a licensed dealer to a governmental entity pursuant to §478.99(e) shall be maintained by the licensed dealer on the licensed premises and shall show the name of the manufacturer, the caliber or gauge, the quantity, the name and address of the entity to which the armor piercing ammunition was transferred, and the date of the transaction. Such information shall be recorded under the format prescribed by §478.122(b). Each licensed dealer disposing of armor piercing ammunition pursuant to §478.99(e) shall also maintain a record showing the date of acquisition of such ammunition which shall be filed in an orderly manner separate from other commercial records maintained and be readily available for inspection. The records required by this paragraph shall be retained on the licensed premises of the licensee for a period not less than two years following the date of the recorded sale or disposition of the armor piercing ammunition.

(d) Commercial records of armor piercing ammunition transactions. When a commercial record is made at the time of sale or other disposition of armor piercing ammunition, and such record contains all information required by the bound record prescribed by paragraph (a) of this section, the licensed collector transferring the armor piercing ammunition may, for a period not exceeding 7 days following the date of such transfer, delay making the required entry into such bound record: Provided, That the commercial record pertaining to the transfer is:

(1) Maintained by the licensed collector separate from other commercial documents maintained by such licensee, and

(2) Is readily available for inspection on the licensed premises until such time as the required entry into the bound record is made.

(e) Firearms receipt and disposition by dealers. Each licensed dealer shall enter into a record each receipt and disposition of firearms. In addition, before commencing or continuing a firearms business, each licensed dealer shall inventory the firearms possessed for such business and shall record same in the record required by this paragraph. The record required by this paragraph shall be maintained in bound form under the format prescribed below. The purchase or other acquisition of a firearm shall, except as provided in paragraph (g) of this section, be recorded not later than the close of the next business day following the date of such purchase or acquisition. The record shall show the date of receipt, the name and address or the name and license number of the person from whom received, the name of the manufacturer and importer (if any), the model, serial number, type, and the caliber or gauge of the firearm. The sale or other disposition of a firearm shall be recorded by the licensed dealer not later than 7 days following the date of such transaction. When such disposition is made to a nonlicensee, the firearms transaction record, Form 4473, obtained by the licensed dealer shall be retained, until the transaction is recorded, separate from the licensee's Form 4473 file and be readily available for inspection. When such disposition is made to a licensee, the commercial record of the transaction shall be retained, until the transaction is recorded, separate from other commercial documents maintained by the licensed dealer, and be readily available for inspection. The record shall show the

date of the sale or other disposition of each firearm, the name and address of the person to whom the firearm is transferred, or the name and license number of the person to whom transferred if such person is a licensee, or the firearms transaction record, Form 4473, serial number if the licensed dealer transferring the firearm serially numbers the Forms 4473 and files them numerically. The format required for the record of receipt and disposition of firearms is as follows: **(See Table 4)**

(f) Firearms receipt and disposition by licensed collectors. (1) Each licensed collector shall enter into a record each receipt and disposition of firearms curios or relics. The record required by this paragraph shall be maintained in bound form under the format prescribed below. The purchase or other acquisition of a curio or relic shall, except as provided in paragraph (g) of this section, be recorded not later than the close of the next business day following the date of such purchase or other acquisition. The record shall show the date of receipt, the name and address or the name and license number of the person from whom received, the name of the manufacturer and importer (if any), the model, serial number, type, and the caliber or gauge of the firearm curio or relic. The sale or other disposition of a curio or relic shall be recorded by the licensed collector not later than 7 days following the date of such transaction. When such disposition is made to a licensee, the commercial record of the transaction shall be retained, until the transaction is recorded, separate from other commercial documents maintained by the licensee, and be readily available for inspection. The record shall show the date of the sale or other disposition of each firearm curio or relic, the name and address of the person to whom the firearm curio or relic is transferred, or the name and license number of the person to whom transferred if such person is a licensee,

TABLE 4: Firearms Acquisition and Disposition Record

Description of firearm					Reciept		Disposition		
Manufacturer and/or importer	Model	Serial No.	Type	Caliber or gauge	Date	Name and address or name and license No.	Date	Name	Address or license No. if licensee, or Form 4473 Serial No. if Forms 4733 filed numerically

and the date of birth of the transferee if other than a licensee. In addition, the licensee shall cause the transferee, if other than a licensee, to be identified in any manner customarily used in commercial transactions (e.g., a driver's license), and note on the record the method used.

(2) The format required for the record of receipt and disposition of firearms by collectors is as follows: **(See Table 5)**

(g) Commercial records of firearms received. When a commercial record is held by a licensed dealer or licensed collector showing the acquisition of a firearm or firearm curio or relic, and such record contains all acquisition information required by the bound record prescribed by paragraphs (e) and (f) of this section, the licensed dealer or licensed collector acquiring such firearm or curio or relic, may, for a period not exceeding 7 days following the date of such acquisition, delay making the required entry into such bound record: Provided, That the commercial record is, until such time as the required entry into the bound record is made, (1) maintained by the licensed dealer or licensed collector separate from other commercial documents maintained by such licensee, and (2) readily available for inspection on the licensed premises: Provided further, That when disposition is made of a firearm or firearm curio or relic not entered in the bound record under the provisions of this paragraph, the licensed dealer or licensed collector making such disposition shall enter all required acquisition information regarding the firearm or firearm curio or relic in the bound record at the time such transfer or disposition is made.

(h) Alternate records. Notwithstanding the provisions of paragraphs (a),

(e), and (f) of this section, the Director of Industry Operations may authorize alternate records to be maintained by a licensed dealer or licensed collector to record the acquisition and disposition of firearms or curios or relics and the disposition of armor piercing ammunition when it is shown by the licensed dealer or the licensed collector that such alternate records will accurately and readily disclose the required information. A licensed dealer or licensed collector who proposes to use alternate records shall submit a letter application, in duplicate, to the Director of Industry Operations and shall describe the proposed alternate records and the need therefor. Such alternate records shall not be employed by the licensed dealer or licensed collector until approval in such regard is received from the Director of Industry Operations.

(i) Requirements for importers and manufacturers. Each licensed importer and licensed manufacturer selling or otherwise disposing of firearms or armor piercing ammunition to nonlicensees shall maintain such records of such transactions as are required of licensed dealers by this section.

(Approved by the Office of Management and Budget under control number 1140–0032)

§ 478.125a Personal firearms collection.

(a) Notwithstanding any other provision of this subpart, a licensed manufacturer, licensed importer, or licensed dealer is not required to comply with the provisions of §478.102 or record on a firearms transaction record, Form 4473, the sale or other disposition of a firearm maintained as part of the licensee's personal firearms collection: **Provided,** That

(1) The licensee has maintained the firearm as part of such collection for 1 year from the date the firearm was transferred from the business inventory into the personal collection or otherwise acquired as a personal firearm,

(2) The licensee recorded in the bound record prescribed by §478.125(e) the receipt of the firearm into the business inventory or other acquisition,

(3) The licensee recorded the firearm as a disposition in the bound record prescribed by §478.125(e) when the firearm was transferred from the business inventory into the personal firearms collection or otherwise acquired as a personal firearm, and

(4) The licensee enters the sale or other disposition of the firearm from the personal firearms collection into a bound record, under the format prescribed below, identifying the firearm transferred by recording the name of the manufacturer and importer (if any), the model, serial number, type, and the caliber or gauge, and showing the date of the sale or other disposition, the name and address of the transferee, or the name and business address of the transferee if such person is a licensee, and the date of birth of the transferee if other than a licensee. In addition, the licensee shall cause the transferee, if other than a licensee, to be identified in any manner customarily used in commercial transactions (e.g., a drivers license). The format required for the disposition record of personal firearms is as follows: **(See Table 6)**

(b) Any licensed manufacturer, licensed importer, or licensed dealer selling or otherwise disposing of a firearm

TABLE 5: Firearms Collectors Acquisition and Disposition Record

Description of firearm					Reciept		Disposition			
Manu-facturer and/or importer	Model	Serial No.	Type	Caliber or gauge	Date	Name and address or name and license No.	Date	Name and address or name and license No.	Date of birth if nonlicensee	Driver's license No. or other identification if nonlicensee

from the licensee's personal firearms collection under this section shall be subject to the restrictions imposed by the Act and this part on the dispositions of firearms by persons other than licensed manufacturers, licensed importers, and licensed dealers.

(Approved by the Office of Management and Budget under control number 1140–0032)

§ 478.126 Furnishing transaction information.

(a) Each licensee shall, when required by letter issued by the Director of Industry Operations, and until notified to the contrary in writing by such officer, submit on Form 5300.5, Report of Firearms Transactions, for the periods and at the times specified in the letter issued by the Director of Industry Operations, all record information required by this subpart, or such lesser record information as the Director of Industry Operations in his letter may specify.

(b) The Director of Industry Operations may authorize the information to be submitted in a manner other than that prescribed in paragraph (a) of this section when it is shown by a licensee that an alternate method of reporting is reasonably necessary and will not unduly hinder the effective administration of this part. A licensee who proposes to use an alternate method of reporting shall submit a letter application, in duplicate, to the Director of Industry Operations and shall describe the proposed alternate method of reporting and the need therefor. An alternate method of reporting shall not be employed by the licensee until approval in such regard is received from the Director of Industry Operations.

(Approved by the Office of Management and Budget under control number 1140–0032)

§ 478.126a Reporting multiple sales or other disposition of pistols and revolvers.

Each licensee shall prepare a report of multiple sales or other disposition when-

ever the licensee sells or otherwise disposes of, at one time or during any five consecutive business days, two or more pistols, or revolvers, or any combination of pistols and revolvers totaling two or more, to an unlicensed person: Provided, That a report need not be made where pistols or revolvers, or any combination thereof, are returned to the same person from whom they were received. The report shall be prepared on Form 3310.4, Report of Multiple Sale or Other Disposition of Pistols and Revolvers. Not later than the close of business on the day that the multiple sale or other disposition occurs, the licensee shall forward two copies of Form 3310.4 to the ATF office specified thereon and one copy to the State police or to the local law enforcement agency in which the sale or other disposition took place. Where the State or local law enforcement officials have notified the licensee that a particular official has been designated to receive Forms 3310.4, the licensee shall forward such forms to that designated official. The licensee shall retain one copy of Form 3310.4 and attach it to the firearms transaction record, Form 4473, executed upon delivery of the pistols or revolvers.

Example 1. A licensee sells a pistol and revolver in a single transaction to an unlicensed person. This is a multiple sale and must be reported not later than the close of business on the date of the transaction.

Example 2. A licensee sells a pistol on Monday and sells a revolver on the following Friday to the same unlicensed person. This is a multiple sale and must be reported not later than the close of business on Friday. If the licensee sells the same unlicensed person another pistol or revolver on the following Monday, this would constitute an additional multiple sale and must also be reported.

Example 3. A licensee maintaining business hours on Monday through Saturday sells a revolver to an unlicensed person on Monday and sells another revolver to the same person on the following Saturday. This does not constitute a multiple sale and need not be reported

since the sales did not occur during five consecutive business days.

(Approved by the Office of Management and Budget under control number 1140–0003)

§ 478.127 Discontinuance of business.

Where a licensed business is discontinued and succeeded by a new licensee, the records prescribed by this subpart shall appropriately reflect such facts and shall be delivered to the successor. Where discontinuance of the business is absolute, the records shall be delivered within 30 days following the business discontinuance to the ATF Out–of–Business Records Center, 244 Needy Road, Martinsburg, West Virginia 25405, or to any ATF office in the division in which the business was located: Provided, however, Where State law or local ordinance requires the delivery of records to other responsible authority, the Chief, Federal Firearms Licensing Center may arrange for the delivery of the records required by this subpart to such authority: Provided further, That where a licensed business is discontinued and succeeded by a new licensee, the records may be delivered within 30 days following the business discontinuance to the ATF Out–of–Business Records Center or to any ATF office in the division in which the business was located.

§ 478.128 False statement or representation.

(a) Any person who knowingly makes any false statement or representation in applying for any license or exemption or relief from disability, under the provisions of the Act, shall be fined not more than $5,000 or imprisoned not more than 5 years, or both.

(b) Any person other than a licensed manufacturer, licensed importer, licensed dealer, or licensed collector who knowingly makes any false statement or representation with respect to any information required by the provisions of the Act or this part to be kept in the records of a person licensed under the Act or

TABLE 6: Disposition Record of Personal Firearms

Description of firearm					Disposition		
Manufacturer and/or importer	Model	Serial No.	Type	Caliber or gauge	Date	Name and address (business address if licensee)	Date of birth of nonlicensee

this part shall be fined not more than $5,000 or imprisoned not more than 5 years, or both.

(c) Any licensed manufacturer, licensed importer, licensed dealer, or licensed collector who knowingly makes any false statement or representation with respect to any information required by the provisions of the Act or this part to be kept in the records of a person licensed under the Act or this part shall be fined not more than $1,000 or imprisoned not more than 1 year, or both.

§ 478.129 Record retention.

(a) *Records prior to Act.* Licensed importers and licensed manufacturers may dispose of records of sale or other disposition of firearms prior to December 16, 1968. Licensed dealers and licensed collectors may dispose of all records of firearms transactions that occurred prior to December 16, 1968.

(b) *Firearms transaction record.* Licensees shall retain each Form 4473 and Form 4473(LV) for a period of not less than 20 years after the date of sale or disposition. Where a licensee has initiated a NICS check for a proposed firearms transaction, but the sale, delivery, or transfer of the firearm is not made, the licensee shall record any transaction number on the Form 4473, and retain the Form 4473 for a period of not less than 5 years after the date of the NICS inquiry. Forms 4473 shall be retained in the licensee's records as provided in §478.124(b): Provided, That Forms 4473 with respect to which a sale, delivery or transfer did not take place shall be separately retained in alphabetical (by name of transferee) or chronological (by date of transferee's certification) order.

(c) *Statement of intent to obtain a handgun, reports of multiple sales or other disposition of pistols and revolvers, and reports of theft or loss of firearms.* Licensees shall retain each Form 5300.35 (Statement of Intent to Obtain a Handgun(s)) for a period of not less than 5 years after notice of the intent to obtain the handgun was forwarded to the chief law enforcement officer, as defined in §478.150(c). Licensees shall retain each copy of Form 3310.4 (Report of Multiple Sale or Other Disposition of Pistols and Revolvers) for a period of not less than 5 years after the date of sale or other disposition. Licensees shall retain each copy of Form 3310.11 (Federal Firearms Licensee Theft/Loss Report) for a period

of not less than 5 years after the date the theft or loss was reported to ATF.

(d) *Records of importation and manufacture.* Licensees will maintain permanent records of the importation, manufacture, or other acquisition of firearms, including ATF Forms 6 and 6A as required by subpart G of this part. Licensed importers' records and licensed manufacturers' records of the sale or other disposition of firearms after December 15, 1968, shall be retained through December 15, 1988, after which records of transactions over 20 years of age may be discarded.

(e) *Records of dealers and collectors under the Act.* The records prepared by licensed dealers and licensed collectors under the Act of the sale or other disposition of firearms and the corresponding record of receipt of such firearms shall be retained through December 15, 1988, after which records of transactions over 20 years of age may be discarded.

(f) *Retention of records of transactions in semiautomatic assault weapons.* The documentation required by §§478.40(c) and 478.132 shall be retained in the licensee's permanent records for a period of not less than 5 years after the date of sale or other disposition.

(Paragraph (b) approved by the Office of Management and Budget under control number 1512–0544; Paragraph (c) approved by the Office of Management and Budget under control numbers 1512–0520, 1512–0006, and 1512–0524; Paragraph (f) approved by the Office of Management and Budget under control number 1512–0526; all other recordkeeping approved by the Office of Management and Budget under control number 1512–0129)

§ 478.131 Firearms transactions not subject to a NICS check.

(a) (1) A licensed importer, licensed manufacturer, or licensed dealer whose sale, delivery, or transfer of a firearm is made pursuant to the alternative provisions of §478.102(d) and is not subject to the NICS check prescribed by §478.102(a) shall maintain the records required by paragraph (a) of this section.

(2) If the transfer is pursuant to a permit or license in accordance with §478.102(d)(1), the licensee shall either retain a copy of the purchaser's permit or license and attach it to the firearms transaction record, Form 4473, or record on the firearms trans-

action record, Form 4473, any identifying number, the date of issuance, and the expiration date (if provided) from the permit or license.

(3) If the transfer is pursuant to a certification by ATF in accordance with §§478.102(d)(3) and 478.150, the licensee shall maintain the certification as part of the records required to be kept under this subpart and for the period prescribed for the retention of Form 5300.35 in §478.129(c).

(b) The requirements of this section shall be in addition to any other recordkeeping requirements contained in this part.

(Approved by the Office of Management and Budget under control number 1140–0045)

§ 478.132 Dispositions of semi-automatic assault weapons and large capacity ammunition feeding devices to law enforcement officers for official use and to employees or contractors of nuclear facilities.

Editor's Note:

Section 478.132 is not applicable on or after September 13, 2004.

§ 478.133 Records of transactions in semiautomatic assault weapons.

The evidence specified in §478.40(c), relating to transactions in semiautomatic assault weapons, shall be retained in the permanent records of the manufacturer or dealer and in the records of the licensee to whom the weapons are transferred.

(Approved by the Office of Management and Budget under control number 1140–0041)

§ 478.134 Sale of firearms to law enforcement officers.

(a) Law enforcement officers purchasing firearms for official use who provide the licensee with a certification on agency letterhead, signed by a person in authority within the agency (other than the officer purchasing the firearm), stating that the officer will use the firearm in official duties and that a records check reveals that the purchasing officer has no convictions for misdemeanor crimes of domestic violence are not required to complete Form 4473 or Form 5300.35. The law enforcement officer purchasing the firearm may purchase a firearm from a licensee in another State, regardless of where the officer resides or where the agency is located.

(b) (1) The following individuals are considered to have sufficient authority to certify that law enforcement officers purchasing firearms will use the firearms in the performance of official duties:

(i) In a city or county police department, the director of public safety or the chief or commissioner of police.

(ii) In a sheriff's office, the sheriff.

(iii) In a State police or highway patrol department, the superintendent or the supervisor in charge of the office to which the State officer or employee is assigned.

(iv) In Federal law enforcement offices, the supervisor in charge of the office to which the Federal officer or employee is assigned.

(2) An individual signing on behalf of the person in authority is acceptable, provided there is a proper delegation of authority.

(c) Licensees are not required to prepare a Form 4473 or Form 5300.35 covering sales of firearm made in accordance with paragraph (a) of this section to law enforcement officers for official use. However, disposition to the officer must be entered into the licensee's permanent records, and the certification letter must be retained in the licensee's files.

Subpart I—Exemptions, Seizures, and Forfeitures

§ 478.141 General.

With the exception of §§478.32(a)(9) and (d)(9) and 478.99(c)(9), the provisions of this part shall not apply with respect to:

(a) The transportation, shipment, receipt, possession, or importation of any firearm or ammunition imported for, sold or shipped to, or issued for the use of, the United States or any department or agency thereof or any State or any department, agency, or political subdivision thereof.

(b) The shipment or receipt of firearms or ammunition when sold or issued by the Secretary of the Army pursuant to section 4308 of Title 10, U.S.C., and the transportation of any such firearm or ammunition carried out to enable a person, who lawfully received such firearm or ammunition from the Secretary of the Army, to engage in military training or in competitions.

(c) The shipment, unless otherwise prohibited by the Act or any other Federal law, by a licensed importer, licensed manufacturer, or licensed dealer to a member of the U.S. Armed Forces on active duty outside the United States or to clubs, recognized by the Department of Defense, whose entire membership is composed of such members of the U.S. Armed Forces, and such members or clubs may receive a firearm or ammunition determined by the Director to be generally recognized as particularly suitable for sporting purposes and intended for the personal use of such member or club. Before making a shipment of firearms or ammunition under the provisions of this paragraph, a licensed importer, licensed manufacturer, or licensed dealer may submit a written request, in duplicate, to the Director for a determination by the Director whether such shipment would constitute a violation of the Act or any other Federal law, or whether the firearm or ammunition is considered by the Director to be generally recognized as particularly suitable for sporting purposes.

(d) The transportation, shipment, receipt, possession, or importation of any antique firearm.

§ 478.142 Effect of pardons and expunctions of convictions.

(a) A pardon granted by the President of the United States regarding a Federal conviction for a crime punishable by imprisonment for a term exceeding 1 year shall remove any disability which otherwise would be imposed by the provisions of this part with respect to that conviction.

(b) A pardon granted by the Governor of a State or other State pardoning authority or by the pardoning authority of a foreign jurisdiction with respect to a conviction, or any expunction, reversal, setting aside of a conviction, or other proceeding rendering a conviction nugatory, or a restoration of civil rights shall remove any disability which otherwise would be imposed by the provisions of this part with respect to the conviction, unless:

(1) The pardon, expunction, setting aside, or other proceeding rendering a conviction nugatory, or restoration of civil rights expressly provides that the person may not ship, transport, possess or receive firearms; or

(2) The pardon, expunction, setting aside, or other proceeding rendering a conviction nugatory, or restoration of civil rights did not fully restore the rights of the person to possess or receive firearms under the law of the jurisdiction where the conviction occurred.

§ 478.143 Relief from disabilities incurred by indictment.

A licensed importer, licensed manufacturer, licensed dealer, or licensed collector who is indicted for a crime punishable by imprisonment for a term exceeding 1 year may, notwithstanding any other provision of the Act, continue operations pursuant to his existing license during the term of such indictment and until any conviction pursuant to the indictment becomes final: Provided, That if the term of the license expires during the period between the date of the indictment and the date the conviction thereunder becomes final, such importer, manufacturer, dealer, or collector must file a timely application for the renewal of his license in order to continue operations. Such application shall show that the applicant is under indictment for a crime punishable by imprisonment for a term exceeding 1 year.

§ 478.144 Relief from disabilities under the Act.

(a) Any person may make application for relief from the disabilities under section 922 (g) and (n) of the Act (see §478.32).

(b) An application for such relief shall be filed, in triplicate, with the Director. It shall include the information required by this section and such other supporting data as the Director and the applicant deem appropriate.

(c) Any record or document of a court or other government entity or official required by this paragraph to be furnished by an applicant in support of an application for relief shall be certified by the court or other government entity or official as a true copy. An application shall include:

(1) In the case of an applicant who is an individual, a written statement from each of 3 references, who are not related to the applicant by blood or marriage and have known the applicant for at least 3 years, recommending the granting of relief;

(2) Written consent to examine and obtain copies of records and to receive statements and information regarding the applicant's background, including records, statements and other information concerning employment, medical history, military service, and criminal record;

(3) In the case of an applicant under indictment, a copy of the indictment or information;

(4) In the case of an applicant having been convicted of a crime punishable by imprisonment for a term exceeding 1 year, a copy of the indictment or information on which the applicant was convicted, the judgment of conviction or record of any plea of nolo contendere or plea of guilty or finding of guilt by the court, and any pardon, expunction, setting aside or other record purporting to show that the conviction was rendered nugatory or that civil rights were restored;

(5) In the case of an applicant who has been adjudicated a mental defective or committed to a mental institution, a copy of the order of a court, board, commission, or other lawful authority that made the adjudication or ordered the commitment, any petition that sought to have the applicant so adjudicated or committed, any medical records reflecting the reasons for commitment and diagnoses of the applicant, and any court order or finding of a court, board, commission, or other lawful authority showing the applicant's discharge from commitment, restoration of mental competency and the restoration of rights;

(6) In the case of an applicant who has been discharged from the Armed Forces under dishonorable conditions, a copy of the applicant's summary of service record (Department of Defense Form 214), charge sheet (Department of Defense Form 458), and final court martial order;

(7) In the case of an applicant who, having been a citizen of the United States, has renounced his or her citizenship, a copy of the formal renunciation of nationality before a diplomatic or consular officer of the United States in a foreign state or

before an officer designated by the Attorney General when the United States was in a state of war (see 8 U.S.C. 1481(a) (5) and (6)); and

(8) In the case of an applicant who has been convicted of a misdemeanor crime of domestic violence, a copy of the indictment or information on which the applicant was convicted, the judgment of conviction or record of any plea of nolo contendere or plea of guilty or finding of guilt by the court, and any pardon, expunction, setting aside or other record purporting to show that the conviction was rendered nugatory or that civil rights were restored.

(d) The Director may grant relief to an applicant if it is established to the satisfaction of the Director that the circumstances regarding the disability, and the applicant's record and reputation, are such that the applicant will not be likely to act in a manner dangerous to public safety, and that the granting of the relief would not be contrary to the public interest. The Director will not ordinarily grant relief if the applicant has not been discharged from parole or probation for a period of at least 2 years. Relief will not be granted to an applicant who is prohibited from possessing all types of firearms by the law of the State where such applicant resides.

(e) In addition to meeting the requirements of paragraph (d) of this section, an applicant who has been adjudicated a mental defective or committed to a mental institution will not be granted relief unless the applicant was subsequently determined by a court, board, commission, or other lawful authority to have been restored to mental competency, to be no longer suffering from a mental disorder, and to have had all rights restored.

(f) Upon receipt of an incomplete or improperly executed application for relief, the applicant shall be notified of the deficiency in the application. If the application is not corrected and returned within 30 days following the date of notification, the application shall be considered as having been abandoned.

(g) Whenever the Director grants relief to any person pursuant to this section, a notice of such action shall be promptly published in the Federal Register, together with the reasons therefor.

(h) A person who has been granted

relief under this section shall be relieved of any disabilities imposed by the Act with respect to the acquisition, receipt, transfer, shipment, transportation, or possession of firearms or ammunition and incurred by reason of such disability.

(i) (1) A licensee who incurs disabilities under the Act (see §478.32(a)) during the term of a current license or while the licensee has pending a license renewal application, and who files an application for removal of such disabilities, shall not be barred from licensed operations for 30 days following the date on which the applicant was first subject to such disabilities (or 30 days after the date upon which the conviction for a crime punishable by imprisonment for a term exceeding 1 year becomes final), and if the licensee files the application for relief as provided by this section within such 30–day period, the licensee may further continue licensed operations during the pendency of the application. A licensee who does not file such application within such 30–day period shall not continue licensed operations beyond 30 days following the date on which the licensee was first subject to such disabilities (or 30 days from the date the conviction for a crime punishable by imprisonment for a term exceeding 1 year becomes final).

(2) In the event the term of a license of a person expires during the 30–day period specified in paragraph (i)(1) of this section, or during the pendency of the application for relief, a timely application for renewal of the license must be filed in order to continue licensed operations. Such license application shall show that the applicant is subject to Federal firearms disabilities, shall describe the event giving rise to such disabilities, and shall state when the disabilities were incurred.

(3) A licensee shall not continue licensed operations beyond 30 days following the date the Director issues notification that the licensee's applications for removal of disabilities has been denied.

(4) When as provided in this paragraph a licensee may no longer continue licensed operations, any application for renewal of license filed by the licensee during the pendency of the application for removal of disabilities shall be denied by the Director of Industry Operations.

§ 478.145 Research organizations.

The provisions of §478.98 with respect to the sale or delivery of destructive devices, machine guns, short–barreled shotguns, and short–barreled rifles shall not apply to the sale or delivery of such devices and weapons to any research organization designated by the Director to receive same. A research organization desiring such designation shall submit a letter application, in duplicate, to the Director. Such application shall contain the name and address of the research organization, the names and addresses of the persons directing or controlling, directly or indirectly, the policies and management of such organization, the nature and purpose of the research being conducted, a description of the devices and weapons to be received, and the identity of the person or persons from whom such devices and weapons are to be received.

§ 478.146 Deliveries by mail to certain persons.

The provisions of this part shall not be construed as prohibiting a licensed importer, licensed manufacturer, or licensed dealer from depositing a firearm for conveyance in the mails to any officer, employee, agent, or watchman who, pursuant to the provisions of section 1715 of title 18, U.S.C., is eligible to receive through the mails pistols, revolvers, and other firearms capable of being concealed on the person, for use in connection with his official duties.

§ 478.147 Return of firearm.

A person not otherwise prohibited by Federal, State or local law may ship a firearm to a licensed importer, licensed manufacturer, or licensed dealer for any lawful purpose, and, notwithstanding any other provision of this part, the licensed manufacturer, licensed importer, or licensed dealer may return in interstate or foreign commerce to that person the firearm or a replacement firearm of the same kind and type. See §478.124(a) for requirements of a Form 4473 prior to return. A person not otherwise prohibited by Federal, State or local law may ship a firearm curio or relic to a licensed collector for any lawful purpose, and, notwithstanding any other provision of this part, the licensed collector may return in interstate or foreign commerce to that person the firearm curio or relic.

§ 478.148 Armor piercing ammunition intended for sporting or industrial purposes.

The Director may exempt certain armor piercing ammunition from the requirements of this part. A person who desires to obtain an exemption under this section for any such ammunition which is primarily intended for sporting purposes or intended for industrial purposes, including charges used in oil and gas well perforating devices, shall submit a written request to the Director. Each request shall be executed under the penalties of perjury and contain a complete and accurate description of the ammunition, the name and address of the manufacturer or importer, the purpose of and use for which it is designed and intended, and any photographs, diagrams, or drawings as may be necessary to enable the Director to make a determination. The Director may require that a sample of the ammunition be submitted for examination and evaluation.

§ 478.149 Armor piercing ammunition manufactured or imported for the purpose of testing or experimentation.

The provisions of §§478.37 and 478.99(d) with respect to the manufacture or importation of armor piercing ammunition and the sale or delivery of armor piercing ammunition by manufacturers and importers shall not apply to the manufacture, importation, sale or delivery of armor piercing ammunition for the purpose of testing or experimentation as authorized by the Director. A person desiring such authorization to receive armor piercing ammunition shall submit a letter application, in duplicate, to the Director. Such application shall contain the name and addresses of the persons directing or controlling, directly or indirectly, the policies and management of the applicant, the nature or purpose of the testing or experimentation, a description of the armor piercing ammunition to be received, and the identity of the manufacturer or importer from whom such ammunition is to be received. The approved application shall be submitted to the manufacturer or importer who shall retain a copy as part of the records required by subpart H of this part.

§ 478.150 Alternative to NICS in certain geographical locations.

(a) The provisions of §478.102(d)(3) shall be applicable when the Director has certified that compliance with the provisions of §478.102(a)(1) is impracticable because:

(1) The ratio of the number of law enforcement officers of the State in which the transfer is to occur to the number of square miles of land area of the State does not exceed 0.0025;

(2) The business premises of the licensee at which the transfer is to occur are extremely remote in relation to the chief law enforcement officer; and

(3) There is an absence of telecommunications facilities in the geographical area in which the business premises are located.

(b) A licensee who desires to obtain a certification under this section shall submit a written request to the Director. Each request shall be executed under the penalties of perjury and contain information sufficient for the Director to make such certification. Such information shall include statistical data, official reports, or other statements of government agencies pertaining to the ratio of law enforcement officers to the number of square miles of land area of a State and statements of government agencies and private utility companies regarding the absence of telecommunications facilities in the geographical area in which the licensee's business premises are located.

(c) For purposes of this section and §478.129(c), the "chief law enforcement officer" means the chief of police, the sheriff, or an equivalent officer or the designee of any such individual.

(Approved by the Office of Management and Budget under control number 1140–0045)

§ 478.151 Semiautomatic rifles or shotguns for testing or experimentation.

(a) The provisions of §478.39 shall not apply to the assembly of semiautomatic rifles or shotguns for the purpose of testing or experimentation as authorized by the Director.

(b) A person desiring authorization to assemble nonsporting semiautomatic rifles or shotguns shall submit a written request, in duplicate, to the Director. Each such request shall be executed under the penalties of perjury and shall contain a complete and accurate description of the firearm to be assembled, and such diagrams or drawings as may be necessary to enable the Director to make a determination. The Director may require the submission of the firearm parts for examination and evaluation. If the submission of the firearm parts is impractical, the person requesting the authorization shall so advise the Director and designate the place where the firearm parts will be available for examination and evaluation.

(Paragraph (b) approved by the Office of Management and Budget under control number 1140–0037)

§ 478.152 Seizure and forfeiture.

(a) Any firearm or ammunition involved in or used in any knowing violation of subsections (a)(4), (a)(6), (f), (g), (h), (i), (j), or (k) of section 922 of the Act, or knowing importation or bringing into the United States or any possession thereof any firearm or ammunition in violation of section 922(l) of the Act, or knowing violation of section 924 of the Act, or willful violation of any other provision of the Act or of this part, or any violation of any other criminal law of the United States, or any firearm or ammunition intended to be used in any offense referred to in paragraph (c) of this section, where such intent is demonstrated by clear and convincing evidence, shall be subject to seizure and forfeiture, and all provisions of the Internal Revenue Code of 1986 relating to the seizure, forfeiture, and disposition of firearms, as defined in section 5845(a) of that Code, shall, so far as applicable, extend to seizures and forfeitures under the provisions of the Act: **Provided**, That upon acquittal of the owner or possessor, or dismissal of the charges against such person other than upon motion of the Government prior to trial, or lapse of or court termination of the restraining order to which he is subject, the seized or relinquished firearms or ammunition shall be returned forthwith to the owner or possessor or to a person delegated by the owner or possessor unless the return of the firearms or ammunition would place the owner or possessor or the delegate of the owner or possessor

in violation of law. Any action or proceeding for the forfeiture of firearms or ammunition shall be commenced within 120 days of such seizure.

(b) Only those firearms or quantities of ammunition particularly named and individually identified as involved in or used in any violation of the provisions of the Act or this part, or any other criminal law of the United States or as intended to be used in any offense referred to in paragraph (c) of this section, where such intent is demonstrated by clear and convincing evidence, shall be subject to seizure, forfeiture and disposition.

(c) The offenses referred to in paragraphs (a) and (b) of this section for which firearms and ammunition intended to be used in such offenses are subject to seizure and forfeiture are:

(1) Any crime of violence, as that term is defined in section 924(c)(3) of the Act;

(2) Any offense punishable under the Controlled Substances Act (21 U.S.C. 801 *et seq.*) or the Controlled Substances Import and Export Act (21 U.S.C. 951 *et seq.*);

(3) Any offense described in section 922(a)(1), 922(a)(3), 922(a)(5), or 922(b)(3) of the Act, where the firearm or ammunition intended to be used in such offense is involved in a pattern of activities which includes a violation of any offense described in section 922(a)(1), 922(a)(3), 922(a)(5), or 922(b)(3) of the Act;

(4) Any offense described in section 922(d) of the Act where the firearm or ammunition is intended to be used in such offense by the transferor of such firearm or ammunition;

(5) Any offense described in section 922(i), 922(j), 922(l), 922(n), or 924(b) of the Act; and

(6) Any offense which may be prosecuted in a court of the United States which involves the exportation of firearms or ammunition.

§478.153 Semiautomatic assault weapons and large capacity ammunition feeding devices manufactured or imported for the purposes of testing or experimentation.

Editor's Note:

Section 478.153 is not applicable on or after September 13, 2004.

Subpart J—[Reserved]

Subpart K—Exportation

§ 478.171 Exportation.

Firearms and ammunition shall be exported in accordance with the applicable provisions of section 38 of the Arms Export Control Act (22 U.S.C. 2778) and regulations thereunder. However, licensed manufacturers, licensed importers, and licensed dealers exporting firearms shall maintain records showing the manufacture or acquisition of the firearms as required by this part and records showing the name and address of the foreign consignee of the firearms and the date the firearms were exported. Licensed manufacturers and licensed importers exporting armor piercing ammunition and semiautomatic assault weapons manufactured after September 13, 1994, shall maintain records showing the name and address of the foreign consignee and the date the armor piercing ammunition or semiautomatic assault weapons were exported.

Editor's Note:

The references to "semiautomatic assault weapons" in section 478.171 are not applicable.

THE NATIONAL FIREARMS ACT

TITLE 26, UNITED STATES CODE, CHAPTER 53
INTERNAL REVENUE CODE

SUBCHAPTER A—TAXES

Part I—Special Occupational Taxes

§ 5801 Imposition of tax.

(a) General rule. On 1st engaging in business and thereafter on or before July 1 of each year, every importer, manufacturer, and dealer in firearms shall pay a special (occupational) tax for each place of business at the following rates:

(1) Importers and manufacturers: $1,000 a year or fraction thereof.

(2) Dealers: $500 a year or fraction thereof.

(b) Reduced rates of tax for small importers and manufacturers

(1) In general. Paragraph (1) of subsection (a) shall be applied by substituting "**$500**" for "**$1,000**" with respect to any taxpayer the gross receipts of which (for the most recent taxable year ending before the 1st day of the taxable period to which the tax imposed by subsection (a) relates) are less than $500,000.

(2) Controlled group rules. All persons treated as 1 taxpayer under section 5061(e)(3) shall be treated as 1 taxpayer for purposes of paragraph (1).

(3) Certain Rules to apply. For purposes of paragraph (1), rules similar to the rules of subparagraphs (B) and (C) of section 448(c)(3) shall apply.

§ 5802 Registration of importers, manufacturers, and dealers.

On first engaging in business and thereafter on or before the first day of July of each year, each importer, manufacturer, and dealer in firearms shall register with the Secretary in each internal revenue district in which such business is to be carried on, his name, including any trade name, and the address of each location in the district where he will conduct such business. An individual required to register under this section shall include a photograph and fingerprints of the individual with the initial application. Where there is a change during the taxable year in the location of, or the trade name used in, such business, the importer, manufacturer, or dealer shall file an application with the Secretary to amend his registration. Firearms operations of an importer, manufacturer, or dealer may not be commenced at the new location or under a new trade name prior to approval by the Secretary of the application.

Part II—Tax on Transferring Firearms.

§ 5811 Transfer tax.

(a) Rate. There shall be levied, collected, and paid on firearms transferred a tax at the rate of $200 for each firearm transferred, except, the transfer tax on any firearm classified as any other weapon under section 5845(e) shall be at the rate of $5 for each such firearm transferred.

(b) By whom paid. The tax imposed by subsection (a) of this section shall be paid by the transferor.

(c) Payment. The tax imposed by subsection (a) of this section shall be payable by the appropriate stamps prescribed for payment by the Secretary.

§ 5812 Transfers.

(a) Application. A firearm shall not be transferred unless **(1)** the transferor of the firearm has filed with the Secretary a written application, in duplicate, for the transfer and registration of the firearm to the transferee on the application form prescribed by the Secretary; **(2)** any tax payable on the transfer is paid as evidenced by the proper stamp affixed to the original application form; **(3)** the transferee is identified in the application form in such manner as the Secretary may by regulations prescribe, except that, if such person is an individual, the identification must include his fingerprints and his photograph; **(4)** the transferor of the firearm is identified in the application form in such manner as the Secretary may by regulations prescribe; **(5)** the firearm is identified in the application form in such manner as the Secretary may by regulations prescribe; and **(6)** the application form shows that the Secretary has approved the transfer and the registration of the firearm to the transferee. Applications shall be denied if the transfer, receipt, or possession of the firearm would place the transferee in violation of law.

(b) Transfer of possession. The transferee of a firearm shall not take possession of the firearm unless the Secretary has approved the transfer and registration of the firearm to the transferee as required by subsection (a) of this section.

Part III—Tax on Making Firearms.

§ 5821 Making tax.

(a) Rate. There shall be levied, collected, and paid upon the making of a firearm a tax at the rate of $200 for each firearm made.

(b) By whom paid. The tax imposed by subsection (a) of this section shall be paid by the person making the firearm.

(c) Payment. The tax imposed by subsection (a) of this section shall be payable by the stamp prescribed for payment by the Secretary.

§ 5822 Making.

No person shall make a firearm unless he has (a) filed with the Secretary a written application, in duplicate, to make and register the firearm on the form prescribed by the Secretary; (b) paid any tax payable on the making and such payment is evidenced by the proper stamp affixed to the original application form; (c) identified the firearm to be made in the application form in such manner as the Secretary may by regulations prescribe; (d) identified himself in the application form in such manner as the Secretary may by regulations prescribe, except that, if such person is an individual, the identification must include his fingerprints and his photograph; and (e) obtained the approval of the Secretary to make and register the firearm and the application form shows such approval. Applications shall be denied if the making or possession of the firearm would place the person making the firearm in violation of law.

SUBCHAPTER B—GENERAL PROVISIONS AND EXEMPTIONS

Part I—General Provisions.

§ 5841 Registration of firearms.

(a) Central registry. The Secretary shall maintain a central registry of all firearms in the United States which are not in the possession or under the control of the United States. This registry shall be known as the National Firearms Registration and Transfer Record. The registry shall include—

(1) identification of the firearm;

(2) date of registration; and

(3) identification and address of person entitled to possession of the firearm.

(b) By whom registered. Each manufacturer, importer, and maker shall register each firearm he manufactures, imports, or makes. Each firearm transferred shall be registered to the transferee by the transferor.

(c) How registered. Each manufacturer shall notify the Secretary of the manufacture of a firearm in such manner as may by regulations be prescribed and such notification shall effect the registration of the firearm required by this section. Each importer, maker, and transferor of a firearm shall, prior to importing, making, or transferring a firearm, obtain authorization in such manner as required by this chapter or regulations issued thereunder to import, make, or transfer the firearm, and such authorization shall effect the registration of the firearm required by this section.

(d) Firearms registered on effective date of the Act. A person shown as possessing a firearm by the records maintained by the Secretary pursuant to the National Firearms Act in force on the day immediately prior to the effective date of the National Firearms Act of 1968 1 shall be considered to have registered under this section the firearms in his possession which are disclosed by that record as being in his possession.

(e) Proof of registration. A person possessing a firearm registered as required by this section shall retain proof of registration which shall be made available to the Secretary upon request.

§ 5842 Identification of firearms.

(a) Identification of firearms other than destructive devices. Each manufacturer and importer and anyone making a firearm shall identify each firearm, other than a destructive device, manufactured, imported, or made by a serial number which may not be readily removed, obliterated, or altered, the name of the manufacturer, importer, or maker, and such other identification as the Secretary may by regulations prescribe.

(b) Firearms without serial number. Any person who possesses a firearm, other than a destructive device, which does not bear the serial number and other information required by subsection (a) of this section shall identify the firearm with a serial number assigned by the Secretary and any other information the Secretary may by regulations prescribe.

(c) Identification of destructive device. Any firearm classified as a destructive device shall be identified in such manner as the Secretary may by regulations prescribe.

§ 5843 Records and returns.

Importers, manufacturers, and dealers shall keep such records of, and render such returns in relation to, the importation, manufacture, making, receipt, and sale, or other disposition, of firearms as the Secretary may by regulations prescribe.

§ 5844 Importation.

No firearm shall be imported or brought into the United States or any territory under its control or jurisdiction unless the importer establishes, under regulations as may be prescribed by the Secretary, that the firearm to be imported or brought in is—

(1) being imported or brought in for

the use of the United States or any department, independent establishment, or agency thereof or any State or possession or any political subdivision thereof; or

(2) being imported or brought in for scientific or research purposes; or

(3) being imported or brought in solely for testing or use as a model by a registered manufacturer or solely for use as a sample by a registered importer or registered dealer;

except that, the Secretary may permit the conditional importation or bringing in of a firearm for examination and testing in connection with classifying the firearm.

§ 5845 Definitions.

For the purpose of this chapter—

(a) Firearm. The term **"firearm"** means **(1)** a shotgun having a barrel or barrels of less than 18 inches in length; **(2)** a weapon made from a shotgun if such weapon as modified has an overall length of less than 26 inches or a barrel or barrels of less than 18 inches in length; **(3)** a rifle having a barrel or barrels of less than 16 inches in length; **(4)** a weapon made from a rifle if such weapon as modified has an overall length of less than 26 inches or a barrel or barrels of less than 16 inches in length; **(5)** any other weapon, as defined in subsection (e); **(6)** a machinegun; **(7)** any silencer (as defined in section 921 of title 18, United States Code); and **(8)** a destructive device. The term **"firearm"** shall not include an antique firearm or any device (other than a machinegun or destructive device) which, although designed as a weapon, the Secretary finds by reason of the date of its manufacture, value, design, and other characteristics is primarily a collector's item and is not likely to be used as a weapon.

(b) Machinegun. The term **"machinegun"** means any weapon which shoots, is designed to shoot, or can be readily restored to shoot, automatically more than one shot, without manual reloading, by a single function of the trigger. The term shall also include the frame or receiver of any such weapon, any part designed and intended solely and exclusively, or combination of parts designed and intended, for use in converting a weapon into a machinegun, and any combination of parts from which a machinegun can be assembled if such parts are in the possession or under the control of a person.

(c) Rifle. The term **"rifle"** means a weapon designed or redesigned, made or remade, and intended to be fired from the shoulder and designed or redesigned and made or remade to use the energy of the explosive in a fixed cartridge to fire only a single projectile through a rifled bore for each single pull of the trigger, and shall include any such weapon which may be readily restored to fire a fixed cartridge.

(d) Shotgun. The term **"shotgun"** means a weapon designed or redesigned, made or remade, and intended to be fired from the shoulder and designed or redesigned and made or remade to use the energy of the explosive in a fixed shotgun shell to fire through a smooth bore either a number of projectiles (ball shot) or a single projectile for each pull of the trigger, and shall include any such weapon which may be readily restored to fire a fixed shotgun shell.

(e) Any other weapon. The term **"any other weapon"** means any weapon or device capable of being concealed on the person from which a shot can be discharged through the energy of an explosive, a pistol or revolver having a barrel with a smooth bore designed or redesigned to fire a fixed shotgun shell, weapons with combination shotgun and rifle barrels 12 inches or more, less than 18 inches in length, from which only a single discharge can be made from either barrel without manual reloading, and shall include any such weapon which may be readily restored to fire. Such term shall not include a pistol or a revolver having a rifled bore, or rifled bores, or weapons designed, made, or intended to be fired from the shoulder and not capable of firing fixed ammunition.

(f) Destructive device. The term **"destructive device"** means **(1)** any explosive, incendiary, or poison gas (A) bomb, **(B)** grenade, **(C)** rocket having a propellent charge of more than four ounces, **(D)** missile having an explosive or incendiary charge of more than one–quarter ounce, **(E)** mine, or **(F)** similar device; **(2)** any type of weapon by whatever name known which will, or which may be readily converted to, expel a projectile by the action of an explosive or other propellant, the barrel or barrels of which have a bore of more than one–half inch in diameter, except a shotgun or shotgun shell which the Secretary finds is generally recognized as particularly suitable for sporting purposes; and **(3)** any combination of parts either designed or intended for use in converting any device into a destructive device as defined in subparagraphs (1) and (2) and from which a destructive device may be readily assembled. The term **"destructive device"** shall not include any device which is neither designed nor redesigned for use as a weapon; any device, although originally designed for use as a weapon, which is redesigned for use as a signaling, pyrotechnic, line throwing, safety, or similar device; surplus ordnance sold, loaned, or given by the Secretary of the Army pursuant to the provisions of section 4684(2), 4685, or 4686 of title 10 of the United States Code; or any other device which the Secretary finds is not likely to be used as a weapon, or is an antique or is a rifle which the owner intends to use solely for sporting purposes.

(g) Antique firearm. The term **"antique firearm"** means any firearm not designed or redesigned for using rim fire or conventional center fire ignition with fixed ammunition and manufactured in or before 1898 (including any matchlock, flintlock, percussion cap, or similar type of ignition system or replica thereof, whether actually manufactured before or after the year 1898) and also any firearm using fixed ammunition manufactured in or before 1898, for which ammunition is no longer manufactured in the United States and is not readily available in the ordinary channels of commercial trade.

(h) Unserviceable firearm. The term **"unserviceable firearm"** means a firearm which is incapable of discharging a shot by means of an explosive and incapable of being readily restored to a firing condition.

(i) Make. The term **"make"**, and the various derivatives of such word, shall include manufacturing (other than by one qualified to engage in such business under this chapter), putting together, altering, any combination of these, or otherwise producing a firearm.

(j) Transfer. The term **"transfer"** and the various derivatives of such word, shall include selling, assigning, pledging, leasing, loaning, giving away, or otherwise disposing of.

(k) Dealer. The term **"dealer"** means any person, not a manufacturer or importer, engaged in the business of selling, renting, leasing, or loaning firearms and shall include pawnbrokers who accept firearms as collateral for loans.

(l) Importer. The term "importer" means any person who is engaged in the business of importing or bringing firearms into the United States.

(m) Manufacturer. The term "manufacturer" means any person who is engaged in the business of manufacturing firearms.

§ 5846 Other laws applicable.

All provisions of law relating to special taxes imposed by chapter 51 and to engraving, issuance, sale, accountability, cancellation, and distribution of stamps for tax payment shall, insofar as not inconsistent with the provisions of this chapter, be applicable with respect to the taxes imposed by sections 5801, 5811, and 5821.

§ 5847 Effect on other laws.

Nothing in this chapter shall be construed as modifying or affecting the requirements of section 414 of the Mutual Security Act of 1954, as amended, with respect to the manufacture, exportation, and importation of arms, ammunition, and implements of war.

§ 5848 Restrictive use of information.

(a) General Rule. No information or evidence obtained from an application, registration, or records required to be submitted or retained by a natural person in order to comply with any provision of this chapter or regulations issued thereunder, shall, except as provided in subsection (b) of this section, be used, directly or indirectly, as evidence against that person in a criminal proceeding with respect to a violation of law occurring prior to or concurrently with the filing of the application or registration, or the compiling of the records containing the information or evidence.

(b) Furnishing false information. Subsection (a) of this section shall not preclude the use of any such information or evidence in a prosecution or other action under any applicable provision of law with respect to the furnishing of false information.

§ 5849 Citation of chapter.

This chapter may be cited as the **"National Firearms Act"** and any reference in any other provision of law to the **"National Firearms Act"** shall be held to refer to the provisions of this chapter.

Part II—Exemptions.

§ 5851 Special (occupational) tax exemption.

(a) Business with United States. Any person required to pay special (occupational) tax under section 5801 shall be relieved from payment of that tax if he establishes to the satisfaction of the Secretary that his business is conducted exclusively with, or on behalf of, the United States or any department, independent establishment, or agency thereof. The Secretary may relieve any person manufacturing firearms for, or on behalf of, the United States from compliance with any provision of this chapter in the conduct of such business.

(b) Application. The exemption provided for in subsection (a) of this section may be obtained by filing with the Secretary an application on such form and containing such information as may by regulations be prescribed. The exemptions must thereafter be renewed on or before July 1 of each year. Approval of the application by the Secretary shall entitle the applicant to the exemptions stated on the approved application.

§ 5852 General transfer and making tax exemption.

(a) Transfer. Any firearm may be transferred to the United States or any department, independent establishment, or agency thereof, without payment of the transfer tax imposed by section 5811.

(b) Making by a person other than a qualified manufacturer. Any firearm may be made by, or on behalf of, the United States, or any department, independent establishment, or agency thereof, without payment of the making tax imposed by section 5821.

(c) Making by a qualified manufacturer. A manufacturer qualified under this chapter to engage in such business may make the type of firearm which he is qualified to manufacture without payment of the making tax imposed by section 5821.

(d) Transfers between special (occupational) taxpayers. A firearm registered to a person qualified under this chapter to engage in business as an importer, manufacturer, or dealer may be transferred by that person without payment of the transfer tax imposed by section 5811 to any other person qualified under this chapter to manufacture, import, or deal in that type of firearm.

(e) Unserviceable firearm. An unserviceable firearm may be transferred as a curio or ornament without payment of the transfer tax imposed by section 5811, under such requirements as the Secretary may by regulations prescribe.

(f) Right to exemption. No firearm may be transferred or made exempt from tax under the provisions of this section unless the transfer or making is performed pursuant to an application in such form and manner as the Secretary may by regulations prescribe.

§ 5853 Transfer and making tax exemption available to certain governmental entities.

(a) Transfer. A firearm may be transferred without the payment of the transfer tax imposed by section 5811 to any State, possession of the United States, any political subdivision thereof, or any official police organization of such a government entity engaged in criminal investigations.

(b) Making. A firearm may be made without payment of the making tax imposed by section 5821 by, or on behalf of, any State, or possession of the United States, any political subdivision thereof, or any official police organization of such a government entity engaged in criminal investigations.

(c) Right to exemption. No firearm may be transferred or made exempt from tax under this section unless the transfer or making is performed pursuant to an application in such form and manner as the Secretary may by regulations prescribe.

§ 5854 Exportation of firearms exempt from transfer tax.

A firearm may be exported without payment of the transfer tax imposed under section 5811 provided that proof of the exportation is furnished in such form and manner as the Secretary may by regulations prescribe.

SUBCHAPTER C—PROHIBITED ACTS

§ 5861 Prohibited acts.

It shall be unlawful for any person—

(a) to engage in business as a manufacturer or importer of, or dealer in, firearms without having paid the special (occupational) tax required by section 5801 for his business or having registered as required by section 5802; or

(b) to receive or possess a firearm transferred to him in violation of the provisions of this chapter; or

(c) to receive or possess a firearm made in violation of the provisions of this chapter; or

(d) to receive or possess a firearm which is not registered to him in the National Firearms Registration and Transfer Record; or

(e) to transfer a firearm in violation of the provisions of this chapter; or

(f) to make a firearm in violation of the provisions of this chapter; or

(g) to obliterate, remove, change, or alter the serial number or other identification of a firearm required by this chapter; or

(h) to receive or possess a firearm having the serial number or other identification required by this chapter obliterated, removed, changed, or altered; or

(i) to receive or possess a firearm which is not identified by a serial number as required by this chapter; or

(j) to transport, deliver, or receive any firearm in interstate commerce which has not been registered as required by this chapter; or

(k) to receive or possess a firearm which has been imported or brought into the United States in violation of section 5844; or

(l) to make, or cause the making of, a false entry on any application, return, or record required by this chapter, knowing such entry to be false.

SUBCHAPTER D—PENALTIES AND FORFEITURES

§ 5871 Penalties.

Any person who violates or fails to comply with any provisions of this chapter shall, upon conviction, be fined not more than $10,000, or be imprisoned not more than ten years, or both.

§ 5872 Forfeitures.

(a) Laws applicable. Any firearm involved in any violation of the provisions of this chapter shall be subject to seizure and forfeiture, and (except as provided in subsection (b)) all the provisions of internal revenue laws relating to searches, seizures, and forfeitures of unstamped articles are extended to and made to apply to the articles taxed under this chapter, and the persons to whom this chapter applies.

(b) Disposal. In the case of the forfeiture of any firearm by reason of a violation of this chapter, no notice of public sale shall be required; no such firearm shall be sold at a public sale; if such firearm is forfeited for a violation of this chapter and there is no remission or mitigation of forfeiture thereof, it shall be delivered by the Secretary to the Administrator of General Services, General Services Administration, who may order such firearm destroyed or may sell it to any State, or possession, or political subdivision thereof, or at the request of the Secretary, may authorize its retention for official use of the Treasury Department, or may transfer it without charge to any executive department or independent establishment of the Government for use by it.

TITLE 27 CFR CHAPTER II

PART 479—MACHINE GUNS, DESTRUCTIVE DEVICES, AND CERTAIN OTHER FIREARMS

(This part was formerly designated as Part 179)

to effect transfer.

Subpart A—Scope of Regulations

§ 479.1 General.

This part contains the procedural and substantive requirements relative to the importation, manufacture, making, exportation, identification and registration of, and the dealing in, machine guns, destructive devices and certain other firearms under the provisions of the National Firearms Act (26 U.S.C. Chapter 53).

Subpart B—Definitions

§ 479.11 Meaning of terms.

When used in this part and in forms prescribed under this part, where not otherwise distinctly expressed or manifestly incompatible with the intent thereof, terms shall have the meanings ascribed in this section. Words in the plural form shall include the singular, and vice versa, and words importing the masculine gender shall include the feminine. The terms "includes" and "including" do not exclude other things not enumerated which are in the same general class or are otherwise within the scope thereof.

Antique firearm. Any firearm not designed or redesigned for using rim fire or conventional center fire ignition with fixed ammunition and manufactured in or before 1898 (including any matchlock, flintlock, percussion cap, or similar type of ignition system or replica thereof, whether actually manufactured before or after the year 1898) and also any firearm using fixed ammunition manufactured in or before 1898, for which ammunition is no longer manufactured in the United States and is not readily available in the ordinary channels of commercial trade.

Any other weapon. Any weapon or device capable of being concealed on the person from which a shot can be discharged through the energy of an explosive, a pistol or revolver having a barrel with a smooth bore designed or redesigned to fire a fixed shotgun shell, weapons with combination shotgun and rifle barrels 12 inches or more, less than 18 inches in length, from which only a single discharge can be made from either barrel without manual reloading, and shall include any such weapon which may be readily restored to fire. Such term shall not include a pistol or a revolver having a rifled bore, or rifled bores, or weapons designed, made, or intended to be fired from the shoulder and not capable of firing fixed ammunition.

ATF officer. An officer or employee of the Bureau of Alcohol, Tobacco, Firearms and Explosives (ATF) authorized to perform any function relating to the administration or enforcement of this part.

Customs officer. Any officer of the U.S. Customs and Border Protection, any commissioned, warrant, or petty officer of the Coast Guard, or any agent or other person authorized by law to perform the duties of a customs officer.

Dealer. Any person, not a manufacturer or importer, engaged in the business of selling, renting, leasing, or loaning firearms and shall include pawnbrokers who accept firearms as collateral for loans.

Destructive device. (a) Any explosive, incendiary, or poison gas **(1)** bomb, **(2)** grenade, **(3)** rocket having a propellent charge of more than 4 ounces, **(4)** missile having an explosive or incendiary charge of more than one–quarter ounce, **(5)** mine, or **(6)** similar device; **(b)** any type of weapon by whatever name known which will, or which may be readily converted to, expel a projectile by the action of an explosive or other propellant, the barrel or barrels of which have a bore of more than one–half inch in diameter, except a shotgun or shotgun shell which the Director finds is generally recognized as particularly suitable for sporting purposes; and **(c)** any combination of parts either designed or intended for use in converting any device into a destructive device as described in paragraphs (a) and (b) of this definition and from which a destructive device may be readily assembled. The term shall not include any device which is neither designed or redesigned for use as a weapon; any device, although originally designed for use as a weapon, which is redesigned for use as a signaling, pyrotechnic, line throwing, safety, or similar device; surplus ordnance sold, loaned, or given by the Secretary of the Army under 10 U.S.C. 4684(2), 4685, or 4686, or any device which the Director finds is not likely to be used as a weapon, or is an antique or is a rifle which the owner intends to use solely for sporting purposes.

Director. The Director, Bureau of Alcohol, Tobacco, Firearms and Explosives, the Department of Justice, Washington, DC.

Director, Industry Operations. The principal regional official responsible for administering regulations in this part.

Director of the Service Center. A director of an Internal Revenue Service Center in an internal revenue region.

District director. A district director of the Internal Revenue Service in an internal revenue district.

Executed under penalties of perjury. Signed with the prescribed declaration under the penalties of perjury as provided on or with respect to the return, form, or other document or, where no form of declaration is prescribed, with the declaration:

> **"I declare under the penalties of perjury that this—(insert type of document, such as, statement, application, request, certificate), including the documents submitted in support thereof, has been examined by me and, to the best of my knowledge and belief, is true, correct, and complete."**

Exportation. The severance of goods from the mass of things belonging to this country with the intention of uniting them to the mass of things belonging to some foreign country.

Exporter. Any person who exports firearms from the United States.

Firearm. (a) A shotgun having a barrel or barrels of less than 18 inches in length; **(b)** a weapon made from a shotgun if such weapon as modified has an overall length of less than 26 inches or a barrel or barrels of less than 18 inches in length; **(c)** a rifle having a barrel or barrels of less than 16 inches in length; **(d)** a weapon made from a rifle if such weapon as modified has an overall length of less than 26 inches or a barrel or barrels of less than 16 inches in length; **(e)** any other weapon, as defined in this subpart; **(f)** a machine gun; **(g)** a muffler or a silencer for any firearm whether or not such firearm is included within this definition; and **(h)** a destructive device. The term shall not include an antique firearm or any device (other than a machine gun or destructive device) which, although designed as a weapon, the Director finds by reason of the date of its manufacture, value, design, and other characteristics is primarily a collector's item and is not likely to be used as a weapon. For purposes of this definition, the length of the barrel having an integral chamber(s) on a shotgun or rifle shall be determined by measuring the distance between the muzzle and the face of the bolt, breech, or breech block when closed and when the shotgun or rifle is cocked. The overall length of a weapon made from a shotgun or rifle is the distance between the extreme ends of the weapon measured along a line parallel to the center line of the bore.

Fixed ammunition. That self–contained unit consisting of the case, primer, propellant charge, and projectile or projectiles.

Frame or receiver. That part of a firearm which provides housing for the hammer, bolt or breechblock and firing mechanism, and which is usually threaded at its forward portion to receive the barrel.

Importation. The bringing of a firearm within the limits of the United States or any territory under its control or jurisdiction, from a place outside thereof (whether such place be a foreign country or territory subject to the jurisdiction of the United States), with intent to unlade. Except that, bringing a firearm from a foreign country or a territory subject to the jurisdiction of the United States into a foreign trade zone for storage pending shipment to a foreign country or subsequent importation into this country, under Title 26 of the United States Code, and this part, shall not be deemed importation.

Importer. Any person who is engaged in the business of importing or bringing firearms into the United States.

Machine gun. Any weapon which shoots, is designed to shoot, or can be readily restored to shoot, automatically more than one shot, without manual reloading, by a single function of the trigger. The term shall also include the frame or receiver of any such weapon, any part designed and intended solely and exclusively, or combination of parts designed and intended, for use in converting a weapon into a machine gun, and any combination of parts from which a machine gun can be assembled if such parts are in the possession or under the control of a person.

Make. This term and the various derivatives thereof shall include manufacturing (other than by one qualified to engage in such business under this part), putting together, altering, any combination of these, or otherwise producing a firearm.

Manual reloading. The inserting of a cartridge or shell into the chamber of a firearm either with the hands or by means of a mechanical device controlled and energized by the hands.

Manufacturer. Any person who is engaged in the business of manufacturing firearms.

Muffler or silencer. Any device for silencing, muffling, or diminishing the report of a portable firearm, including any combination of parts, designed or redesigned, and intended for the use in assembling or fabricating a firearm silencer or firearm muffler, and any part intended only for use in such assembly or fabrication.

Person. A partnership, company, association, trust, estate, or corporation, as well as a natural person.

Pistol. A weapon originally designed, made, and intended to fire a projectile (bullet) from one or more barrels when held in one hand, and having **(a)** a chamber(s) as an integral part(s) of, or permanently aligned with, the bore(s); and **(b)** a short stock designed to be gripped by one hand and at an angle to and extending below the line of the bore(s).

Revolver. A projectile weapon, of the pistol type, having a breechloading chambered cylinder so arranged that the cocking of the hammer or movement of the trigger rotates it and brings the next cartridge in line with the barrel for firing.

Rifle. A weapon designed or redesigned, made or remade, and intended to be fired from the shoulder and designed or redesigned and made or remade to use the energy of the explosive in a fixed cartridge to fire only a single projectile through a rifled bore for each single pull of the trigger, and shall include any such weapon which may be readily restored to fire a fixed cartridge.

Shotgun. A weapon designed or redesigned, made or remade, and intended to be fired from the shoulder and designed or redesigned and made or remade to use the energy of the explosive in a fixed shotgun shell to fire through a smooth bore either a number of projectiles (ball shot) or a single projectile for each pull of the trigger, and shall include any such weapon which may be readily restored to fire a fixed shotgun shell.

Transfer. This term and the various derivatives thereof shall include selling, assigning, pledging, leasing, loaning, giving away, or otherwise disposing of.

United States. The States and the District of Columbia.

U.S.C. The United States Code.

Unserviceable firearm. A firearm which is incapable of discharging a shot by means of an explosive and incapable of being readily restored to a firing condition.

Subpart C—Administrative and Miscellaneous Provisions

§ 479.21 Forms prescribed.

(a) The Director is authorized to prescribe all forms required by this part. All of the information called for in each form shall be furnished as indicated by the headings on the form and the instructions on or pertaining to the form. In addition, information called for in each form shall be furnished as required by this part. Each form requiring that it be executed under penalties of perjury shall be executed under penalties of perjury.

(b) Requests for forms should be submitted to the ATF Distribution Center (http://www.atf.gov) or by calling (703) 870-7526 or (703) 870-7528..

§ 479.22 Right of entry and examination.

Any ATF officer or employee of the Bureau of Alcohol, Tobacco, Firearms and Explosives duly authorized to perform any function relating to the administration or enforcement of this part may enter during business hours the premises (including places of storage) of any importer or manufacturer of or dealer in firearms, to examine any books, papers, or records required to be kept pursuant to this part, and any firearms kept by such importer, manufacturer or dealer on such premises, and may require the production of any books, papers, or records necessary to determine any liability for tax under 26 U.S.C. Chapter 53, or the observance of 26 U.S.C. Chapter 53, and this part.

§ 479.23 Restrictive use of required information.

No information or evidence obtained from an application, registration, or record required to be submitted or retained by a natural person in order to comply with any provision of 26 U.S.C. Chapter 53, or this part or section 207 of the Gun Control Act of 1968 shall be used, directly or indirectly, as evidence against that person in a criminal proceeding with respect to a violation of law occurring prior to or concurrently with the filing of the application or registration, or the compiling of the record containing the information or evidence: **Provided, however,** That the provisions of this section shall not preclude the use of any such information or evidence in a prosecution or other action under any applicable provision of law with respect to the furnishing of false information.

§ 479.24 Destructive device determination.

The Director shall determine in accordance with 26 U.S.C. 5845(f), whether a device is excluded from the definition of a destructive device. A person who desires to obtain a determination under that provision of law for any device which he believes is not likely to be used as a weapon shall submit a written request, in triplicate, for a ruling thereon to the Director. Each such request shall be executed under the penalties of perjury and contain a complete and accurate description of the device, the name and address of the manufacturer or importer thereof, the purpose of and use for which it is intended, and such photographs, diagrams, or drawings as may be necessary to enable the Director to make his determination. The Director may require the submission to him, of a sample of such device for examination and evaluation. If the submission of such device is impracticable, the person requesting the ruling shall so advise the Director and designate the place where the device will be available for examination and evaluation.

§ 479.25 Collector's items.

The Director shall determine in accordance with 26 U.S.C. 5845(a), whether a firearm or device, which although originally designed as a weapon, is by reason of the date of its manufacture, value, design, and other characteristics primarily a collector's item and is not likely to be used as a weapon. A person who desires to obtain a determination under that provision of law shall follow the procedures prescribed in §479.24 relating to destructive device determinations, and shall include information as to date of manufacture, value, design and other characteristics which would sustain a finding that the firearm or device is primarily a collector's item and is not likely to be used as a weapon.

§ 479.26 Alternate methods or procedures; emergency variations from requirements.

(a) Alternate methods or procedures. Any person subject to the provisions of this part, on specific approval by the Director as provided in this paragraph, may use an alternate method or procedure in lieu of a method or procedure specifically prescribed in this part. The Director may approve an alternate method or procedure, subject to stated conditions, when it is found that:

(1) Good cause is shown for the use of the alternate method or procedure;

(2) The alternate method or procedure is within the purpose of, and consistent with the effect intended by, the specifically prescribed method or procedure and that the alternate method or procedure is substantially equivalent to that specifically prescribed method or procedure; and

(3) The alternate method or procedure will not be contrary to any provision of law and will not result in an increase in cost to the Government or hinder the effective administration of this part. Where such person desires to employ an alternate method or procedure, a written application shall be submitted to the appropriate Director, Industry Operations, for transmittal to the Director. The application shall specifically describe the proposed alternate method or procedure and shall set forth the reasons for it. Alternate methods or procedures may not be employed until the application is approved by the Director. Such person shall, during the period of authorization of an alternate method or procedure, comply with the terms of the approved application. Authorization of any alternate method or procedure may be withdrawn whenever, in the judgment of the Director, the effective administration of this part is hindered by the continuation of the authorization.

(b) Emergency variations from requirements. The Director may approve a method of operation other than as specified in this part, where it is found that an emergency exists and the proposed variation from the specified requirements are necessary and the proposed variations (1) will not hinder the effective administration of this part, and (2) will not be contrary to any provisions of law. Variations from requirements granted under this paragraph are conditioned on compliance with the procedures, conditions, and limitations set forth in the approval of the application. Failure to comply in good faith with the procedures, conditions, and limitations shall automatically terminate the authority for the variations, and the person granted the variance shall fully comply with the prescribed requirements of regulations from which the variations were authorized. Authority for any variation may be withdrawn whenever, in the judgment of the Director, the effective administration of this part is hindered by the continuation of the variation. Where a person desires to employ an emergency variation, a written application shall be submitted to the appropriate Director, Industry Operations for transmittal to the Director. The application shall describe the proposed variation and set forth the reasons for it. Variations may not be employed until the application is approved.

(c) Retention of approved variations. The person granted the variance shall retain and make available for examination by ATF officers any application approved by the Director under this section.

Subpart D—Special (Occupational) Taxes

§ 479.31 Liability for tax.

(a) General. Every person who engages in the business of importing, manufacturing, or dealing in (including pawnbrokers) firearms in the United States shall pay a special (occupational) tax at a rate specified by §479.32. The tax shall be paid on or before the date of commencing the taxable business, and thereafter every year on or before July 1. Special (occupational) tax shall not be prorated. The tax shall be computed for the entire tax year (July 1 through June 30), regardless of the portion of the year during which the taxpayer engages in business. Persons commencing business at any time after July 1 in any year are liable for the special (occupational) tax for the entire tax year.

(b) Each place of business taxable. An importer, manufacturer, or dealer in firearms incurs special tax liability at each place of business where an occupation subject to special tax is conducted. A place of business means the entire office, plant or area of the business in any one location under the same proprietorship. Passageways, streets, highways, rail crossings, waterways, or partitions dividing the premises are not sufficient separation to require additional special tax, if the divisions of the premises are otherwise contiguous. See also §§479.38–479.39.

§ 479.32 Special (occupational) tax rates.

(a) Prior to January 1, 1988, the special (occupational) tax rates were as follows:

	Per year or fraction thereof
Class 1 – Importer of firearms	$500.
Class 2 – Manufacturer of firearms	$500.
Class 3 – Dealer in firearms	$200.
Class 4 – Importer only of weapons classified as "any other weapon"	$25.
Class 5 – Manufacturer only of weapons classified as "any other weapon"	$25.
Class 6 – Dealer only in weapons classified as "any other weapon"	$10.

(b) Except as provided in §479.32a, the special (occupational) tax rates effective January 1, 1988, are as follows:

	Per year or fraction thereof
Class 1 – Importer of firearms (including an importer only of weapons classified as "any other weapon")	$1000.
Class 2 – Manufacturer of firearms (including an manufacturer only of weapons classified as "any other weapon")	$1000.
Class 3 – Dealer of firearms (including an dealer only of weapons classified as "any other weapon")	$500.

(c) A taxpayer who was engaged in a business on January 1, 1988, for which a special (occupational) tax was paid for a taxable period which began before January 1, 1988, and included that date, shall pay an increased special tax for the period January 1, 1988, through June 30, 1988. The increased tax shall not exceed one–half the excess (if any) of **(1)** the rate of special tax in effect on January 1, 1988, over **(2)** the rate of such tax in effect on December 31, 1987. The increased special tax shall be paid on or before April 1, 1988.

§ 479.32a Reduced rate of tax for small importers and manufacturers.

(a) General. Effective January 1, 1988, 26 U.S.C. 5801(b) provides for a reduced rate of special tax with respect to any importer or manufacturer whose gross receipts (for the most recent taxable year ending before the first day of the taxable period to which the special tax imposed by §479.32 relates) are less than $500,000. The rate of tax for such an importer or manufacturer is $500 per year or fraction thereof. The **"taxable year"** to be used for determining gross receipts is the taxpayer's income tax year. All gross receipts of the taxpayer shall be included, not just the gross receipts of the business subject to special tax. Proprietors of new businesses that have not yet begun a taxable year, as well as proprietors of existing businesses that have not yet ended a taxable year, who commence a new activity subject to special tax, quality for the reduced special (occupational) tax rate, unless the business is a member of a **"controlled group"**; in that case, the rules of paragraph (b) of this section shall apply.

(b) Controlled group. All persons treated as one taxpayer under 26 U.S.C. 5061(e)(3) shall be treated as one taxpayer for the purpose of determining gross receipts under paragraph (a) of this section. **"Controlled group"** means a controlled group of corporations, as defined in 26 U.S.C. 1563 and implementing regulations in 26 CFR 1.1563–1 through 1.1563–4, except that the words "at least 80 percent" shall be replaced by the words "more than 50 percent" in each place they appear in subsection (a) of 26 U.S.C. 1563, as well as in the implementing regulations. Also, the rules for a **"controlled group of corporations"** apply in a similar fashion to groups which include partnerships and/or sole proprietorships. If one entity maintains more than 50% control over a group consisting of corporations and one, or more, partnerships and/or sole proprietorships, all of the members of the controlled group are one taxpayer for the purpose of this section.

(c) Short taxable year. Gross receipts for any taxable year of less than 12 months shall be annualized by multiplying the gross receipts for the short period by 12 and dividing the result by the number of months in the short period, as required by 26 U.S.C. 448(c)(3).

(d) Returns and allowances. Gross receipts for any taxable year shall be reduced by returns and allowances made during that year under 26 U.S.C. 448(c)(3).

§ 479.33 Special exemption.

(a) Any person required to pay special (occupational) tax under this part shall be relieved from payment of that tax if he establishes to the satisfaction of the Director that his business is conducted exclusively with, or on behalf of, the United States or any department, independent establishment, or agency thereof. The Director may relieve any person manufacturing firearms for or on behalf of the United States from compliance with any provision of this part in the conduct of the business with respect to such firearms.

(b) The exemption in this section may be obtained by filing with the Director an application, in letter form, setting out the manner in which the applicant conducts his business, the type of firearm to be manufactured, and proof satisfactory to the Director of the existence of the contract with the United States, department, independent establishment, or agency thereof, under which the applicant intends to operate.

§ 479.34 Special tax registration and return.

(a) General. Special tax shall be paid by return. The prescribed return is ATF Form 5630.7, Special Tax Registration and Return. Special tax returns, with payment of tax, shall be filed with ATF in accordance with instructions on the form. Properly completing, signing, and timely filing of a return (Form 5630.7) constitutes compliance with 26 U.S.C. 5802.

(b) Preparation of ATF Form 5630.7. All of the information called for on Form 5630.7 shall be provided, including:

(1) The true name of the taxpayer.

(2) The trade name(s) (if any) of the business(es) subject to special tax.

(3) The employer identification number (see §479.35).

(4) The exact location of the place of business, by name and number of building or street, or if these do not exist, by some description in addition to the post office address. In the case of one return for two or more locations, the address to be shown shall be the taxpayer's principal place of business (or principal office, in the case of a corporate taxpayer).

(5) The class(es) of special tax to which the taxpayer is subject.

(6) Ownership and control information: That is, the name, position, and residence address of every owner of the business and of every person having power to control its management and policies with respect to the activity subject to special tax. "Owner of the business" shall include every partner, if the taxpayer is a partnership, and every person owning 10% or more of its stock, if the taxpayer is a corporation. However, the ownership and control information required by this paragraph need not be stated if the same information has been previously provided to ATF in connection with a license application under Part 478 of this chapter, and if the information previously provided is still current.

(c) Multiple locations and/or classes of tax. A taxpayer subject to special tax for the same period at more than one location or for more than one class of tax shall—

(1) File one special tax return, ATF Form 5630.7, with payment of tax, to cover all such locations and classes of tax; and

(2) Prepare, in duplicate, a list identified with the taxpayer's name, address (as shown on ATF Form 5630.7), employer identification number, and period covered by the return. The list shall show, by States, the name, address, and tax class of each location for which special tax is being paid. The original of the list shall be filed with ATF in accordance with instructions on the return, and the copy shall be retained at the taxpayer's principal place of business (or principal office, in the case of a corporate taxpayer) for not less than 3 years.

(d) Signing of ATF Forms 5630.7—(1) Ordinary returns. The return of an individual proprietor shall be signed by the individual. The return of a partnership shall be signed by a general partner. The return of a corporation shall be signed by any officer. In each case, the person signing the return shall designate his or her capacity as "individual owner," "member of firm," or, in the case of a corporation, the title of the officer.

(2) Fiduciaries. Receivers, trustees, assignees, executors, administrators, and other legal representatives who continue the business of a bankrupt, insolvent, deceased person, etc., shall indicate the fiduciary capacity in which they act.

(3) Agent or attorney in fact. If a return is signed by an agent or attorney in fact, the signature shall be preceded by the name of the principal and followed by the title of the agent or attorney in fact. A return signed by a person as agent will not be accepted unless there is filed, with the ATF office with which the return is required to be filed, a power of attorney authorizing the agent to perform the act.

(4) Perjury statement. ATF Forms 5630.7 shall contain or be verified by a written declaration that the return has been executed under the penalties of perjury.

(e) Identification of taxpayer. If the taxpayer is an individual, with the initial return such person shall securely attach to Form 5630.7 a photograph of the individual 2 × 2 inches in size, clearly showing a full front view of the features of the individual with head bare, with the distance from the top of the head to the point of the chin approximately 1 1⁄4 inches, and which shall have been taken within 6 months prior to the date of completion of the return. The individual shall also attach to the return a properly completed FBI Form FD–258 (Fingerprint Card). The fingerprints must be clear for accurate classification and should be taken by someone properly equipped to take them: Provided, That the provisions of this paragraph shall not apply to individuals who have filed with ATF a properly executed Application for License under 18 U.S.C. Chapter 44, Firearms, ATF Form 7 (5310.12), as specified in §478.44(a).

§ 479.35 Employer identification number.

(a) Requirement. The employer identification number (defined in 26 CFR 301.7701–12) of the taxpayer who has been assigned such a number shall be shown on each special tax return, including amended returns, filed under this subpart. Failure of the taxpayer to include the employer identification number may result in the imposition of the penalty specified in §70.113 of this chapter.

(b) Application for employer identification number. Each taxpayer who files a special tax return, who has not already been assigned an employer identification number, shall file IRS Form SS–4 to apply for one. The taxpayer shall apply for and be assigned only one employer identification number, regardless of the number of places of business for which the taxpayer is required to file a special tax return. The employer identification number shall be applied for no later than 7 days after the filing of the taxpayer's first special tax return. IRS Form SS–4 may be obtained from the director of an IRS service center or from any IRS district director.

(c) Preparation and filing of IRS Form SS–4. The taxpayer shall prepare and file IRS Form SS–4, together with any supplementary statement, in accordance with the instructions on the form or issued in respect to it.

§ 479.36 The special tax stamp, receipt for special (occupational) taxes.

Upon filing a properly completed and executed return (Form 5630.7) accompanied by remittance of the full amount due, the taxpayer will be issued a special tax stamp as evidence of payment of the special (occupational) tax.

§ 479.37 Certificates in lieu of stamps lost or destroyed.

When a special tax stamp has been lost or destroyed, such fact should be reported immediately to the Chief, National Firearms Act Branch who issued the stamp. A certificate in lieu of the lost or destroyed stamp will be issued to the taxpayer upon the submission of an affidavit showing to the satisfaction of the Chief, National Firearms Act Branch that the stamp was lost or destroyed.

§ 479.38 Engaging in business at more than one location.

A person shall pay the special (occupational) tax for each location where he engages in any business taxable under 26 U.S.C. 5801. However, a person paying a special (occupational) tax covering his principal place of business may utilize other locations solely for storage of firearms without incurring special (occupational) tax liability at such locations. A manufacturer, upon the single payment of the appropriate special (occupational) tax, may sell firearms, if such firearms are of his own manufacture, at the place of manufacture and at his principal office or place of business if no such firearms, except samples, are kept at such office or place of business. When a person changes the location of a business for which he has paid the special (occupational) tax, he will be liable for another such tax unless the change is properly registered with the Chief, National Firearms Act Branch for the region in which the special tax stamp was issued, as provided in §479.46.

§ 479.39 Engaging in more than one business at the same location.

If more than one business taxable under 26 U.S.C. 5801, is carried on at the same location during a taxable year, the special (occupational) tax imposed on each such business must be paid. This section does not require a qualified manufacturer or importer to qualify as a dealer if such manufacturer or importer also engages in business on his qualified premises as a dealer. However, a qualified manufacturer who engages in business as an importer must also qualify as an importer. Further, a qualified dealer is not entitled to engage in business as a manufacturer or importer.

§ 479.40 Partnership liability.

Any number of persons doing business in partnership at any one location shall be required to pay but one special (occupational) tax.

§ 479.41 Single sale.

A single sale, unattended by circumstances showing the one making the sale to be engaged in business, does not create special (occupational) tax liability.

Change of Ownership

§ 479.42 Changes through death of owner.

Whenever any person who has paid special (occupational) tax dies, the surviving spouse or child, or executors or administrators, or other legal representatives, may carry on this business for the remainder of the term for which tax has been paid and at the place (or places) for which the tax was paid, without any additional payment, subject to the following conditions. If the surviving spouse or child, or executor or administrator, or other legal representative of the deceased taxpayer continues the business, such person shall, within 30 days after the date on which the successor begins to carry on the business, file a new return, Form 5630.7, with ATF in accordance with the instructions on the form. The return thus executed shall show the name of the original taxpayer, together with the basis of the succession. (As to liability in case of failure to register, see §479.49.)

§ 479.43 Changes through bankruptcy of owner.

A receiver or referee in bankruptcy may continue the business under the stamp issued to the taxpayer at the place and for the period for which the tax was paid. An assignee for the benefit of creditors may continue business under his assignor's special tax stamp without incurring additional special (occupational) tax liability. In such cases, the change shall be registered with ATF in a manner similar to that required by §479.42.

§ 479.44 Change in partnership or unincorporated association.

When one or more members withdraw from a partnership or an unincorporated association, the remaining member, or members, may, without incurring additional special (occupational) tax liability, carry on the same business at the same location for the balance of the taxable period for which special (occupational) tax was paid, provided any such change shall be registered in the same manner as required by §479.42. Where new member(s) are taken into a partnership or an unincorporated association, the new firm so constituted may not carry on business under the special tax stamp of the old firm. The new firm must file a return, pay the special (occupational) tax and register in the same manner as a person who first engages in business is required to do under §479.34 even

though the name of the new firm may be the same as that of the old. Where the members of a partnership or an unincorporated association, which has paid special (occupational) tax, form a corporation to continue the business, a new special tax stamp must be taken out in the name of the corporation.

§ 479.45 Changes in corporation.

Additional special (occupational) tax is not required by reason of a mere change of name or increase in the capital stock of a corporation if the laws of the State of incorporation provide for such change or increase without the formation of a new corporation. A stockholder in a corporation who after its dissolution continues the business, incurs new special (occupational) tax liability.

Change of Business Location

§ 479.46 Notice by taxpayer.

Whenever during the taxable year a taxpayer intends to remove his business to a location other than specified in his last special (occupational) tax return (see §479.34), he shall file with ATF (a) a return, Form 5630.7, bearing the notation "Removal Registry," and showing the new address intended to be used, (b) his current special tax stamp, and (c) a letter application requesting the amendment of his registration. The Chief, National Firearms Act Branch, upon approval of the application, shall return the special tax stamp, amended to show the new business location. Firearms operations shall not be commenced at the new business location by the taxpayer prior to the required approval of his application to so change his business location.

Change of Trade Name

§ 479.47 Notice by taxpayer.

Whenever during the taxable year a taxpayer intends to change the name of his business, he shall file with ATF (a) a return, Form 5630.7, bearing the notation "Amended," and showing the trade name intended to be used, (b) his current special tax stamp, and (c) a letter application requesting the amendment of his registration. The Chief, National Firearms Act Branch, upon approval of the application, shall return the special tax stamp, amended to show the new trade name. Firearms operations shall not be commenced under the new trade name by the taxpayer prior to the required approval of his application to so change the trade name.

Penalties and Interest

§ 479.48 Failure to pay special (occupational) tax.

Any person who engages in a business taxable under 26 U.S.C. 5801, without timely payment of the tax imposed with respect to such business (see §479.34) shall be liable for such tax, plus the interest and penalties thereon (see 26 U.S.C. 6601 and 6651). In addition, such person may be liable for criminal penalties under 26 U.S.C. 5871.

§ 479.49 Failure to register change or removal.

Any person succeeding to and carrying on a business for which special (occupational) tax has been paid without registering such change within 30 days thereafter, and any taxpayer removing his business with respect to which special (occupational) tax has been paid to a place other than that for which tax was paid without obtaining approval therefor (see §479.46), will incur liability to an additional payment of the tax, addition to tax and interest, as provided in sections 5801, 6651, and 6601, respectively, I.R.C., for failure to make return (see §479.50) or pay tax, as well as criminal penalties for carrying on business without payment of special (occupational) tax (see section 5871 I.R.C.).

§ 479.50 Delinquency.

Any person liable for special (occupational) tax under section 5801, I.R.C., who fails to file a return (Form 5630.7), as prescribed, will be liable for a delinquency penalty computed on the amount of tax due unless a return (Form 5630.7) is later filed and failure to file the return timely is shown to the satisfaction of the Chief, National Firearms Act Branch, to be due to reasonable cause. The delinquency penalty to be added to the tax is 5 percent if the failure is for not more than 1 month, with an additional 5 percent for each additional month or fraction thereof during which failure continues, not to exceed 25 percent in the aggregate (section 6651, I.R.C.). However, no delinquency penalty is assessed where the 50 percent addition to tax is assessed for fraud (see §479.51).

§ 479.51 Fraudulent return.

If any part of any underpayment of tax required to be shown on a return is due to fraud, there shall be added to the tax an amount equal to 50 percent of the underpayment, but no delinquency penalty shall be assessed with respect to the same underpayment (section 6653, I.R.C.).

Application of State Laws

§ 479.52 State regulations.

Special tax stamps are merely receipts for the tax. Payment of tax under Federal law confers no privilege to act contrary to State law. One to whom a special tax stamp has been issued may still be punishable under a State law prohibiting or controlling the manufacture, possession or transfer of firearms. On the other hand, compliance with State law confers no immunity under Federal law. Persons who engage in the business of importing, manufacturing or dealing in firearms, in violation of the law of a State, are nevertheless required to pay special (occupational) tax as imposed under the internal revenue laws of the United States. For provisions relating to restrictive use of information furnished to comply with the provisions of this part see §479.23.

Subpart E—Tax on Making Firearms

§ 479.61 Rate of tax.

Except as provided in this subpart, there shall be levied, collected, and paid upon the making of a firearm a tax at the rate of $200 for each firearm made. This tax shall be paid by the person making the firearm. Payment of the tax on the making of a firearm shall be represented by a $200 adhesive stamp bearing the words "National Firearms Act." The stamps are maintained by the Director.

Application to Make a Firearm

§ 479.62 Application to make.

No person shall make a firearm unless the person has filed with the Director a written application on Form 1 (Firearms), Application to Make and Register a Firearm, in duplicate, executed under the penalties of perjury, to make and register the firearm and has received the approval of the Director to make the firearm which approval shall effectuate registration of the weapon to the applicant. The application shall identify the firearm to be made by serial number, type, model, caliber or gauge, length of barrel, other marks of identification, and the name and address of original manufacturer (if the applicant is not the original

manufacturer). The applicant must be identified on the Form 1 (Firearms) by name and address and, if other than a natural person, the name and address of the principal officer or authorized representative and the employer identification number and, if an individual, the identification must include the date and place of birth and the information prescribed in §479.63. Each applicant shall identify the Federal firearms license and special (occupational) tax stamp issued to the applicant, if any. The applicant shall also show required information evidencing that making or possession of the firearm would not be in violation of law. If the making is taxable, a remittance in the amount of $200 shall be submitted with the application in accordance with the instructions on the form. If the making is taxable and the application is approved, the Director will affix a National Firearms Act stamp to the original application in the space provided therefor and properly cancel the stamp (see §479.67). The approved application will be returned to the applicant. If the making of the firearm is tax exempt under this part, an explanation of the basis of the exemption shall be attached to the Form 1 (Firearms).

§ 479.63 Identification of applicant.

If the applicant is an individual, the applicant shall securely attach to each copy of the Form 1 (Firearms), in the space provided on the form, a photograph of the applicant 2 × 2 inches in size, clearly showing a full front view of the features of the applicant with head bare, with the distance from the top of the head to the point of the chin approximately 1¼ inches, and which shall have been taken within 1 year prior to the date of the application. The applicant shall attach two properly completed FBI Forms FD–258 (Fingerprint Card) to the application. The fingerprints must be clear for accurate classification and should be taken by someone properly equipped to take them. A certificate of the local chief of police, sheriff of the county, head of the State police, State or local district attorney or prosecutor, or such other person whose certificate may in a particular case be acceptable to the Director, shall be completed on each copy of the Form 1 (Firearms). The certificate shall state that the certifying official is satisfied that the fingerprints and photograph accompanying the application are those of the applicant and that the certifying official has no information indicating that posses-

sion of the firearm by the maker would be in violation of State or local law or that the maker will use the firearm for other than lawful purposes.

§ 479.64 Procedure for approval of application.

The application to make a firearm, Form 1 (Firearms), must be forwarded directly, in duplicate, by the maker of the firearm to the Director in accordance with the instructions on the form. The Director will consider the application for approval or disapproval. If the application is approved, the Director will return the original thereof to the maker of the firearm and retain the duplicate. Upon receipt of the approved application, the maker is authorized to make the firearm described therein. The maker of the firearm shall not, under any circumstances, make the firearm until the application, satisfactorily executed, has been forwarded to the Director and has been approved and returned by the Director with the National Firearms Act stamp affixed. If the application is disapproved, the original Form 1 (Firearms) and the remittance submitted by the applicant for the purchase of the stamp will be returned to the applicant with the reason for disapproval stated on the form.

§ 479.65 Denial of application.

An application to make a firearm shall not be approved by the Director if the making or possession of the firearm would place the person making the firearm in violation of law.

§ 479.66 Subsequent transfer of firearms.

Where a firearm which has been made in compliance with 26 U.S.C. 5821, and the regulations contained in this part, is to be transferred subsequently, the transfer provisions of the firearms laws and regulations must be complied with. (See subpart F of this part).

§ 479.67 Cancellation of stamp.

The person affixing to a Form 1 (Firearms) a "National Firearms Act" stamp shall cancel it by writing or stamping thereon, in ink, his initials, and the day, month and year, in such manner as to render it unfit for reuse. The cancellation shall not so deface the stamp as to prevent its denomination and genuineness from being readily determined.

Exceptions to Tax on Making Firearms

§ 479.68 Qualified manufacturer.

A manufacturer qualified under this part to engage in such business may make firearms without payment of the making tax. However, such manufacturer shall report and register each firearm made in the manner prescribed by this part.

§ 479.69 Making a firearm for the United States.

A firearm may be made by, or on behalf of, the United States or any department, independent establishment, or agency thereof without payment of the making tax. However, if a firearm is to be made on behalf of the United States, the maker must file an application, in duplicate, on Form 1 (Firearms) and obtain the approval of the Director in the manner prescribed in §479.62.

§ 479.70 Certain government entities.

A firearm may be made without payment of the making tax by, or on behalf of, any State, or possession of the United States, any political subdivision thereof, or any official police organization of such a government entity engaged in criminal investigations. Any person making a firearm under this exemption shall first file an application, in duplicate, on Form 1 (Firearms) and obtain the approval of the Director as prescribed in §479.62.

Registration

§ 479.71 Proof of registration.

The approval by the Director of an application, Form 1 (Firearms), to make a firearm under this subpart shall effectuate registration of the firearm described in the Form 1 (Firearms) to the person making the firearm. The original Form 1 (Firearms) showing approval by the Director shall be retained by the maker to establish proof of his registration of the firearm described therein, and shall be made available to any ATF officer on request.

Subpart F—Transfer Tax

§ 479.81 Scope of tax.

Except as otherwise provided in this part, each transfer of a firearm in the United States is subject to a tax to be represented by an adhesive stamp of the proper denomination bearing the words

"National Firearms Act" to be affixed to the Form 4 (Firearms), Application for Transfer and Registration of Firearm, as provided in this subpart.

§ 479.82 Rate of tax.

The transfer tax imposed with respect to firearms transferred within the United States is at the rate of $200 for each firearm transferred, except that the transfer tax on any firearm classified as "any other weapon" shall be at the rate of $5 for each such firearm transferred. The tax imposed on the transfer of the firearm shall be paid by the transferor.

§ 479.83 Transfer tax in addition to import duty.

The transfer tax imposed by section 5811, I.R.C., is in addition to any import duty.

Application and Order for Transfer of Firearm

§ 479.84 Application to transfer.

Except as otherwise provided in this subpart, no firearm may be transferred in the United States unless an application, Form 4 (Firearms), Application for Transfer and Registration of Firearm, in duplicate, executed under the penalties of perjury to transfer the firearm and register it to the transferee has been filed with and approved by the Director. The application, Form 4 (Firearms), shall be filed by the transferor and shall identify the firearm to be transferred by type; serial number; name and address of the manufacturer and importer, if known; model; caliber, gauge or size; in the case of a short–barreled shotgun or a short–barreled rifle, the length of the barrel; in the case of a weapon made from a rifle or shotgun, the overall length of the weapon and the length of the barrel; and any other identifying marks on the firearm. In the event the firearm does not bear a serial number, the applicant shall obtain a serial number from the Special Agent in Charge and shall stamp (impress) or otherwise conspicuously place such serial number on the firearm in a manner not susceptible of being readily obliterated, altered or removed. The application, Form 4 (Firearms), shall identify the transferor by name and address; shall identify the transferor's Federal firearms license and special (occupational) Chapter tax stamp, if any; and if the transferor is other than a natural person, shall show the title or status of the person executing the application.

The application also shall identify the transferee by name and address, and, if the transferee is a natural person not qualified as a manufacturer, importer or dealer under this part, he shall be further identified in the manner prescribed in §479.85. The application also shall identify the special (occupational) tax stamp and Federal firearms license of the transferee, if any. Any tax payable on the transfer must be represented by an adhesive stamp of proper denomination being affixed to the application, Form 4 (Firearms), properly cancelled.

§ 479.85 Identification of transferee.

If the transferee is an individual, such person shall securely attach to each copy of the application, Form 4 (Firearms), in the space provided on the form, a photograph of the applicant 2 × 2 inches in size, clearly showing a full front view of the features of the applicant with head bare, with the distance from the top of the head to the point of the chin approximately 1¼ inches, and which shall have been taken within 1 year prior to the date of the application. The transferee shall attach two properly completed FBI Forms FD–258 (Fingerprint Card) to the application. The fingerprints must be clear for accurate classification and should be taken by someone properly equipped to take them. A certificate of the local chief of police, sheriff of the county, head of the State police, State or local district attorney or prosecutor, or such other person whose certificate may in a particular case be acceptable to the Director, shall be completed on each copy of the Form 4 (Firearms). The certificate shall state that the certifying official is satisfied that the fingerprints and photograph accompanying the application are those of the applicant and that the certifying official has no information indicating that the receipt or possession of the firearm would place the transferee in violation of State or local law or that the transferee will use the firearm for other than lawful purposes.

§ 479.86 Action on application.

The Director will consider a completed and properly executed application, Form 4 (Firearms), to transfer a firearm. If the application is approved, the Director will affix the appropriate National Firearms Act stamp, cancel it, and return the original application showing approval to the transferor who may then transfer the firearm to the transferee along with the approved application. The approval of an application, Form 4 (Firearms), by the

Director will effectuate registration of the firearm to the transferee. The transferee shall not take possession of a firearm until the application, Form 4 (Firearms), for the transfer filed by the transferor has been approved by the Director and registration of the firearm is effectuated to the transferee. The transferee shall retain the approved application as proof that the firearm described therein is registered to the transferee, and shall make the approved Form 4 (Firearms) available to any ATF officer on request. If the application, Form 4 (Firearms), to transfer a firearm is disapproved by the Director, the original application and the remittance for purchase of the stamp will be returned to the transferor with reasons for the disapproval stated on the application. An application, Form 4 (Firearms), to transfer a firearm shall be denied if the transfer, receipt, or possession of a firearm would place the transferee in violation of law. In addition to any other records checks that may be conducted to determine whether the transfer, receipt, or possession of a firearm would place the transferee in violation of law, the Director shall contact the National Instant Criminal Background Check System.

§ 479.87 Cancellation of stamp.

The method of cancellation of the stamp required by this subpart as prescribed in §479.67 shall be used.

Exemptions Relating to Transfers of Firearms

§ 479.88 Special (occupational) taxpayers.

(a) A firearm registered to a person qualified under this part to engage in business as an importer, manufacturer, or dealer may be transferred by that person without payment of the transfer tax to any other person qualified under this part to manufacture, import, or deal in firearms.

(b) The exemption provided in paragraph (a) of this section shall be obtained by the transferor of the firearm filing with the Director an application, Form 3 (Firearms), Application for Tax–exempt Transfer of Firearm and Registration to Special (Occupational) Taxpayer, in duplicate, executed under the penalties of perjury. The application, Form 3 (Firearms), shall (1) show the name and address of the transferor and of the transferee, (2) identify the Federal firearms license and special (occupa-

tional) tax stamp of the transferor and of the transferee, (3) show the name and address of the manufacturer and the importer of the firearm, if known, (4) show the type, model, overall length (if applicable), length of barrel, caliber, gauge or size, serial number, and other marks of identification of the firearm, and (5) contain a statement by the transferor that he is entitled to the exemption because the transferee is a person qualified under this part to manufacture, import, or deal in firearms. If the Director approves an application, Form 3 (Firearms), he shall return the original Form 3 (Firearms) to the transferor with the approval noted thereon. Approval of an application, Form 3 (Firearms), by the Director shall remove registration of the firearm reported thereon from the transferor and shall effectuate the registration of that firearm to the transferee. Upon receipt of the approved Form 3 (Firearms), the transferor shall deliver same with the firearm to the transferee. The transferor shall not transfer the firearm to the transferee until his application, Form 3 (Firearms), has been approved by the Director and the original thereof has been returned to the transferor. If the Director disapproves the application, Form 3 (Firearms), he shall return the original Form 3 (Firearms) to the transferor with the reasons for the disapproval stated thereon.

(c) The transferor shall be responsible for establishing the exempt status of the transferee before making a transfer under the provisions of this section. Therefore, before engaging in transfer negotiations with the transferee, the transferor should satisfy himself as to the claimed exempt status of the transferee and the bona fides of the transaction. If not fully satisfied, the transferor should communicate with the Director, report all circumstances regarding the proposed transfer, and await the Director's advice before making application for the transfer. An unapproved transfer or a transfer to an unauthorized person may subject the transferor to civil and criminal liabilities. (See 26 U.S.C. 5852, 5861, and 5871.)

§ 479.89 Transfers to the United States.

A firearm may be transferred to the United States or any department, independent establishment or agency thereof without payment of the transfer tax. However, the procedures for the transfer of a firearm as provided in §479.90 shall be followed in a tax–exempt transfer of a firearm under this section, un-

less the transferor is relieved of such requirement under other provisions of this part.

§ 479.90 Certain government entities.

(a) A firearm may be transferred without payment of the transfer tax to or from any State, possession of the United States, any political subdivision thereof, or any official police organization of such a governmental entity engaged in criminal investigations.

(b) The exemption provided in paragraph (a) of this section shall be obtained by the transferor of the firearm filing with the Director an application, Form 5 (Firearms), Application for Tax–exempt Transfer and Registration of Firearm, in duplicate, executed under the penalties of perjury. The application shall (1) show the name and address of the transferor and of the transferee, (2) identify the Federal firearms license and special (occupational) tax stamp, if any, of the transferor and of the transferee, (3) show the name and address of the manufacturer and the importer of the firearm, if known, (4) show the type, model, overall length (if applicable), length of barrel, caliber, gauge or size, serial number, and other marks of identification of the firearm, and (5) contain a statement by the transferor that the transferor is entitled to the exemption because either the transferor or the transferee is a governmental entity coming within the purview of paragraph (a) of this section. In the case of a transfer of a firearm by a governmental entity to a transferee who is a natural person not qualified as a manufacturer, importer, or dealer under this part, the transferee shall be further identified in the manner prescribed in §479.85. If the Director approves an application, Form 5 (Firearms), the original Form 5 (Firearms) shall be returned to the transferor with the approval noted thereon. Approval of an application, Form 5 (Firearms), by the Director shall effectuate the registration of that firearm to the transferee. Upon receipt of the approved Form 5 (Firearms), the transferor shall deliver same with the firearm to the transferee. The transferor shall not transfer the firearm to the transferee until the application, Form 5 (Firearms), has been approved by the Director and the original thereof has been returned to the transferor. If the Director disapproves the application, Form 5 (Firearms), the original Form 5 (Firearms) shall be returned to the transferor with the reasons for the disapproval stated thereon. An application by a governmental entity to transfer a firearm shall be denied if the transfer, receipt, or possession of a firearm would place the transferee in violation of law.

(c) The transferor shall be responsible for establishing the exempt status of the transferee before making a transfer under the provisions of this section. Therefore, before engaging in transfer negotiations with the transferee, the transferor should satisfy himself of the claimed exempt status of the transferee and the bona fides of the transaction. If not fully satisfied, the transferor should communicate with the Director, report all circumstances regarding the proposed transfer, and await the Director's advice before making application for transfer. An unapproved transfer or a transfer to an unauthorized person may subject the transferor to civil and criminal liabilities. (See 26 U.S.C. 5852, 5861, and 5871.)

§ 479.91 Unserviceable firearms.

An unserviceable firearm may be transferred as a curio or ornament without payment of the transfer tax. However, the procedures for the transfer of a firearm as provided in §479.90 shall be followed in a tax–exempt transfer of a firearm under this section, except a statement shall be entered on the transfer application, Form 5 (Firearms), by the transferor that he is entitled to the exemption because the firearm to be transferred is unserviceable and is being transferred as a curio or ornament. An unapproved transfer, the transfer of a firearm under the provisions of this section which is in fact not an unserviceable firearm, or the transfer of an unserviceable firearm as something other than a curio or ornament, may subject the transferor to civil and criminal liabilities. (See 26 U.S.C. 5811, 5852, 5861, and 5871.)

§ 479.92 Transportation of firearms to effect transfer.

Notwithstanding any provision of §478.28 of this chapter, it shall not be required that authorization be obtained from the Director for the transportation in interstate or foreign commerce of a firearm in order to effect the transfer of a firearm authorized under the provisions of this subpart.

Other Provisions

§ 479.93 Transfers of firearms to certain persons.

Where the transfer of a destructive device, machine gun, short–barreled shotgun, or short–barreled rifle is to be made by a person licensed under the provisions of Title I of the Gun Control Act of 1968 (82 Stat. 1213) to a person not so licensed, the sworn statement required by §478.98 of this chapter shall be attached to and accompany the transfer application required by this subpart.

Subpart G—Registration and Identification of Firearms

§ 479.101 Registration of firearms.

(a) The Director shall maintain a central registry of all firearms in the United States which are not in the possession of or under the control of the United States. This registry shall be known as the National Firearms Registration and Transfer Record and shall include:

(1) Identification of the firearm as required by this part;

(2) Date of registration; and

(3) Identification and address of person entitled to possession of the firearm as required by this part.

(b) Each manufacturer, importer, and maker shall register each firearm he manufactures, imports, or makes in the manner prescribed by this part. Each firearm transferred shall be registered to the transferee by the transferor in the manner prescribed by this part. No firearm may be registered by a person unlawfully in possession of the firearm except during an amnesty period established under section 207 of the Gun Control Act of 1968 (82 Stat. 1235).

(c) A person shown as possessing firearms by the records maintained by the Director pursuant to the National Firearms Act (26 U.S.C. Chapter 53) in force on October 31, 1968, shall be considered to have registered the firearms in his possession which are disclosed by that record as being in his possession on October 31, 1968.

(d) The National Firearms Registration and Transfer Record shall include firearms registered to the possessors thereof under the provisions of section 207 of the Gun Control Act of 1968.

(e) A person possessing a firearm registered to him shall retain proof of registration which shall be made available to any ATF officer upon request.

(f) A firearm not identified as required by this part shall not be registered.

§ 479.102 How must firearms be identified?

(a) You, as a manufacturer, importer, or maker of a firearm, must legibly identify the firearm as follows:

(1) By engraving, casting, stamping (impressing), or otherwise conspicuously placing or causing to be engraved, cast, stamped (impressed) or placed on the frame or receiver thereof an individual serial number. The serial number must be placed in a manner not susceptible of being readily obliterated, altered, or removed, and must not duplicate any serial number placed by you on any other firearm. For firearms manufactured, imported, or made on and after January 30, 2002, the engraving, casting, or stamping (impressing) of the serial number must be to a minimum depth of .003 inch and in a print size no smaller than 1⁄16 inch; and

(2) By engraving, casting, stamping (impressing), or otherwise conspicuously placing or causing to be engraved, cast, stamped (impressed), or placed on the frame, receiver, or barrel thereof certain additional information. This information must be placed in a manner not susceptible of being readily obliterated, altered or removed. For firearms manufactured, imported, or made on and after January 30, 2002, the engraving, casting, or stamping (impressing) of this information must be to a minimum depth of .003 inch. The additional information includes:

(i) The model, if such designation has been made;

(ii) The caliber or gauge;

(iii) Your name (or recognized abbreviation) and also, when applicable, the name of the foreign manufacturer or maker;

(iv) In the case of a domestically made firearm, the city and State (or recognized abbreviation thereof) where you as the manufacturer maintain your place of business, or where you, as the maker, made the firearm; and

(v) In the case of an imported firearm, the name of the country in which it was manufactured and the city and State (or recognized abbreviation thereof) where you as the importer maintain your place of business. For additional requirements relating to imported firearms, see Customs regulations at 19 CFR part 134.

(b) The depth of all markings required by this section will be measured from the flat surface of the metal and not the peaks or ridges. The height of serial numbers required by paragraph (a)(1) of this section will be measured as the distance between the latitudinal ends of the character impression bottoms (bases).

(c) The Director may authorize other means of identification upon receipt of a letter application from you, submitted in duplicate, showing that such other identification is reasonable and will not hinder the effective administration of this part.

(d) In the case of a destructive device, the Director may authorize other means of identifying that weapon upon receipt of a letter application from you, submitted in duplicate, showing that engraving, casting, or stamping (impressing) such a weapon would be dangerous or impractical.

(e) A firearm frame or receiver that is not a component part of a complete weapon at the time it is sold, shipped, or otherwise disposed of by you must be identified as required by this section.

(f)(1) Any part defined as a machine gun, muffler, or silencer for the purposes of this part that is not a component part of a complete firearm at the time it is sold, shipped, or otherwise disposed of by you must be identified as required by this section.

(2) The Director may authorize other means of identification of parts defined as machine guns other than frames or receivers and parts defined as mufflers or silencers upon receipt of a letter application from you, submitted in duplicate, showing that such other identification is reasonable and will not hinder the effective administration of this part.

(Approved by the Office of Management and Budget under control number 1140–0050)

§ 479.103 Registration of firearms manufactured.

Each manufacturer qualified under this part shall file with the Director an accurate notice on Form 2 (Firearms), Notice of Firearms Manufactured or Imported, executed under the penalties of perjury, to show his manufacture of firearms. The notice shall set forth the name and address of the manufacturer, identify his special (occupational) tax stamp and Federal firearms license, and show the date of manufacture, the type, model, length of barrel, overall length, caliber, gauge or size, serial numbers, and other marks of identification of the firearms he manufactures, and the place where the manufactured firearms will be kept. All firearms manufactured by him during a single day shall be included on one notice, Form 2 (Firearms), filed by the manufacturer no later than the close of the next business day. The manufacturer shall prepare the notice, Form 2 (Firearms), in duplicate, file the original notice as prescribed herein and keep the copy with the records required by subpart I of this part at the premises covered by his special (occupational) tax stamp. Receipt of the notice, Form 2 (Firearms), by the Director shall effectuate the registration of the firearms listed on that notice. The requirements of this part relating to the transfer of a firearm are applicable to transfers by qualified manufacturers.

§ 479.104 Registration of firearms by certain governmental entities.

Any State, any political subdivision thereof, or any official police organization of such a government entity engaged in criminal investigations, which acquires for official use a firearm not registered to it, such as by abandonment or by forfeiture, will register such firearm with the Director by filing Form 10 (Firearms), Registration of Firearms Acquired by Certain Governmental Entities, and such registration shall become a part of the National Firearms Registration and Transfer Record. The application shall identify the applicant, describe each firearm covered by the application, show the location where each firearm usually will be kept, and, if the firearm is unserviceable, the application shall show how the firearm was made unserviceable. This section shall not apply to a firearm merely being held for use as evidence in a criminal proceeding. The Form 10 (Firearms) shall be executed in duplicate in accordance with the instructions thereon. Upon reg-

istering the firearm, the Director shall return the original Form 10 (Firearms) to the registrant with notification thereon that registration of the firearm has been made. The registration of any firearm under this section is for official use only and a subsequent transfer will be approved only to other governmental entities for official use.

Machine Guns

§479.105 Transfer and possession of machine guns.

(a) **General.** As provided by 26 U.S.C. 5812 and 26 U.S.C. 5822, an application to make or transfer a firearm shall be denied if the making, transfer, receipt, or possession of the firearm would place the maker or transferee in violation of law. Section 922(o), Title 18, U.S.C., makes it unlawful for any person to transfer or possess a machine gun, except a transfer to or by, or possession by or under the authority of, the United States or any department or agency thereof or a State, or a department, agency, or political subdivision thereof; or any lawful transfer or lawful possession of a machine gun that was lawfully possessed before May 19, 1986. Therefore, notwithstanding any other provision of this part, no application to make, transfer, or import a machine gun will be approved except as provided by this section.

(b) **Machine guns lawfully possessed prior to May 19, 1986.** A machine gun possessed in compliance with the provisions of this part prior to May 19, 1986, may continue to be lawfully possessed by the person to whom the machine gun is registered and may, upon compliance with the provisions of this part, be lawfully transferred to and possessed by the transferee.

(c) **Importation and manufacture.** Subject to compliance with the provisions of this part, importers and manufacturers qualified under this part may import and manufacture machine guns on or after May 19, 1986, for sale or distribution to any department or agency of the United States or any State or political subdivision thereof, or for use by dealers qualified under this part as sales samples as provided in paragraph (d) of this section. The registration of such machine guns under this part and their subsequent transfer shall be conditioned upon and restricted to the sale or distribution of such weapons for the official use of Federal, State or local governmental entities. Subject to compliance with the provisions of this part, manufacturers qualified under this part may manufacture machine guns on or after May 19, 1986, for exportation in compliance with the Arms Export Control Act (22 U.S.C. 2778) and regulations prescribed thereunder by the Department of State.

(d) Dealer sales samples. Subject to compliance with the provisions of this part, applications to transfer and reg-ister a machine gun manufactured or imported on or after May 19, 1986, to dealers qualified under this part will be approved if it is established by specific information the expected governmental customers who would require a demon-stration of the weapon, information as to the availability of the machine gun to fill subsequent orders, and letters from governmental entities expressing a need for a particular model or interest in seeing a demonstration of a particular weapon. Applications to transfer more than one machine gun of a particular model to a dealer must also establish the dealer's need for the quantity of samples sought to be transferred.

(e) The making of machine guns on or after May 19, 1986. Subject to compliance with the provisions of this part, applications to make and regis-ter machine guns on or after May 19, 1986, for the benefit of a Federal, State or local governmental entity (e.g., an invention for possible future use of a governmental entity or the making of a weapon in connection with research and development on behalf of such an entity) will be approved if it is estab-lished by specific information that the machine gun is particularly suitable for use by Federal, State or local govern-mental entities and that the making of the weapon is at the request and on behalf of such an entity.

(f) Discontinuance of business. Since section 922(o), Title 18, U.S.C., makes it unlawful to transfer or possess a machine gun except as provided in the law, any qualified manufacturer, im-porter, or dealer intending to discontin-ue business shall, prior to going out of business, transfer in compliance with the provisions of this part any machine gun manufactured or imported after May 19, 1986, to a Federal, State or local gov-ernmental entity, qualified manufacturer, qualified importer, or, subject to the pro-visions of paragraph (d) of this section, dealer qualified to possess such, ma-chine gun.

Subpart H—Importation and Exportation

Importation

§ 479.111 Procedure.

(a) No firearm shall be imported or brought into the United States or any territory under its control or jurisdiction unless the person importing or bringing in the firearm establishes to the satisfaction of the Director that the firearm to be imported or brought in is being imported or brought in for:

(1) The use of the United States or any department, independent establishment, or agency thereof or any State or possession or any political subdivision thereof; or

(2) Scientific or research purposes; or

(3) Testing or use as a model by a registered manufacturer or solely for use as a sample by a registered importer or registered dealer.

The burden of proof is affirmatively on any person importing or bringing the firearm into the United States or any territory under its control or jurisdiction to show that the firearm is being imported or brought in under one of the above paragraphs. Any person desiring to import or bring a firearm into the United States under this paragraph shall file with the Director an application on Form 6 (Firearms), Application and Permit for Importation of Firearms, Ammunition and Implements of War, in triplicate, executed under the penalties of perjury. The application shall show the information required by subpart G of Part 478 of this chapter. A detailed explanation of why the importation of the firearm falls within the standards set out in this paragraph shall be attached to the application. The person seeking to import or bring in the firearm will be notified of the approval or disapproval of his application. If the application is approved, the original Form 6 (Firearms) will be returned to the applicant showing such approval and he will present the approved application, Form 6 (Firearms), to the Customs officer at the port of importation. The approval of an application to import a firearm shall be automatically terminated at the expiration of two years from the date of approval unless, upon request, it is further extended by the Director. If the firearm described in the approved application is not imported prior to the expiration

of the approval, the Director shall be so notified. Customs officers will not permit release of a firearm from Customs custody, except for exportation, unless covered by an application which has been approved by the Director and which is currently effective. The importation or bringing in of a firearm not covered by an approved application may subject the person responsible to civil and criminal liabilities. (26 U.S.C. 5861, 5871, and 5872.)

(b) Part 478 of this chapter also contains requirements and procedures for the importation of firearms into the United States. A firearm may not be imported into the United States under this part unless those requirements and procedures are also complied with by the person importing the firearm.

(c) The provisions of this subpart shall not be construed as prohibiting the return to the United States or any territory under its control or jurisdiction of a firearm by a person who can establish to the satisfaction of Customs that (1) the firearm was taken out of the United States or any territory under its control or jurisdiction by such person, (2) the firearm is registered to that person, and (3) if appropriate, the authorization required by Part 478 of this chapter for the transportation of such a firearm in interstate or foreign commerce has been obtained by such person.

§ 479.112 Registration of imported firearms.

(a) Each importer shall file with the Director an accurate notice on Form 2 (Firearms), Notice of Firearms Manufactured or Imported, executed under the penalties of perjury, showing the importation of a firearm. The notice shall set forth the name and address of the importer, identify the importer's special (occupational) tax stamp and Federal firearms license, and show the import permit number, the date of release from Customs custody, the type, model, length of barrel, overall length, caliber, gauge or size, serial number, and other marks of identification of the firearm imported, and the place where the imported firearm will be kept. The Form 2 (Firearms) covering an imported firearm shall be filed by the importer no later than fifteen (15) days from the date the firearm was released from Customs custody. The importer shall prepare the notice, Form 2 (Firearms), in duplicate, file the original return as prescribed herein, and keep the copy with the records required by subpart I of this part at the premises covered by the special (occupational) tax stamp. The timely receipt by the Director of the notice, Form 2 (Firearms), and the timely receipt by the Director of the copy of Form 6A (Firearms), Release and Receipt of Imported Firearms, Ammunition and Implements of War, required by §478.112 of this chapter, covering the weapon reported on the Form 2 (Firearms) by the qualified importer, shall effectuate the registration of the firearm to the importer.

(b) The requirements of this part relating to the transfer of a firearm are applicable to the transfer of imported firearms by a qualified importer or any other person.

(c) Subject to compliance with the provisions of this part, an application, Form 6 (Firearms), to import a firearm by an importer or dealer qualified under this part, for use as a sample in connection with sales of such firearms to Federal, State or local governmental entities, will be approved if it is established by specific information attached to the application that the firearm is suitable or potentially suitable for use by such entities. Such information must show why a sales sample of a particular firearm is suitable for such use and the expected governmental customers who would require a demonstration of the firearm. Information as to the availability of the firearm to fill subsequent orders and letters from governmental entities expressing a need for a particular model or interest in seeing a demonstration of a particular firearm would establish suitability for governmental use. Applications to import more than one firearm of a particular model for use as a sample by an importer or dealer must also establish the importer's or dealer's need for the quantity of samples sought to be imported.

(d) Subject to compliance with the provisions of this part, an application, Form 6 (Firearms), to import a firearm by an importer or dealer qualified under this part, for use as a sample in connection with sales of such firearms to Federal, State or local governmental entities, will be approved if it is established by specific information attached to the application that the firearm is particularly suitable for use by such entities. Such information must show why a sales sample of a particular firearm is suitable for such use and the expected governmental customers who would require a demonstration of the firearm.

Information as to the availability of the firearm to fill subsequent orders and letters from governmental entities expressing a need for a particular model or interest in seeing a demonstration of a particular firearm would establish suitability for governmental use. Applications to import more than one firearm of a particular model for use as a sample by an importer or dealer must also establish the importer's or dealer's need for the quantity of samples sought to be imported.

§ 479.113 Conditional importation.

The Director shall permit the conditional importation or bringing into the United States of any firearm for the purpose of examining and testing the firearm in connection with making a determination as to whether the importation or bringing in of such firearm will be authorized under this subpart. An application under this section shall be filed on Form 6 (Firearms), in triplicate, with the Director. The Director may impose conditions upon any importation under this section including a requirement that the firearm be shipped directly from Customs custody to the Director and that the person importing or bringing in the firearm must agree to either export the weapon or destroy it if a final determination is made that it may not be imported or brought in under this subpart. A firearm so imported or brought into the United States may be released from Customs custody in the manner prescribed by the conditional authorization of the Director.

Exportation

§ 479.114 Application and permit for exportation of firearms.

Any person desiring to export a firearm without payment of the transfer tax must file with the Director an application on Form 9 (Firearms), Application and Permit for Exportation of Firearms, in quadruplicate, for a permit providing for deferment of tax liability. Part 1 of the application shall show the name and address of the foreign consignee, number of firearms covered by the application, the intended port of exportation, a complete description of each firearm to be exported, the name, address, State Department license number (or date of application if not issued), and identification of the special (occupational) tax stamp of the transferor. Part 1 of the application shall be executed under the penalties of perjury by the transferor and shall be supported by a certified copy of a writ-

ten order or contract of sale or other evidence showing that the firearm is to be shipped to a foreign designation. Where it is desired to make a transfer free of tax to another person who in turn will export the firearm, the transferor shall likewise file an application supported by evidence that the transfer will start the firearm in course of exportation, except, however, that where such transferor and exporter are registered special–taxpayers the transferor will not be required to file an application on Form 9 (Firearms).

§ 479.115 Action by Director.

If the application is acceptable, the Director will execute the permit, Part 2 of Form 9 (Firearms), to export the firearm described on the form and return three copies thereof to the applicant. Issuance of the permit by the Director will suspend assertion of tax liability for a period of six (6) months from the date of issuance. If the application is disapproved, the Director will indicate thereon the reason for such action and return the forms to the applicant.

§ 479.116 Procedure by exporter.

Shipment may not be made until the permit, Form 9 (Firearms), is received from the Director. If exportation is to be made by means other than by parcel post, two copies of the form must be addressed to the District Director of Customs at the port of exportation, and must precede or accompany the shipment in order to permit appropriate inspection prior to lading. If exportation is to be made by parcel post, one copy of the form must be presented to the postmaster at the office receiving the parcel who will execute Part 4 of such form and return the form to the exporter for transmittal to the Director. In the event exportation is not effected, all copies of the form must be immediately returned to the Director for cancellation.

§ 479.117 Action by Customs.

Upon receipt of a permit, Form 9 (Firearms), in duplicate, authorizing the exportation of firearms, the District Director of Customs may order such inspection as deemed necessary prior to lading of the merchandise. If satisfied that the shipment is proper and the information contained in the permit to export is in agreement with information shown in the shipper's export declaration, the District Director of Customs will, after the merchandise has been duly exported, execute the certificate of exportation (Part 3 of Form 9 (Firearms)).

One copy of the form will be retained with the shipper's export declaration and the remaining copy thereof will be transmitted to the Director.

§ 479.118 Proof of exportation.

Within a six–month's period from date of issuance of the permit to export firearms, the exporter shall furnish or cause to be furnished to the Director (a) the certificate of exportation (Part 3 of Form 9 (Firearms)) executed by the District Director of Customs as provided in §479.117, or (b) the certificate of mailing by parcel post (Part 4 of Form 9 (Firearms)) executed by the postmaster of the post office receiving the parcel containing the firearm, or (c) a certificate of landing executed by a Customs officer of the foreign country to which the firearm is exported, or (d) a sworn statement of the foreign consignee covering the receipt of the firearm, or (e) the return receipt, or a reproduced copy thereof, signed by the addressee or his agent, where the shipment of a firearm was made by insured or registered parcel post. Issuance of a permit to export a firearm and furnishing of evidence establishing such exportation under this section will relieve the actual exporter and the person selling to the exporter for exportation from transfer tax liability. Where satisfactory evidence of exportation of a firearm is not furnished within the stated period, the transfer tax will be assessed.

§ 479.119 Transportation of firearms to effect exportation.

Notwithstanding any provision of §478.28 of this chapter, it shall not be required that authorization be obtained from the Director for the transportation in interstate or foreign commerce of a firearm in order to effect the exportation of a firearm authorized under the provisions of this subpart.

§ 479.120 Refunds.

Where, after payment of tax by the manufacturer, a firearm is exported, and satisfactory proof of exportation (see §479.118) is furnished, a claim for refund may be submitted on Form 843 (see §479.172). If the manufacturer waives all claim for the amount to be refunded, the refund shall be made to the exporter. A claim for refund by an exporter of tax paid by a manufacturer should be accompanied by waiver of the manufacturer and proof of tax payment by the latter.

§ 479.121 Insular possessions.

Transfers of firearms to persons in the insular possessions of the United States are exempt from transfer tax, provided title in cases involving change of title (and custody or control, in cases not involving change of title), does not pass to the transferee or his agent in the United States. However, such exempt transactions must be covered by approved permits and supporting documents corresponding to those required in the case of firearms exported to foreign countries (see §§479.114 and 479.115), except that the Director may vary the requirements herein set forth in accordance with the requirements of the governing authority of the insular possession. Shipments to the insular possessions will not be authorized without compliance with the requirements of the governing authorities thereof. In the case of a nontaxable transfer to a person in such insular possession, the exemption extends only to such transfer and not to prior transfers.

Arms Export Control Act

§ 479.122 Requirements.

(a) Persons engaged in the business of importing firearms are required by the Arms Export Control Act (22 U.S.C. 2778) to register with the Director. (See Part 447 of this chapter.)

(b) Persons engaged in the business of exporting firearms caliber .22 or larger are subject to the requirements of a license issued by the Secretary of State. Application for such license should be made to the Office of Munitions Control, Department of State, Washington, DC 20502, prior to exporting firearms.

Subpart I—Records and Returns

§ 479.131 Records.

For the purposes of this part, each manufacturer, importer, and dealer in firearms shall keep and maintain such records regarding the manufacture, importation, acquisition (whether by making, transfer, or otherwise), receipt, and disposition of firearms as are prescribed, and in the manner and place required, by part 478 of this chapter. In addition, each manufacturer, importer, and dealer shall maintain, in chronological order, at his place of business a separate record consisting of the documents required by this part showing the registration of any firearm to him. If firearms owned or

possessed by a manufacturer, importer, or dealer are stored or kept on premises other than the place of business shown on his special (occupational) tax stamp, the record establishing registration shall show where such firearms are stored or kept. The records required by this part shall be readily accessible for inspection at all reasonable times by ATF officers.

(Approved by the Office of Management and Budget under control number 1140–0032)

Subpart J—Stolen or Lost Firearms or Documents

§ 479.141 Stolen or lost firearms.

Whenever any registered firearm is stolen or lost, the person losing possession thereof will, immediately upon discovery of such theft or loss, make a report to the Director showing the following: (a) Name and address of the person in whose name the firearm is registered, (b) kind of firearm, (c) serial number, (d) model, (e) caliber, (f) manufacturer of the firearm, (g) date and place of theft or loss, and (h) complete statement of facts and circumstances surrounding such theft or loss.

§ 479.142 Stolen or lost documents.

When any Forms 1, 2, 3, 4, 5, 6A, or 10 (Firearms) evidencing possession of a firearm is stolen, lost, or destroyed, the person losing possession will immediately upon discovery of the theft, loss, or destruction report the matter to the Director. The report will show in detail the circumstances of the theft, loss, or destruction and will include all known facts which may serve to identify the document. Upon receipt of the report, the Director will make such investigation as appears appropriate and may issue a duplicate document upon such conditions as the circumstances warrant.

Subpart K—Examination of Books and Records

§ 479.151 Failure to make returns: Substitute returns.

If any person required by this part to make returns shall fail or refuse to make any such return within the time prescribed by this part or designated by the Director, then the return shall be made by an ATF officer upon inspection of the books, but the making of such return by an ATF officer shall not relieve the person from any default or penalty incurred by reason of failure to make such return.

§ 479.152 Penalties (records and returns).

Any person failing to keep records or make returns, or making, or causing the making of, a false entry on any application, return or record, knowing such entry to be false, is liable to fine and imprisonment as provided in section 5871, I.R.C.

Subpart L—Distribution and Sale of Stamps

§ 479.161 National Firearms Act stamps.

"National Firearms Act" stamps evidencing payment of the transfer tax or tax on the making of a firearm are maintained by the Director. The remittance for purchase of the appropriate tax stamp shall be submitted with the application. Upon approval of the application, the Director will cause the appropriate tax to be paid by affixing the appropriate stamp to the application.

§ 479.162 Stamps authorized.

Adhesive stamps of the $5 and $200 denomination, bearing the words "National Firearms Act," have been prepared and only such stamps shall be used for the payment of the transfer tax and for the tax on the making of a firearm.

§ 479.163 Reuse of stamps prohibited.

A stamp once affixed to one document cannot lawfully be removed and affixed to another. Any person willfully reusing such a stamp shall be subject to the penalty prescribed by 26 U.S.C. 7208.

Subpart M—Redemption of or Allowance for Stamps or Refunds

§ 479.171 Redemption of or allowance for stamps.

Where a National Firearms Act stamp is destroyed, mutilated or rendered useless after purchase, and before liability has been incurred, such stamp may be redeemed by giving another stamp in lieu thereof. Claim for redemption of the stamp should be filed on ATF Form 2635 (5620.8) with the Director. Such claim shall be accompanied by the stamp or by a satisfactory explanation of the reasons why the stamp cannot be returned, and shall be filed within 3 years after the purchase of the stamp.

§ 479.172 Refunds.

As indicated in this part, the transfer tax or tax on the making of a firearm is ordinarily paid by the purchase and affixing of stamps, while special tax stamps are issued in payment of special (occupational) taxes. However, in exceptional cases, transfer tax, tax on the making of firearms, and/or special (occupational) tax may be paid pursuant to assessment. Claims for refunds of such taxes, paid pursuant to assessment, shall be filed on ATF Form 2635 (5620.8) within 3 years next after payment of the taxes. Such claims shall be filed with the Chief, National Firearms Act Branch serving the region in which the tax was paid. (For provisions relating to hand–carried documents and manner of filing, see 26 CFR 301.6091–1(b) and 301.6402–2(a).) When an applicant to make or transfer a firearm wishes a refund of the tax paid on an approved application where the firearm was not made pursuant to an approved Form 1 (Firearms) or transfer of the firearm did not take place pursuant to an approved Form 4 (Firearms), the applicant shall file a claim for refund of the tax on ATF Form 2635 (5620.8) with the Director. The claim shall be accompanied by the approved application bearing the stamp and an explanation why the tax liability was not incurred. Such claim shall be filed within 3 years next after payment of the tax.

Subpart N—Penalties and Forfeitures

§ 479.181 Penalties.

Any person who violates or fails to comply with the requirements of 26 U.S.C. Chapter 53 shall, upon conviction, be subject to the penalties imposed under 26 U.S.C. 5871.

§ 479.182 Forfeitures.

Any firearm involved in any violation of the provisions of 26 U.S.C. Chapter 53, shall be subject to seizure, and forfeiture under the internal revenue laws: Provided, however, That the disposition of forfeited firearms shall be in conformance with the requirements of 26 U.S.C. 5872. In addition, any vessel, vehicle or aircraft used to transport, carry, convey or conceal or possess any firearm with respect to which there has been committed any violation of any provision of 26 U.S.C. Chapter 53, or the regulations in this part issued pursuant thereto, shall be subject to seizure and forfeiture under the Customs laws, as provided by the act of August 9, 1939 (49 U.S.C. App., Chapter 11).

Subpart O—Other Laws Applicable

§ 479.191 Applicability of other provisions of internal revenue laws.

All of the provisions of the internal revenue laws not inconsistent with the provisions of 26 U.S.C. Chapter 53 shall be applicable with respect to the taxes imposed by 26 U.S.C. 5801, 5811, and 5821 (see 26 U.S.C. 5846).

§ 479.192 Commerce in firearms and ammunition.

For provisions relating to commerce in firearms and ammunition, including the movement of destructive devices, machine guns, short–barreled shotguns, or short–barreled rifles, see 18 U.S.C. Chapter 44, and Part 478 of this chapter issued pursuant thereto.

§ 479.193 Arms Export Control Act.

For provisions relating to the registration and licensing of persons engaged in the business of manufacturing, importing or exporting arms, ammunition, or implements of war, see the Arms Export Control Act (22 U.S.C. 2778), and the regulations issued pursuant thereto. (See also Part 447 of this chapter.)

TITLE 22, UNITED STATES CODE, § 2778

Editor's Note:

With respect to Section 38 of the Arms Export Control Act of 1976 (22 U.S.C. 2778), only the importation provisions are administered by ATF. Export provisions are administered by the Department of State. Importation regulations issued under this law are in 27 CFR Part 447, and are included in this publication.

§ 2778. Control of arms exports and imports

(a) Presidential control of exports and imports of defense articles and services, guidance of policy, etc.; designation of United States Munitions List; issuance of export licenses; condition for export; negotiations information.

(1) In furtherance of world peace and the security and foreign policy of the United States, the President is authorized to control the import and the export of defense articles and defense services and to provide foreign policy guidance to persons of the United States involved in the export and import of such articles and services. The President is authorized to designate those items which shall be considered as defense articles and defense services for the purposes of this section and to promulgate regulations for the import and export of such articles and services. The items so designated shall constitute the United States Munitions List.

(2) Decisions on issuing export licenses under this section shall take into account whether the export of an article would contribute to an arms race, aid in the development of weapons of mass destruction, support international terrorism, increase the possibility of outbreak or escalation of conflict, or prejudice the development of bilateral or multilateral arms control or nonproliferation agreements or other arrangements.

(3) In exercising the authorities conferred by this section, the President may require that any defense article or defense service be sold under this chapter as a condition of its eligibility for export, and may require that persons engaged in the negotiation for the export of defense articles and services keep the President fully and currently informed of the progress and future prospects of such negotiations.

(b) Registration and licensing requirements for manufacturers, exporters, or importers of designated defense articles and defense services; exceptions.

(1) (A) (i) As prescribed in regulations issued under this section, every person (other than an officer or employee of the United States Government acting in an official capacity) who engages in the business of manufacturing, exporting, or importing any defense articles or defense services designated by the President under subsection (a)(1) of this section shall register with the United States Government agency charged with the administration of this section, and shall pay a registration fee which shall be prescribed by such regulations. Such regulations shall prohibit the return to the United States for sale in the United States (other than for the Armed Forces of the United States and its allies or for any State or local law enforcement agency) of any military firearms or ammunition of United States manufacture furnished to foreign governments by the United States under this chapter or any other foreign assistance or sales program of the United States, whether or not enhanced in value or improved in condition in a foreign country. This prohibition shall not extend to similar firearms that have been so substantially transformed as to become, in effect, articles of foreign manufacture.

(ii)(I) As prescribed in regulations issued under this section, every person (other than an officer or employee of the United States Government acting in official capacity) who engages in the business of brokering activities with respect to the manufacture, export, import, or transfer of any defense article or defense service designated by the President under subsection (a)(1) of this section, or in the business of brokering activities with respect to the manufacture, export, import, or transfer of any foreign defense article or defense service (as defined in subclause (IV)), shall register with the United States Government agency charged with the administration of this section, and shall pay a registration fee which shall be prescribed by such regulations.

(II) Such brokering activities shall include the financing, transportation, freight forwarding, or taking of any other action that facilitates the manufacture, export, or import of a defense article or defense service.

(III) No person may engage in the business of brokering activities described in subclause (I) without a license, issued in accordance with this chapter, except that no license shall be required for such activities undertaken by or for an agency of the United States Government—

(aa) for use by an agency of the United States Government; or

(bb) for carrying out any foreign assistance or sales program authorized by law and subject to the control of the President by other means.

(IV) For purposes of this clause, the term **"foreign defense article or defense service"** includes any non–United States defense article or defense service of a nature described on the United States Munitions List regardless of whether such article or service is of United States origin or whether such article or service contains United States origin components.

(B) The prohibition under such regulations required by the second sentence of subparagraph (A) shall not extend to any military firearms (or ammunition, components, parts, accessories, and attachments for such firearms) of United States manufacture furnished to any foreign government by the United States under this chapter or any other foreign assistance or sales program of the United States if—

(i) such firearms are among those firearms that the Secretary of the Treasury is, or was at any time, required to authorize the importation of by reason of the provisions of section 925(e) of title 18 (including the requirement for the listing of such firearms as curios or relics under section 921(a)(13) of that title); and

(ii) such foreign government certifies to the United States Government that such firearms are owned by such foreign government.

(C) A copy of each registration made under this paragraph shall be transmitted to the Secretary of the Treasury for review regarding law enforcement concerns. The Secretary shall report to the President regarding such concerns as necessary.

(2) Except as otherwise specifically provided in regulations issued under subsection (a)(1) of this section, no defense articles or defense services designated by the President under subsection (a)(1) of this section may be exported or imported without a license for such export or import, issued in accordance with this chapter and regulations issued under this chapter, except that no license shall be required for exports or imports

made by or for an agency of the United States Government (A) for official use by a department or agency of the United States Government, or (B) for carrying out any foreign assistance or sales program authorized by law and subject to the control of the President by other means.

(3)(A) For each of the fiscal years 1988 and 1989, $250,000 of registration fees collected pursuant to paragraph (1) shall be credited to a Department of State account, to be available without fiscal year limitation. Fees credited to that account shall be available only for the payment of expenses incurred for—

(i) contract personnel to assist in the evaluation of munitions control license applications, reduce processing time for license applications, and improve monitoring of compliance with the terms of licenses; and

(ii) the automation of munitions control functions and the processing of munitions control license applications, including the development, procurement, and utilization of computer equipment and related software.

(B) The authority of this paragraph may be exercised only to such extent or in such amounts as are provided in advance in appropriation Acts.

(c) Criminal violations; punishment.

Any person who willfully violates any provision of this section, section 2779 of this title, a treaty referred to in subsection (j)(1)(C)(i), or any rule or regulation issued under this section or section 2779 of this title, including any rule or regulation issued to implement or enforce a treaty referred to in subsection (j)(1)(C)(i) or an implementing arrangement pursuant to such treaty, or who willfully, in a registration or license application or required report, makes any untrue statement of a material fact or omits to state a material fact required to be stated therein or necessary to make the statements therein not misleading, shall upon conviction be fined for each violation not more than $1,000,000 or imprisoned not more than 20 years, or both.

(d) [Repealed]

(e) Enforcement powers of President.

In carrying out functions under this section with respect to the export of defense articles and defense services, including defense articles and defense services exported or imported pursuant to a treaty referred to in subsection (j)(1)(C)(i), the President is authorized to exercise the same powers concerning violations and enforcement which are conferred upon departments, agencies and officials by subsections (c), (d), (e), and (g) of section 11 of the Export Administration Act of 1979 [50 U.S.C. App. 2410(c), (d), (e), and (g)], and by subsections (a) and (c) of section 12 of such Act [50 U.S.C. App. 2411(a) and (c)], subject to the same terms and conditions as are applicable to such powers under such Act [50 U.S.C. App. 2401 et seq.], except that section 11(c)(2)(B) of such Act shall not apply, and instead, as prescribed in regulations issued under this section, the Secretary of State may assess civil penalties for violations of this chapter and regulations prescribed thereunder and further may commence a civil action to recover such civil penalties, and except further that the names of the countries and the types and quantities of defense articles for which licenses are issued under this section shall not be withheld from public disclosure unless the President determines that the release of such information would be contrary to the national interest. Nothing in this subsection shall be construed as authorizing the withholding of information from the Congress. Notwithstanding section 11(c) of the Export Administration Act of 1979, the civil penalty for each violation involving controls imposed on the export of defense articles and defense services under this section may not exceed $500,000.

(f) Periodic review of items on the munitions list; notification regarding exemption from licensing requirements for export of defense items.

(1) The President shall periodically review the items on the United States Munitions List to determine what items, if any, no longer warrant export controls under this section. The results of such reviews shall be reported to the Speaker of the House of Representatives and to the Committee on Foreign Relations and the Committee on Banking, Housing, and Urban Affairs of the Senate. The President may not remove any item from the Munitions List until 30 days after the date on which the President has provided notice of the proposed removal to the Committee on International Relations of the House of Representatives and to the Committee on Foreign Relations of the Senate in accordance with the procedures applicable to reprogramming notifications under section 2394–1(a) of this title. Such notice shall describe the nature of any controls to be imposed on that item under any other provision of law.

(2) The President may not authorize an exemption for a foreign country from the licensing requirements of this chapter for the export of defense items under subsection (j) of this section or any other provision of this chapter until 30 days after the date on which the President has transmitted to the Committee on International Relations of the House of Representatives and the Committee on Foreign Relations of the Senate a notification that includes—

(A) a description of the scope of the exemption, including a detailed summary of the defense articles, defense services, and related technical data covered by the exemption; and

(B) a determination by the Attorney General that the bilateral agreement concluded under subsection (j) of this section requires the compilation and maintenance of sufficient documentation relating to the export of United States defense articles, defense services, and related technical data to facilitate law enforcement efforts to detect, prevent, and prosecute criminal violations of any provision of this chapter, including the efforts on the part of countries and factions engaged in international terrorism to illicitly acquire sophisticated United States defense items.

(3) Paragraph (2) shall not apply with respect to an exemption for Canada from the licensing requirements of this chapter for the export of defense items.

TITLE 27 CFR CHAPTER II

PART 447—IMPORTATION OF ARMS, AMMUNITION AND IMPLEMENTS OF WAR
(This section was formerly designated as Part 47)

Subpart A—Scope

§ 447.1 General.

The regulations in this part relate to that portion of section 38 of the Arms Export Control Act of 1976, as amended, authorizing the President to designate defense articles and defense services as part of the United States Munitions List (USML) for purposes of import and export controls. To distinguish the list of defense articles and defense services controlled in this part for purposes of permanent import from the list of defense articles and defense services controlled by the Secretary of State for purposes of export and temporary import, this part shall refer to the defense articles and defense services controlled for purposes of permanent import as the U.S. Munitions Import List (USMIL) and shall refer to the export and temporary import control list set out by the Department of State in its International Traffic in Arms Regulations as the USML. Part 447 contains the USMIL and includes procedural and administrative requirements relating to registration of importers, permits, articles in transit, import certification, delivery verification, import restrictions applicable to certain countries, exemptions, U.S. military firearms and ammunition, penalties, seizures, and forfeitures. The President's delegation of permanent import control authorities to the Attorney General provides the Attorney General the authority to assess whether controls are justified, but in designating the defense articles and defense services set out in the USMIL the Attorney General shall be guided by the views of the Secretary of State on matters affecting world peace and the external security and foreign policy of the United States. All designations and changes in designations of defense articles and defense services subject to permanent import control under this part must have the concurrence of the Secretary of State and the Secretary of Defense, with notice given to the Secretary of Commerce.

§ 447.2 Relation to other laws and regulations.

(a) All of those items on the U.S. Munitions Import List (see §447.21) which are **"firearms"** or **"ammunition"** as defined in 18 U.S.C. 921(a) are subject to the interstate and foreign commerce controls contained in Chapter 44 of Title 18 U.S.C. and 27 CFR Part 478 and if they are **"firearms"** within the definition set out in 26 U.S.C. 5845(a) are also subject to the provisions of 27 CFR Part 479. Any person engaged in the business of importing firearms or ammunition as defined in 18 U.S.C. 921(a) must obtain a license under the provisions of 27 CFR Part 478, and if he imports firearms which fall within the definition of 26 U.S.C. 5845(a) must also register and pay special tax pursuant to the provisions of 27 CFR Part 479. Such licensing, registration and special tax requirements are in addition to registration under subpart D of this part.

(b) The permit procedures of subpart E of this part are applicable to all importations of articles on the U.S. Munitions Import List not subject to controls under 27 CFR Part 478 or 479. U.S. Munitions Import List articles subject to controls under 27 CFR Part 478 or 27 CFR Part 479 are subject to the import permit procedures of those regulations if imported into the United States (within the meaning of 27 CFR Parts 478 and 479).

(c) Articles on the U.S. Munitions Import List imported for the United States or any State or political subdivision thereof are exempt from the import controls of 27 CFR Part 478 but are not exempt from control under Section 38, Arms Export Control Act of 1976, unless imported by the United States or any agency thereof. All such importations not imported by the United States or any agency thereof shall be subject to the import permit procedures of subpart E of this part.

(d) For provisions requiring the registration of persons engaged in the business of brokering activities with respect to the importation of any defense article or defense service, see Department of State regulations in 22 CFR part 129.

Subpart B—Definitions

§ 447.11 Meaning of terms.

When used in this part and in forms prescribed under this part, where not otherwise distinctly expressed or manifestly incompatible with the intent thereof, terms shall have the meanings ascribed in this section. Words in the plural form shall include the singular, and vice versa, and words imparting the masculine gender shall include the feminine. The terms **"includes"** and **"including"** do not exclude other things not enumerated which are in the same general class or are otherwise within the scope thereof.

Appropriate ATF officer. An officer or employee of the Bureau of Alcohol, Tobacco, Firearms and Explosives (ATF) specified by ATF Order 1130.34, Delegation of the Director's Authorities in 27 CFR Part 447, Importation of Arms, Ammunition and Implements of War.

Article. Any of the defense articles enumerated in the U.S. Munitions Import List (USMIL).

Bureau. Bureau of Alcohol, Tobacco, Firearms and Explosives, the Department of Justice.

Carbine. A short-barrelled rifle whose barrel is generally not longer than 22 inches and is characterized by light weight.

CFR. The Code of Federal Regulations.

Chemical agent. A substance useful in war which, by its ordinary and direct chemical action, produces a powerful physiological effect.

Defense articles. Any item designated in §447.21 or §447.22. This term includes models, mockups, and other such items which reveal technical data directly relating to §447.21 or §447.22.

Defense services. (a) The furnishing of assistance, including training, to foreign persons in the design, engineering, development, production, processing, manufacture, use, operation, overhaul, repair, maintenance, modification, or reconstruction of defense articles, whether in the United States or abroad; or

(b) The furnishing to foreign persons of any technical data, whether in the United States or abroad.

Director. The Director, Bureau of Alcohol, Tobacco, Firearms and Explosives, the Department of Justice, Washington, DC 20226.

Executed under the penalties of perjury. Signed with the prescribed declaration under the penalties of perjury as provided on or with respect to the application, form, or other document or, where no form of declaration is prescribed, with the declaration: **"I declare under the penalties of perjury that this _____ (insert type of document such as statement, certificate, application, or other document), including the documents submitted in support thereof, has been examined by me and, to best of my knowledge and belief, is true, correct, and complete."**

Firearms. A weapon, and all components and parts therefor, not over.50 caliber which will or is designed to or may be readily converted to expel a projectile by the action of an explosive, but shall not include BB and pellet guns, and muzzle loading (black powder) firearms (including any firearm with a matchlock, flintlock, percussion cap, or similar type of ignition system) or firearms covered by Category I(a) established to have been manufactured in or before 1898.

Import or importation. Bringing into the United States from a foreign country any of the articles on the Import List, but

shall not include intransit, temporary import or temporary export transactions subject to Department of State controls under Title 22, Code of Federal Regulations.

Import List. The list of articles contained in § 447.21 and identified therein as **"The U.S. Munitions Import List"**.

Machinegun. A **"machinegun,"** **"machine pistol,"** **"submachinegun,"** or **"automatic rifle"** is a firearm originally designed to fire, or capable of being fired fully automatically by a single pull of the trigger.

Permit. The same as **"license"** for purposes of 22 U.S.C. 1934(c).

Person. A partnership, company, association, or corporation, as well as a natural person.

Pistol. A hand-operated firearm having a chamber integral with, or permanently aligned with, the bore.

Revolver. A hand-operated firearm with a revolving cylinder containing chambers for individual cartridges.

Rifle. A shoulder firearm discharging bullets through a rifled barrel at least 16 inches in length, including combination and drilling guns.

Sporting type sight including optical. A telescopic sight suitable for daylight use on a rifle, shotgun, pistol, or revolver for hunting or target shooting.

This chapter. Title 27, Code of Federal Regulations, Chapter II (27 CFR Chapter II).

United States. When used in the geographical sense, includes the several States, the Commonwealth of Puerto Rico, the insular possessions of the United States, the District of Columbia, and any territory over which the United States exercises any powers of administration, legislation, and jurisdiction.

(26 U.S.C. 7805 (68A Stat. 917), 27 U.S.C. 205 (49 Stat. 981 as amended), 18 U.S.C. 926 (82 Stat. 959), and sec. 38, Arms Export Control Act (22 U.S.C. 2778, 90 Stat. 744))

Subpart C—The U.S. Munitions Import List

§ 447.21 The U.S. Munitions Import List.

The following defense articles and defense services, designated pursuant to section 38(a) of the Arms Export Control Act, 22 U.S.C. 2778(a), and E.O. 13637 are subject to controls under this part. For purposes of this part, the list shall be known as the U.S. Munitions Import List (USMIL):

THE U.S. MUNITIONS IMPORT LIST CATEGORY I—FIREARMS

(a) Nonautomatic and semiautomatic firearms, to caliber .50 inclusive, combat shotguns, and shotguns with barrels less than 18 inches in length, and all components and parts for such firearms.

(b) Automatic firearms and all components and parts for such firearms to caliber .50 inclusive.

(c) Insurgency-counterinsurgency type firearms of other weapons having a special military application (e.g. close assault weapons systems) regardless of caliber and all components and parts for such firearms.

(d) Firearms silencers and suppressors, including flash suppressors.

(e) [Reserved]

Note: Rifles, carbines, revolvers, and pistols, to caliber .50 inclusive, combat shotguns, and shotguns with barrels less than 18 inches in length are included under Category I(a). Machineguns, submachineguns, machine pistols and fully automatic rifles to caliber .50 inclusive are included under Category I(b).

CATEGORY II—ARTILLERY PROJECTORS

(a) Guns over caliber .50, howitzers, mortars, and recoiless rifles.

(b) Military flamethrowers and projectors.

(c) Components, parts, accessories, and attachments for the articles in paragraphs (a) and (b) of this category, including but not limited to mounts and carriages for these articles.

CATEGORY III—AMMUNITION

(a) Ammunition for the arms in Categories I and II of this section.

(b) Components, parts, accessories, and attachments for articles in paragraph (a) of this category, including but not limited to cartridge cases, powder bags, bullets, jackets, cores, shells (excluding shotgun shells), projectiles, boosters, fuzes and components therefor, primers, and other detonating devices for such ammunition.

(c) [Reserved]

(d) [Reserved]

Note: Cartridge and shell casings are included under Category III unless, prior to their importation, they have been rendered useless beyond the possibility of restoration for use as a cartridge or shell casing by means of heating, flame treatment, mangling, crushing, cutting, or popping.

CATEGORY IV—LAUNCH VEHICLES, GUIDED MISSILES, BALLISTIC MISSILES, ROCKETS, TORPEDOES, BOMBS AND MINES

(a) Rockets (including but not limited to meteorological and other sounding rockets), bombs, grenades, torpedoes, depth charges, land and naval mines, as well as launchers for such defense articles, and demolition blocks and blasting caps.

(b) Launch vehicles and missile and anti-missile systems including but not limited to guided, tactical and strategic missiles, launchers, and systems.

(c) Apparatus, devices, and materials for the handling, control, activation, monitoring, detection, protection, discharge, or detonation of the articles in paragraphs (a) and (b) of this category. Articles in this category include, but are not limited to, the following: Fuses and components for the items in this category, bomb racks and shackles, bomb shackle release units, bomb ejectors, torpedo tubes, torpedo and guided missile boosters, guidance system equipment and parts, launching racks and projectors, pistols (exploders), igniters, fuze arming devices, intervalometers, guided missile launchers and specialized handling equipment, and hardened missile launching facilities.

(d) Missile and space vehicle powerplants.

(e) Military explosive excavating devices.

(f) [Reserved]

(g) Non/nuclear warheads for rockets and guided missiles.

(h) All specifically designed components or modified components, parts, accessories, attachments, and associated equipment for the articles in this category.

Note: Military demolition blocks and blasting caps referred to in Category IV(a) do not include the following articles:

(a) Electric squibs.
(b) No. 6 and No. 8 blasting caps, including electric ones.
(c) Delay electric blasting caps (including No. 6 and No. 8 millisecond ones).
(d) Seismograph electric blasting caps (including SSS, Static-Master, Vibrocap SR, and SEISMO SR).
(e) Oil well perforating devices.

CATEGORY V [Reserved]

CATEGORY VI—VESSELS OF WAR AND SPECIAL NAVAL EQUIPMENT

(a) Vessels of War, if they are armed and equipped with offensive or defensive weapon systems, including but not limited to amphibious warfare vessels, landing craft, mine warfare vessels, patrol vessels, auxiliary vessels, service craft, experimental types of naval ships, and any vessels specifically designed or modified for military purposes or other surface vessels equipped with offensive or defensive military systems.

(b) Turrets and gun mounts, special weapons systems, protective systems, and other components, parts, attachments, and accessories specifically designed or modified for such articles on combatant vessels.

(c)-(d)[Reserved]

(e) Naval nuclear propulsion plants, their land prototypes and special facilities for their construction, support and maintenance. This includes any machinery, device, component, or equipment specifically developed or designed or modified for use in such plants or facilities.

Note: The term **"vessels of war"** includes, but is not limited to, the following, if armed and equipped with offensive or defensive weapons systems:".

(a) Combatant vessels:
(1) Warships (including nuclear-powered versions):
(i) Aircraft carriers (CV, CVN)
(ii) Battleships (BB)
(iii) Cruisers (CA, CG, CGN)
(iv) Destroyers (DD, DDG)
(v) Frigates (FF, FFG)

(vi) Submarines (SS, SSN, SSBN, SSG, SSAG).

(2) Other Combatant Classifications:

(i) Patrol Combatants (PC, PHM)

(ii) Amphibious Helicopter/Landing Craft Carriers (LHA, LPD, LPH)

(iii) Amphibious Landing Craft Carriers (LKA, LPA, LSD, LST)

(iv) Amphibious Command Ships (LCC)

(v) Mine Warfare Ships (MSO).

(b) Auxiliaries:

(1) Mobile Logistics Support:

(i) Under way Replenishment (AD, AF, AFS, AO, AOE, AOR)

(ii) Material Support (AD, AR, AS).

(2) Support Ships:

(i) Fleet Support Ships (ARS, ASR, ATA, ATF, ATS)

(ii) Other Auxiliaries (AG, AGDS, AGF, AGM, AGOR, AGOS, AGS, AH, AK, AKR, AOG, AOT, AP, APB, ARC, ARL, AVM, AVT).

(c) Combatant Craft:

(1) Patrol Craft:

(i) Coastal Patrol Combatants (PB, PCF, PCH, PTF)

(ii) River, Roadstead Craft (ATC, PBR).

(2) Amphibious Warfare Craft:

(i) Landing Craft (AALC, LCAC, LCM, LCPL, LCPR, LCU, LWT, SLWT)

(ii) Special Warfare Craft (LSSC, MSSC, SDV, SWCL, SWCM).

(3) Mine Warfare Craft:

(i) Mine Countermeasures Craft (MSB, MSD, MSI, MSM, MSR).

(d) Support and Service Craft:

(1) Tugs (YTB, YTL, YTM).

(2) Tankers (YO, YOG, YW)

(3) Lighters (YC, YCF, YCV, YF, YFN, YFNB, YFNX, YFR, YFRN, YFU, YG, YGN, YOGN, YON, YOS, YSR, YWN)

(4) Floating Dry Docks (AFDB, AFDL, AFDM, ARD, ARDM, YFD)

(5) Miscellaneous (APL, DSRV, DSV, IX, NR, YAG, YD, YDT, YFB, YFND, YEP, YFRT, YHLC, YM, YNG, YP, YPD, YR, YRB, YRBN, YRDH, YRDM, YRR, YRST, YSD).

(e) Coast Guard Patrol and Service Vessels and Craft:

(1) Coast Guard Cutters (CGC, WHEC, WMEC)

(2) Patrol Craft (WPB)

(3) Icebreakers (WAGB)

(4) Oceanography Vessels (WAGO)

(5) Special Vessels (WIX)

(6) Buoy Tenders (WLB, WLM, WLI, WLR, WLIC)

(7) Tugs (WYTM, WYTL)

(8) Light Ships (WLV).

CATEGORY VII—TANKS AND MILITARY VEHICLES

(a) Military type armed or armored vehicles, military railway trains, and vehicles specifically designed or modified to accommodate mountings for arms or other specialized military equipment or fitted with such items.

(b) Military tanks, combat engineer vehicles, bridge launching vehicles, halftracks and gun carriers.

(c) Self-propelled guns and howitzers.

(d)-(e) [Reserved]

(f) Amphibious vehicles.

(g) Reserved]

(h) Tank and military vehicle parts, components, accessories, attachments, and associated equipment for offensive or defensive systems for the articles in this category, as follows:

(1) Armored hulls, armored turrets and turret support rings;

(2) Active protection systems (i.e., defensive systems that actively detect and track incoming threats and launch a ballistic, explosive, energy or electromagnetic countermeasure(s) to neutralize the threat prior to contact with a vehicle);

(3) Composite armor parts and components;

(4) Spaced armor components and parts, including slat armor parts and components;

(5) Reactive armor and components;

(6) Electromagnetic armor parts and components, including pulsed power;

(7) Gun mount, stabilization, turret drive, and automatic elevating systems;

(8) Kits specifically designed to convert a vehicle in this category into either an unmanned or a driver-optional vehicle. For a kit to be controlled by this paragraph it must include all of the following:

(i) Remote or autonomous steering;

(ii) Acceleration and braking; and

(iii) A control system;

(9) Fire control computers, stored management systems, armaments control processors, vehicle weapon interface units and computers;

(10) Electro-optical sighting systems; and

(11) Laser rangefinder or target designating devices.

(i) Other ground vehicles having all of the following:

(1) Manufactured or fitted with materials or components to provide ballistic protection to level III (NIJ 0108.01, September 1985) or better;

(2) A transmission to provide drive to both front and rear wheels simultaneously, including those vehicles having additional wheels for load bearing purposes whether driven or not;

(3) Gross Vehicle Weight Rating (GVWR) greater than 4,500 kg; and

(4) Designed or modified for off-road use.

Note: An **"amphibious vehicle"** in Category VII(f) is a vehicle or chassis that is equipped to meet special military requirements, and that is designed or adapted for operation on or under water, as well as on land.

Note: Engines and engine parts are not included in paragraph (h) of Category VII.

Note: Paragraph (i) of Category VII does not apply to civil vehicles designed or modified for transporting money or valuables.

CATEGORY VIII—AIRCRAFT AND ASSOCIATED EQUIPMENT

(a) Aircraft, including but not limited to helicopters, non-expansive balloons, drones and lighter-than-air aircraft, which are specifically designed, modified, or equipped for military purposes. This includes but is not limited to the following military purposes: gunnery, bombing, rocket or missile launching, electronic and other surveillance, reconnaissance, refueling, aerial mapping, military liaison, cargo carrying or dropping, personnel dropping, airborne warning and control, and military training.

(b) [Reserved]

Note: In Category VIII, "aircraft" means aircraft designed, modified, or equipped for a military purpose, including aircraft described as "demilitarized." All aircraft bearing an original military designation are included in Category VIII. However, the following aircraft are not so included so long as they have not been specifically equipped, reequipped, or modified for military operations:

(a) Cargo aircraft bearing "C" designations and numbered C-45 through C-118 inclusive, and C-121 through C-125 inclusive, and C-131, using reciprocating engines only.

(b) Trainer aircraft bearing "T" designations and using reciprocating engines or turboprop engines with less than 600 horsepower (s.h.p.).

(c) Utility aircraft bearing "U" designations and using reciprocating engines only.

(d) All liaison aircraft bearing an "L" designation.

(e) All observation aircraft bearing "O" designations and using reciprocating engines.

CATEGORIES IX—XIII [Reserved]

CATEGORY XIV—TOXICOLOGICAL AGENTS AND EQUIPMENT AND RADIOLOGICAL EQUIPMENT

(a) Chemical agents, including but not limited to lung irritants, vesicants, lachrymators, and tear gases (except tear gas formulations containing 1% or less CN or CS), sternutators and irritant smoke, and nerve gases and incapacitating agents.

(b) [Reserved]

(c) All specifically designed or modified equipment, including components, parts, accessories, and attachments for disseminating the articles in paragraph (a) of this category.

(d)-(e) [Reserved]

Note: A chemical agent in Category XIV(a) is a substance having military application which by its ordinary and direct chemical action produces a powerful physiological effect. The term "chemical agent" includes, but is not limited to, the following chemical compounds:

(a) Lung irritants:
(1) Diphenylcyanoarsine (DC).
(2) Fluorine (but not fluorene).
(3) Trichloronitro methane (chloropicrin PS).
(b) Vesicants:
(1) B-Chlorovinyldichloroarsine (Lewisite, L).

(2) Bis(dichlorethyl) sulphide (Mustard Gas, HD or H).
(3) Ethyldichloroarsine (ED).
(4) Methyldichloroarsine (MD).
(c) Lachrymators and tear gases:
(1) A-Brombenzyl cyanide (BBC).
(2) Chloroacetophenone (CN).
(3) Dibromodimethyl ether.
(4) Dichlorodimethyl ether (ClCi).
(5) Ethyldibromoarsine.
(6) Phenylcarbylamine chloride.
(7) Tear gas solutions (CNB and CNS).
(8) Tear gas orthochlorobenzalmalononitrile (CS).
(d) Sternutators and irritant smokes:
(1) Diphenylamine chloroarsine (Adamsite, DM).
(2) Diphenylchloroarsine (BA).
(3) Liquid pepper.
(e) Nerve agents, gases, and aerosols. These are toxic compounds which affect the nervous system, such as:
(1) Dimethylaminoethoxycyanophosphine oxide (GA).
(2) Methylisopropoxyfluorophosphine oxide (GB).
(3) Methylpinacolyloxyfluoriphosphine oxide (GD).
(f) Antiplant chemicals, such as: Butyl 2-chloro-4-fluorophenoxyacetate (LNF).

CATEGORY XV [Reserved]

CATEGORY XVI—NUCLEAR WEAPONS DESIGN AND TEST EQUIPMENT

(a) [Reserved]

(b) Modeling or simulation tools that model or simulate the environments generated by nuclear detonations or the effects of these environments on systems, subsystems, components, structures, or humans.

Note: Category XVI does not include equipment, technical data, or services controlled by the Department of Energy pursuant to the Atomic Energy Act of 1954, as amended, and the Nuclear Non-Proliferation Act of 1978, as amended, or are government transfers authorized pursuant to these Acts.

CATEGORY XVII—XIX [Reserved]

CATEGORY XX—SUBMERSIBLE VESSELS, OCEANOGRAPHIC AND ASSOCIATED EQUIPMENT

(a) Submersible vessels, manned and unmanned, designed or modified for military purposes or having independent capability to maneuver vertically or horizontally at depths below 1,000 feet, or powered by nuclear propulsion plants.

(b) Submersible vessels, manned or unmanned, designed or modified in whole or in part from technology developed by or for the U.S. Armed Forces.

(c) Any of the articles in Category VI and elsewhere in this part specifically designed or modified for use with submersible vessels, and oceanographic or associated equipment assigned a military designation.

(d) Equipment, components, parts, accessories, and attachments specifically designed for any of the articles in paragraphs (a) and (b) of this category.

CATEGORY XXI—MISCELLANEOUS ARTICLES

Any defense article or defense service not specifically enumerated in the other categories of the USMIL that has substantial military applicability and that has been specifically designed or modified for military purposes. The decision as to whether any article may be included in this category shall be made by the Attorney General with the concurrence of the Secretary of State and the Secretary of Defense.

§ 447.22 Forgings, castings, and machined bodies.

Articles on the U.S. Munitions Import List include articles in a partially completed state (such as forgings, castings, extrusions, and machined bodies) which have reached a stage in manufacture where they are clearly identifiable as defense articles. If the end-item is an article on the U.S. Munitions Import List, (including components, accessories, attachments and parts) then the particular forging, casting, extrusion, machined body, etc., is considered a defense article subject to the controls of this part, except for such items as are in normal commercial use.

Subpart D—Registration

§ 447.31 Registration requirement.

Persons engaged in the business, in the United States, of importing articles enumerated on the U.S. Munitions Import List must register by making an application on ATF Form 4587.

§ 447.32 Application for registration and refund of fee.

(a) Application for registration must be filed on ATF Form 4587 and must be accompanied by the registration fee at the rate prescribed in this section. The appropriate ATF officer will approve the application and return the original to the applicant.

1 year	$250
2 years	$500
2 years	$700
4 years	$850
5 years	$1,000

(b) Registration may be effected for periods of from 1 to 5 years at the option of the registrant by identifying on Form 4587 the period of registration desired. The registration fees are as follows:

(c) Fees paid in advance for whole future years of a multiple year registration will be refunded upon request if the registrant ceases to engage in importing articles on the U.S. Munitions Import List. A request for a refund must be submitted to the appropriate ATF officer at the Bureau of Alcohol, Tobacco, Firearms and Explosives, Martinsburg, WV 25405, prior to the beginning of any year for which a refund is claimed.

(Approved by the Office of Management and Budget under control number 1140-0009)

§ 447.33 Notification of changes in information furnished by registrants.

Registered persons shall notify the appropriate ATF officer in writing, in duplicate, of significant changes in the information set forth in their registration

(Approved by the Office of Management and Budget under control number 1140-0009)

§ 447.34 Maintenance of records by persons required to register as importers of Import List articles.

(a) Registrants under this part engaged in the business of importing articles subject to controls under 27 CFR Parts 478 and 479 shall maintain records in accordance with the applicable provisions of those parts.

(b) Registrants under this part engaged in importing articles on the U.S. Munitions Import List subject to the permit procedures of subpart E of this part must maintain for a period of 6 years records bearing on such articles imported, including records concerning their acquisition and disposition, including Forms 6 and 6A. The appropri-

ate ATF officer may prescribe a longer or shorter period in individual cases as such officer deems necessary. See §478.129 of this chapter for articles subject to import control under part 478 of this chapter.

(Approved by the Office of Management and Budget under control number 1140-0032)

§ 447.35 Forms prescribed.

(a) The appropriate ATF officer is authorized to prescribe all forms required by this part. All of the information called for in each form shall be furnished as indicated by the headings on the form and the instructions on or pertaining to the form. In addition, information called for in each form shall be furnished as required by this part. The form will be filed in accordance with the instructions for the form.

(b) Forms may be requested from the ATF Distribution Center (http://www.atf.gov)or by calling (703) 870-7526 or (703) 870-7528.

Subpart E—Permits

§ 447.41 Permit requirement.

(a) Articles on the U.S. Munitions Import List will not be imported into the United States except pursuant to a permit under this subpart. For articles subject to control under parts 478 or 479 of this chapter, a separate permit is not necessary.

(b) Articles on the U.S. Munitions Import List intended for the United States or any State or political subdivision thereof, or the District of Columbia, which are exempt from import controls of 27 CFR 478.115 shall not be imported into the United States, except by the United States or agency thereof, without first obtaining a permit under this subpart.

(c) A permit is not required for the importation of—

(1)(i) The U.S. Munitions Import List articles from Canada, except articles enumerated in Categories I, II, III, IV, VI(e), VIII(a), XVI, and XX; and

(ii) Nuclear weapons strategic delivery systems and all specifically designed components, parts, accessories, attachments, and associated equipment thereof (see Category XXI); or

(2) Minor components and parts for Category I(a) and I(b) firearms, except barrels, cylinders, receivers

(frames) or complete breech mechanisms, when the total value does not exceed $100 wholesale in any single transaction.

§ 447.42 Application for permit.

(a)(1) Persons required to obtain a permit as provided in §447.41 must file a Form 6—Part I. The application must be signed and dated and must contain the information requested on the form, including:

(i) The name, address, telephone number, license and registration number, if any (including expiration date) of the importer;

(ii) The country from which the defense article is to be imported;

(iii) The name and address of the foreign seller and foreign shipper;

(iv) A description of the defense article to be imported, including—

(A) The name and address of the manufacturer;

(B) The type (e.g., rifle, shotgun, pistol, revolver, aircraft, vessel, and in the case of ammunition only, ball, wadcutter, shot, etc.);

(C) The caliber, gauge, or size;

(D) The model;

(E) The length of barrel, if any (in inches);

(F) The overall length, if a firearm (in inches);

(G) The serial number, if known;

(H) Whether the defense article is new or used;

(I) The quantity;

(J) The unit cost of the firearm, firearm barrel, ammunition, or other defense article to be imported;

(K) The category of U.S. Munitions Import List under which the article is regulated;

(v) The specific purpose of importation, including final recipient information if different from the importer; and

(vi) Certification of origin.

(2) (i) If the appropriate ATF officer approves the application, such approved application will serve as the permit to import the defense article described therein, and importation of such defense article may continue to be made by the licensed/registered importer (if applicable) under the approved application (permit) during the period specified thereon. The appropriate ATF officer will furnish the approved application (permit) to the applicant and retain two copies thereof for administrative use.

(ii) If the Director disapproves the application, the licensed/registered importer (if applicable) will be notified of the basis for the disapproval.

(b) For additional requirements relating to the importation of plastic explosives into the United States on or after April 24, 1997, see §555.183 of this title.

(Approved by the Office of Management and Budget under control number 1140-0005)

§ 447.43 Terms of permit.

(a) Import permits issued under this subpart are valid for two years from their issuance date unless a different period of validity is stated thereon. They are not transferable.

(b) If shipment cannot be completed during the period of validity of the permit, another application must be submitted for permit to cover the unshipped balance. Such an application shall make reference to the previous permit and may include materials in addition to the unshipped balance.

(c) No amendments or alteration of a permit may be made, except by the appropriate ATF officer.

§ 447.44 Permit denial, revocation or suspension.

(a) Import permits under this subpart may be denied, revoked, suspended or revised without prior notice whenever the appropriate ATF officer finds the proposed importation to be inconsistent with the purpose or in violation of section 38, Arms Export Control Act of 1976 or the regulations in this part.

(b) Whenever, after appropriate consideration, a permit application is denied or an outstanding permit is revoked,

suspended, or revised, the applicant or permittee shall be promptly advised in writing of the appropriate ATF officer's decision and the reasons therefor.

(c) Upon written request made within 30 days after receipt of an adverse decision, the applicant or permittee shall be accorded an opportunity to present additional information and to have a full review of his case by the appropriate ATF officer.

(d) Unused, expired, suspended, or revoked permits must be returned immediately to the appropriate ATF officer.

§ 447.45 Importation.

(a) Articles subject to the import permit procedures of this subpart imported into the United States may be released from Customs custody to the person authorized to import same upon his showing that he has a permit for the importation of the article or articles to be released. For articles in Categories I and III imported by a registered importer, the importer will also submit to Customs a copy of the export license authorizing the export of the article or articles from the exporting country. If the exporting country does not require issuance of an export license, the importer must submit a certification, under penalty of perjury, to that effect.

(1) In obtaining the release from Customs custody of an article imported pursuant to a permit, the permit holder will prepare and file Form 6A according to its instructions.

(2) The ATF Form 6A must contain the information requested on the form, including:

(i) The name, address, and license number (if any) of the importer;

(ii) The name of the manufacturer of the defense article;

(iii) The country of manufacture;

(iv) The type;

(v) The model;

(vi) The caliber, gauge, or size;

(vii) The serial number in the case of firearms, if known; and

(viii) The number of defense articles released.

(b) Within 15 days of the date of their

release from Customs custody, the importer of the articles released will forward to the address specified on the form a copy of Form 6A on which will be reported any error or discrepancy appearing on the Form 6A certified by Customs and serial numbers if not previously provided on ATF Form 6A.

(Approved by the Office of Management and Budget under control number 1140-0007)

§ 447.46 Articles in transit.

Articles subject to the import permit procedures of this subpart which enter the United States for temporary deposit pending removal therefrom and such articles which are temporarily taken out of the United States for return thereto shall be regarded as in transit and will be considered neither imported nor exported under this part. Such transactions are subject to the Intransit or Temporary Export License procedures of the Department of State (see 22 CFR Part 123).

Subpart F—Miscellaneous Provisions

§ 447.51 Import certification and delivery verification.

Pursuant to agreement with the United States, certain foreign countries are entitled to request certification of legality of importation of articles on the U.S. Munitions Import List. Upon request of a foreign government, the appropriate ATF officer will certify the importation, on Form ITA-645P/ATF-4522/DSP53, for the U.S. importer. Normally, the U.S. importer will submit this form at the time he applies for an import permit. This document will serve as evidence to the government of the exporting company that the U.S. importer has complied with import regulations of the U.S. Government and is prohibited from diverting, transshipping, or reexporting the material described therein without the approval of the U.S. Government. Foreign governments may also require documentation attesting to the delivery of the material into the United States. When such delivery certification is requested by a foreign government, the U.S. importer may obtain directly from the U.S. District Director of Customs the authenticated Delivery Verification Certificate (U.S. Department of Commerce Form ITA-647P) for this purpose.

(Approved by the Office of Management and Budget under control number 0625-0064)

§ 447.52 Import restrictions applicable to certain countries.

(a) It is the policy of the United States to deny licenses and other approvals with respect to defense articles and defense services originating in certain countries or areas. This policy applies to Afghanistan, Belarus (one of the states composing the former Soviet Union), Cuba, Iran, Iraq, Libya, Mongolia, North Korea, Sudan, Syria, and Vietnam. This policy applies to countries or areas with respect to which the United States maintains an arms embargo (e.g., Burma, China, the Democratic Republic of the Congo, Haiti, Liberia, Rwanda, Somalia, Sudan, and UNITA (Angola)). It also applies when an import would not be in furtherance of world peace and the security and foreign policy of the United States.

Note: Changes in foreign policy may result in additions to and deletions from the above list of countries. The ATF will publish changes to this list in the Federal Register. Contact the Firearms and Explosives Imports Branch at (304) 616-4550 for current information.

(b) Notwithstanding paragraph (a) of this section, the appropriate ATF officer shall deny applications to import into the United States the following firearms and ammunition:

(1) Any firearm located or manufactured in Georgia, Kazakstan, Kyrgyzstan, Moldova, Russian Federation, Turkmenistan, Ukraine, or Uzbekistan, and any firearm previously manufactured in the Soviet Union, that is not one of the models listed below:

(i) Pistols/Revolvers:

(A) German Model P08 Pistol.

(B) IZH 34M, .22 caliber Target Pistol.

(C) IZH 35M, .22 caliber Target Pistol.

(D) Mauser Model 1896 Pistol.

(E) MC-57-1 Pistol.

(F) MC-1-5 Pistol.

(G) Polish Vis Model 35 Pistol.

(H) Soviet Nagant Revolver.

(I) TOZ 35, .22 caliber Target Pistol.

(ii) Rifles:

(A) BARS-4 Bolt Action Carbine.

(B) Biathlon Target Rifle, .22LR caliber.

(C) British Enfield Rifle.

(D) CM2, .22 caliber Target Rifle (also known as SM2, 22 caliber).

(E) German Model 98K Rifle.

(F) German Model G41 Rifle.

(G) German Model G43 Rifle.

(H) IZH-94.

(I) LOS-7 Bolt Action Rifle.

(J) MC-7-07.

(K) MC-18-3.

(L) MC-19-07.

(M) MC-105-01.

(N) MC-112-02.

(O) MC-113-02.

(P) MC-115-1.

(Q) MC-125/127.

(R) MC-126.

(S) MC-128.

(T) Saiga Rifle.

(U) Soviet Model 38 Carbine.

(V) Soviet Model 44 Carbine.

(W) Soviet Model 91/30 Rifle.

(X) TOZ 18, .22 caliber Bolt Action Rifle.

(Y) TOZ 55.

(Z) TOZ 78.

(AA) Ural Target Rifle, .22LR caliber.

(BB) VEPR Rifle.

(CC) Winchester Model 1895, Russian Model Rifle;

Editor's Note:

This list has been modified. The regulation will be amended to formally incorporate the revised list. The current list of affected firearms is available at www.ATF.gov.

(2) Ammunition located or manufactured in Georgia, Kazakstan, Kyrgyzstan, Moldova, Russian Federation, Turkmenistan, Ukraine, or Uzbekistan, and ammunition previously manufactured in the Soviet Union, that is 7.62×25mm caliber (also known as 7.63×25mm caliber or .30 Mauser); or

(3) A type of firearm the manufacture of which began after February 9, 1996.

(c) The provisions of paragraph (b) of this section shall not affect the fulfillment of contracts with respect to firearms or ammunition entered or withdrawn from warehouse for consumption in the United States on or before February 9, 1996.

(d) A defense article authorized for importation under this part may not be shipped on a vessel, aircraft or other means or conveyance which is owned or operated by, or leased to or from, any of the countries or areas covered by paragraph (a) of this section.

(e) Applications for permits to import articles that were manufactured in, or have been in, a country or area proscribed under this section may be approved where the articles are covered by Category I(a) of the Import List (other than those subject to the provisions of 27 CFR Part 479), are importable as curios or relics under the provisions of 27 CFR 478.118, and meet the following criteria:

(1) The articles were manufactured in a proscribed country or area prior to the date, as established by the Department of State, the country or area became proscribed, or, were manufactured in a non-proscribed country or area; and

(2) The articles have been stored for the five year period immediately prior to importation in a non-proscribed country or area.

(f) Applicants desiring to import articles claimed to meet the criteria specified in paragraph (e) of this section shall explain, and certify to, how the firearms meet the criteria. The certification statement will be prepared in letter form, executed under the penalties of perjury, and should be submitted with the application for an import permit. The certification statement must be accompanied by documentary information on the country or area of original manufacture and on the country or area of storage for the five year period immediately prior to importation. Such information may, for example, include a verifiable statement in the English language of a government official or any other person having knowledge of the date and place of manufacture and/or the place of storage; a warehouse receipt or other document which provides the required history of storage; and any other document that the applicant believes substantiates the place and date of manufacture and the place of storage. The appropriate ATF officer, however, reserves the right to determine whether documentation is acceptable. Applicants shall, when required by the appropriate ATF officer, furnish additional documentation as may be necessary to determine whether an import permit application should be approved.

§ 447.53 Exemptions.

(a) The provisions of this part are not applicable to:

(1) Importations by the United States or any agency thereof;

(2) Importation of components for items being manufactured under contract for the Department of Defense; or

(3) Importation of articles (other than those which would be **"firearms"** as defined in 18 U.S.C. 921(a)(3) manufactured in foreign countries for persons in the United States pursuant to Department of State approval.

(b) Any person seeking to import articles on the U.S. Munitions Import List as exempt under paragraph (a)(2) or (3) of this section may obtain release of such articles from Customs custody by submitting, to the Customs officer with authority to release, a statement claiming the exemption accompanied by satisfactory proof of eligibility. Such proof may be in the form of a letter from the Department of Defense or State, as the case may be, confirming that the conditions of the exemption are met.

§ 447.54 Administrative procedures inapplicable.

The functions conferred under section 38, Arms Export Control Act of 1976, as amended, are excluded from the operation of Chapter 5, Title 5, United States Code, with respect to Rule Making and Adjudication, 5 U.S.C. 553 and 554.

§ 447.55 Departments of State and Defense consulted.

The administration of the provisions of this part will be subject to the guidance of the Secretaries of State and Defense on matters affecting world peace and the external security and foreign policy of the United States.

§ 447.56 Authority of Customs officers.

(a) Officers of the U.S. Customs Service are authorized to take appropriate action to assure compliance with this part and with 27 CFR Parts 478 and 479 as to the importation or attempted importation of articles on the U.S. Munitions Import List, whether or not authorized by permit.

(b) Upon the presentation to him of a permit or written approval authorizing importation of articles on the U.S. Munitions Import List, the Customs officer who has authority to release same may require, in addition to such documents as may be required by Customs regulations, the production of other relevant documents relating to the proposed importation, including, but not limited to, invoices, orders, packing lists, shipping documents, correspondence, and instructions.

§ 447.57 U.S. military defense articles.

(a)(1) Notwithstanding any other provision of this part or of parts 478 or 479 of this chapter, no military defense article of United States manufacture may be imported into the United States if such article was furnished to a foreign government under a foreign assistance or foreign military sales program of the United States.

(2) The restrictions in paragraph (a)(1) of this section cover defense articles which are advanced in value or improved in condition in a foreign country, but do not include those which have been substantially transformed as to become, in effect, articles of foreign manufacture.

(b) Paragraph (a) of this section will not apply if:

(1) The applicant submits with the ATF Form 6—Part I application written authorization from the Department of State to import the defense article; and

(2) In the case of firearms, such firearms are curios or relics under 18 U.S.C. 925(e) and the person seeking to import such firearms provides a certification of a foreign government that the firearms were furnished to such government under a foreign assistance or foreign military sales program of the United States and that the firearms are owned by such foreign government. (See §478.118 of this chapter providing for the importation of certain curio or relic handguns, rifles and shotguns.)

(c) For the purpose of this section, the term **"military defense article"** includes all defense articles furnished to foreign governments under a foreign assistance or foreign military sales program of the United States as set forth in paragraph (a) of this section.

(Approved by the Office of Management and Budget under OMB Control No. 1140-0005)

§ 447.58 Delegations of the Director.

The regulatory authorities of the Director contained in this part are delegated to appropriate ATF officers. These ATF officers are specified in ATF O 1130.34, Delegation of the Director's Authorities in 27 CFR Part 447. ATF delegation orders, such as ATF O 1130.34, are available to any interested party by submitting a request to the ATF Distribution Center (http://www.atf.gov) or by calling (703) 870-7526 or (703) 870-7528

Subpart G—Penalties, Seizures and Forfeitures

§ 447.61 Unlawful importation.

Any person who willfully:

(a) Imports articles on the U.S. Munitions Import List without a permit;

(b) Engages in the business of importing articles on the U.S. Munitions Import List without registering under this part; or

(c) Otherwise violates any provisions of this part;

Shall upon conviction be fined not more than $1,000,000 or imprisoned not more than 10 years, or both.

§ 447.62 False statements or concealment of facts.

Any person who willfully, in a registration or permit application, makes any untrue statement of a material fact or fails to state a material fact required to be stated therein or necessary to make the statements therein not misleading, shall upon conviction be fined not more than $1,000,000, or imprisoned not more than 10 years, or both.

§ 447.63 Seizure and forfeiture.

Whoever knowingly imports into the United States contrary to law any article on the U.S. Munitions Import List; or receives, conceals, buys, sells, or in any manner facilitates its transportation, concealment, or sale after importation, knowing the same to have been imported contrary to law, shall be fined not more than $10,000 or imprisoned not more than 5 years, or both; and the merchandise so imported, or the value thereof shall be forfeited to the United States.

TITLE 28 CFR CHAPTER I

PART 25—DEPARTMENT OF JUSTICE INFORMATION SYSTEMS

§ 25.1 Purpose and authority.

The purpose of this subpart is to establish policies and procedures implementing the Brady Handgun Violence Prevention Act (Brady Act), Public Law 103–159, 107 Stat. 1536. The Brady Act requires the Attorney General to establish a National Instant Criminal Background Check System (NICS) to be contacted by any licensed importer, licensed manufacturer, or licensed dealer of firearms for information as to whether the transfer of a firearm to any person who is not licensed under 18 U.S.C. 923 would be in violation of Federal or state law. The regulations in this subpart are issued pursuant to section 103(h) of the Brady Act, 107 Stat. 1542 (18 U.S.C. 922 note), and include requirements to ensure the privacy and security of the NICS and appeals procedures for persons who have been denied the right to obtain a firearm as a result of a NICS background check performed by the Federal Bureau of Investigation (FBI) or a state or local law enforcement agency.

§ 25.2 Definitions.

Appeal means a formal procedure to challenge the denial of a firearm transfer.

ARI means a unique Agency Record Identifier assigned by the agency submitting records for inclusion in the NICS Index.

ATF means the Bureau of Alcohol, Tobacco, and Firearms of the Department of Treasury.

Audit log means a chronological record of system (computer) activities that enables the reconstruction and examination of the sequence of events and/or changes in an event.

Business day means a 24–hour day (beginning at 12:01 a.m.) on which state offices are open in the state in which the proposed firearm transaction is to take place.

Control Terminal Agency means a state or territorial criminal justice agency recognized by the FBI as the agency responsible for providing state–or territory–wide service to criminal justice users of NCIC data.

Data source means an agency that provided specific information to the NICS.

Delayed means the response given to the FFL indicating that the transaction is in an **"Open"** status and that more research is required prior to a NICS **"Proceed"** or **"Denied"** response. A **"Delayed"** response to the FFL indicates that it would be unlawful to transfer the firearm until receipt of a follow–up **"Proceed"** response from the NICS or the expiration of three business days, whichever occurs first.

Denied means denial of a firearm transfer based on a NICS response indicating one or more matching records were found providing information demonstrating that receipt of a firearm by a prospective transferee would violate 18 U.S.C. 922 or state law.

Denying agency means a POC or the NICS Operations Center, whichever determines that information in the NICS indicates that the transfer of a firearm to a person would violate Federal or state law, based on a background check.

Dial–up access means any routine access through commercial switched circuits on a continuous or temporary basis.

Federal agency means any authority of the United States that is an **"Agency"** under 44 U.S.C. 3502(1), other than those considered to be independent regulatory agencies, as defined in 44 U.S.C. 3502(10).

FFL (federal firearms licensee) means a person licensed by the ATF as a manufacturer, dealer, or importer of firearms.

Firearm has the same meaning as in 18 U.S.C. 921(a)(3).

Licensed dealer means any person defined in 27 CFR 178.11.

Licensed importer has the same meaning as in 27 CFR 178.11.

Licensed manufacturer has the same meaning as in 27 CFR 178.11.

NCIC (National Crime Information Center) means the nationwide computerized information system of criminal justice data established by the FBI as a service to local, state, and Federal criminal justice agencies.

NICS means the National Instant Criminal Background Check System, which an FFL must, with limited exceptions, contact for information on whether receipt of a firearm by a person who is not licensed under 18 U.S.C. 923 would violate Federal or state law.

NICS Index means the database, to be managed by the FBI, containing information provided by Federal and state agencies about persons prohibited under Federal law from receiving or possessing a firearm. The NICS Index is separate and apart from the NCIC and the Interstate Identification Index (III).

NICS operational day means the period during which the NICS Operations Center has its daily regular business hours.

NICS Representative means a person who receives telephone inquiries to the NICS Operations Center from FFLs requesting background checks and provides a response as to whether the receipt or transfer of a firearm may proceed or is delayed.

NRI (NICS Record Identifier) means the system–generated unique number associated with each record in the NICS Index.

NTN (NICS Transaction Number) means the unique number that will be assigned to each valid background check inquiry received by the NICS. Its primary purpose will be to provide a means of associating inquiries to the NICS with the responses provided by the NICS to the FFLs.

Open means those non–canceled transactions where the FFL has not been notified of the final determination. In cases of **"open"** responses, the NICS continues researching potentially prohibiting records regarding the transferee and, if definitive information is obtained, communicates to the FFL the final determination that the check resulted in a proceed or a deny. An **"open"** response does not prohibit an FFL from transferring a firearm after three business days have elapsed since the FFL provided to the system the identifying information about the prospective transferee.

ORI (Originating Agency Identifier) means a nine–character identifier assigned by the FBI to an agency that has met the established qualifying criteria for ORI assignment to identify the agency in transactions on the NCIC System.

Originating Agency means an agency that provides a record to a database checked by the NICS.

POC (Point of Contact) means a state or local law enforcement agency serving as an intermediary between an FFL and the federal databases checked by the NICS. A POC will receive NICS background check requests from FFLs, check state or local record systems, perform NICS inquiries, determine whether matching records provide information demonstrating that an individual is disqualified from possessing a firearm under Federal or state law, and respond to FFLs with the results of a NICS background check. A POC will be an agency with express or implied authority to perform POC duties pursuant to state statute, regulation, or executive order.

Proceed means a NICS response indicating that the information available to the system at the time of the response did not demonstrate that transfer of the firearm would violate federal or state law. A **"Proceed"** response would not relieve an FFL from compliance with other provisions of Federal or state law that may be applicable to firearms transfers. For example, under 18 U.S.C. 922(d), an FFL may not lawfully transfer a firearm if he or she knows or has reasonable cause to believe that the prospective recipient is prohibited by law from receiving or possessing a firearm.

Record means any item, collection, or grouping of information about an individual that is maintained by an agency, including but not limited to information that disqualifies the individual from receiving a firearm, and that contains his or her name or other personal identifiers.

STN (State–Assigned Transaction Number) means a unique number that may be assigned by a POC to a valid background check inquiry.

System means the National Instant Criminal Background Check System (NICS).

§ 25.3 System information.

(a) There is established at the FBI a National Instant Criminal Background Check System.

(b) The system will be based at the Federal Bureau of Investigation, 1000 Custer Hollow Road, Clarksburg, West Virginia 26306–0147.

(c) The system manager and address are: Director, Federal Bureau of Investigation, J. Edgar Hoover F.B.I. Building, 935 Pennsylvania Avenue, NW, Washington, D.C. 20535.

§ 25.4 Record source categories.

It is anticipated that most records in the NICS Index will be obtained from Federal agencies. It is also anticipated that a limited number of authorized state and local law enforcement agencies will voluntarily contribute records to the NICS Index. Information in the NCIC and III systems that will be searched during a background check has been or will be contributed voluntarily by Federal, state, local, and international criminal justice agencies.

§ 25.5 Validation and data integrity of records in the system.

(a) The FBI will be responsible for maintaining data integrity during all NICS operations that are managed and carried out by the FBI. This responsibility includes:

(1) Ensuring the accurate adding, canceling, or modifying of NICS Index records supplied by Federal agencies;

(2) Automatically rejecting any attempted entry of records into the NICS Index that contain detectable invalid data elements;

(3) Automatic purging of records in the NICS Index after they are on file for a prescribed period of time; and

(4) Quality control checks in the form of periodic internal audits by FBI personnel to verify that the information provided to the NICS Index remains valid and correct.

(b) Each data source will be responsible for ensuring the accuracy and validity of the data it provides to the NICS Index and will immediately correct any record determined to be invalid or incorrect.

§ 25.6 Accessing records in the system.

(a) FFLs may initiate a NICS background check only in connection with a proposed firearm transfer as required by the Brady Act. FFLs are strictly prohibited from initiating a NICS background check for any other purpose. The process of accessing the NICS for the purpose of conducting a NICS background check is initiated by an FFL's contacting the FBI NICS Operations Center (by telephone or electronic dial–up access) or a POC. FFLs in each state will be advised by the ATF whether they are required to initiate NICS background checks with the NICS Operations Center or a POC and how they are to do so.

(b) Access to the NICS through the FBI NICS Operations Center. FFLs may contact the NICS Operations Center by use of a toll–free telephone number, only during its regular business hours. In addition to telephone access, toll–free electronic dial–up access to the NICS will be provided to FFLs after the beginning of the NICS operation. FFLs with electronic dial–up access will be able to contact the NICS 24 hours each day, excluding scheduled and unscheduled downtime.

(c)(1) The FBI NICS Operations Center, upon receiving an FFL telephone or electronic dial–up request for a background check, will:

    (i) Verify the FFL Number and code word;

    (ii) Assign a NICS Transaction Number (NTN) to a valid inquiry and provide the NTN to the FFL;

    (iii) Search the relevant databases (i.e., NICS Index, NCIC, III) for any matching records; and

    (iv) Provide the following NICS responses based upon the consolidated NICS search results to the FFL that requested the background check:

        (A) **"Proceed"** response, if no disqualifying information was found in the NICS Index, NCIC, or III.

        (B) **"Delayed"** response, if the NICS search finds a record that requires more research to determine whether the prospective transferee is disqualified from possessing a firearm by Federal or state law. A **"Delayed"** response to the FFL indicates that the firearm transfer should not proceed pending receipt of a follow–up **"Proceed"** response from the NICS or the expiration of three business days (exclusive of the day on which the query is made), whichever occurs first. (Example: An FFL requests a NICS check on a prospective firearm transferee at 9:00 a.m. on Friday and shortly thereafter

receives a **"Delayed"** response from the NICS. If state offices in the state in which the FFL is located are closed on Saturday and Sunday and open the following Monday, Tuesday, and Wednesday, and the NICS has not yet responded with a **"Proceed"** or **"Denied"** response, the FFL may transfer the firearm at 12:01 a.m. Thursday.)

        (C) **"Denied"** response, when at least one matching record is found in either the NICS Index, NCIC, or III that provides information demonstrating that receipt of a firearm by the prospective transferee would violate 18 U.S.C. 922 or state law. The **"Denied"** response will be provided to the requesting FFL by the NICS Operations Center during its regular business hours.

    (2) None of the responses provided to the FFL under paragraph (c)(1) of this section will contain any of the underlying information in the records checked by the system.

(d) Access to the NICS through POCs. In states where a POC is designated to process background checks for the NICS, FFLs will contact the POC to initiate a NICS background check. Both ATF and the POC will notify FFLs in the POC's state of the means by which FFLs can contact the POC. The NICS will provide POCs with electronic access to the system virtually 24 hours each day through the NCIC communication network. Upon receiving a request for a background check from an FFL, a POC will:

    (1) Verify the eligibility of the FFL either by verification of the FFL number or an alternative POC–verification system;

    (2) Enter a purpose code indicating that the query of the system is for the purpose of performing a NICS background check in connection with the transfer of a firearm; and (3) Transmit the request for a background check via the NCIC interface to the NICS.

(e) Upon receiving a request for a NICS background check, POCs may also conduct a search of available files in state and local law enforcement and other relevant record systems, and may provide a unique State–Assigned

Transaction Number (STN) to a valid inquiry for a background check.

(f) When the NICS receives an inquiry from a POC, it will search the relevant databases (i.e., NICS Index, NCIC, III) for any matching record(s) and will provide an electronic response to the POC. This response will consolidate the search results of the relevant databases and will include the NTN. The following types of responses may be provided by the NICS to a state or local agency conducting a background check:

    (1) No record response, if the NICS determines, through a complete search, that no matching record exists.

    (2) Partial response, if the NICS has not completed the search of all of its records. This response will indicate the databases that have been searched (i.e., III, NCIC, and/or NICS Index) and the databases that have not been searched. It will also provide any potentially disqualifying information found in any of the databases searched. A follow–up response will be sent as soon as all the relevant databases have been searched. The follow–up response will provide the complete search results.

    (3) Single matching record response, if all records in the relevant databases have been searched and one matching record was found.

    (4) Multiple matching record response, if all records in the relevant databases have been searched and more than one matching record was found.

(g) Generally, based on the response(s) provided by the NICS, and other information available in the state and local record systems, a POC will:

    (1) Confirm any matching records; and

    (2) Notify the FFL that the transfer may proceed, is delayed pending further record analysis, or is denied. **"Proceed"** notifications made within three business days will be accompanied by the NTN or STN traceable to the NTN. The POC may or may not provide a transaction number (NTN or STN) when notifying the FFL of a **"Denied"** response.

(h) POC Determination Messages. POCs shall transmit electronic NICS transaction determination messages to the FBI for the following transactions: open transactions that are not resolved before the end of the operational day on which the check is requested; denied transactions; transactions reported to the NICS as open and later changed to proceed; and denied transactions that have been overturned. The FBI shall provide POCs with an electronic capability to transmit this information. These electronic messages shall be provided to the NICS immediately upon communicating the POC determination to the FFL. For transactions where a determination has not been communicated to the FFL, the electronic messages shall be communicated no later than the end of the operational day on which the check was initiated. With the exception of permit checks, newly created POC NICS transactions that are not followed by a determination message (deny or open) before the end of the operational day on which they were initiated will be assumed to have resulted in a proceed notification to the FFL. The information provided in the POC determination messages will be maintained in the NICS Audit Log described in §25.9(b). The NICS will destroy its records regarding POC determinations in accordance with the procedures detailed in §25.9(b).

(i) Response recording. FFLs are required to record the system response, whether provided by the FBI NICS Operations Center or a POC, on the appropriate ATF form for audit and inspection purposes, under 27 CFR part 178 recordkeeping requirements. The FBI NICS Operations Center response will always include an NTN and associated **"Proceed," "Delayed,"** or **"Denied"** determination. POC responses may vary as discussed in paragraph (g) of this section. In these instances, FFLs will record the POC response, including any transaction number and/or determination.

(j) Access to the NICS Index for purposes unrelated to NICS background checks required by the Brady Act. Access to the NICS Index for purposes unrelated to NICS background checks pursuant to 18 U.S.C. 922(t) shall be limited to uses for the purpose of:

(1) Providing information to Federal, state, or local criminal justice agencies in connection with the issuance of a firearm–related or explosives–related permit or license, including permits or licenses to possess, acquire, or transfer a firearm, or to carry a concealed firearm, or to import, manufacture, deal in, or purchase explosives; or

(2) Responding to an inquiry from the ATF in connection with a civil or criminal law enforcement activity relating to the Gun Control Act (18 U.S.C. Chapter 44) or the National Firearms Act (26 U.S.C. Chapter 53).

§ 25.7 Querying records in the system.

(a) The following search descriptors will be required in all queries of the system for purposes of a background check:

(1) Name;

(2) Sex;

(3) Race;

(4) Complete date of birth; and

(5) State of residence.

(b) A unique numeric identifier may also be provided to search for additional records based on exact matches by the numeric identifier. Examples of unique numeric identifiers for purposes of this system are: Social Security number (to comply with Privacy Act requirements, a Social Security number will not be required by the NICS to perform any background check) and miscellaneous identifying numbers (e.g., military number or number assigned by Federal, state, or local authorities to an individual's record). Additional identifiers that may be requested by the system after an initial query include height, weight, eye and hair color, and place of birth. At the option of the querying agency, these additional identifiers may also be included in the initial query of the system.

§ 25.8 System safeguards.

(a) Information maintained in the NICS Index is stored electronically for use in an FBI computer environment. The NICS central computer will reside inside a locked room within a secure facility. Access to the facility will be restricted to authorized personnel who have identified themselves and their need for access to a system security officer.

(b) Access to data stored in the NICS is restricted to duly authorized agencies. The security measures listed in paragraphs (c) through (f) of this section are the minimum to be adopted by all POCs and data sources having access to the NICS.

(c) State or local law enforcement agency computer centers designated by a Control Terminal Agency as POCs shall be authorized NCIC users and shall observe all procedures set forth in the NCIC Security Policy of 1992 when processing NICS background checks. The responsibilities of the Control Terminal Agencies and the computer centers include the following:

(1) The criminal justice agency computer site must have adequate physical security to protect against any unauthorized personnel gaining access to the computer equipment or to any of the stored data.

(2) Since personnel at these computer centers can have access to data stored in the NICS, they must be screened thoroughly under the authority and supervision of a state Control Terminal Agency. This authority and supervision may be delegated to responsible criminal justice agency personnel in the case of a satellite computer center being serviced through a state Control Terminal Agency. This screening will also apply to non–criminal justice maintenance or technical personnel.

(3) All visitors to these computer centers must be accompanied by staff personnel at all times.

(4) POCs utilizing a state/NCIC terminal to access the NICS must have the proper computer instructions written and other built–in controls to prevent data from being accessible to any terminals other than authorized terminals.

(5) Each state Control Terminal Agency shall build its data system around a central computer, through which each inquiry must pass for screening and verification.

(d) Authorized state agency remote terminal devices operated by POCs and having access to the NICS must meet the following requirements:

(1) POCs and data sources having terminals with access to the NICS must physically place these terminals in secure locations within the authorized agency;

(2) The agencies having terminals with access to the NICS must screen terminal operators and must restrict access to the terminals to a minimum number of authorized employees; and

(3) Copies of NICS data obtained from terminal devices must be afforded appropriate security to prevent any unauthorized access or use.

(e) FFL remote terminal devices may be used to transmit queries to the NICS via electronic dial–up access. The following procedures will apply to such queries:

(1) The NICS will incorporate a security authentication mechanism that performs FFL dial–up user authentication before network access takes place;

(2) The proper use of dial–up circuits by FFLs will be included as part of the periodic audits by the FBI; and

(3) All failed authentications will be logged by the NICS and provided to the NICS security administrator.

(f) FFLs may use the telephone to transmit queries to the NICS, in accordance with the following procedures:

(1) FFLs may contact the NICS Operations Center during its regular business hours by a telephone number provided by the FBI;

(2) FFLs will provide the NICS Representative with their FFL Number and code word, the type of sale, and the name, sex, race, date of birth, and state of residence of the prospective buyer; and

(3) The NICS will verify the FFL Number and code word before processing the request.

(g) The following precautions will be taken to help ensure the security and privacy of NICS information when FFLs contact the NICS Operations Center:

(1) Access will be restricted to the initiation of a NICS background check in connection with the proposed transfer of a firearm.

(2) The NICS Representative will only provide a response of **"Proceed"** or **"Delayed"** (with regard to the prospective firearms transfer), and will not provide the details of any record information about the transferee. In cases where potentially disqualifying information is found in response to an FFL query, the NICS Representative will provide a **"Delayed"** response to the FFL. Follow–up **"Proceed"** or **"Denied"** responses will be provided by the NICS Operations Center during its regular business hours.

(3) The FBI will periodically monitor telephone inquiries to ensure proper use of the system.

(h) All transactions and messages sent and received through electronic access by POCs and FFLs will be automatically logged in the NICS Audit Log described in §25.9(b). Information in the NICS Audit Log will include initiation and termination messages, failed authentications, and matching records located by each search transaction.

(i) The FBI will monitor and enforce compliance by NICS users with the applicable system security requirements outlined in the NICS POC Guidelines and the NICS FFL Manual (available from the NICS Operations Center, Federal Bureau of Investigation, 1000 Custer Hollow Road, Clarksburg, West Virginia 26306–0147).

§ 25.9 Retention and destruction of records in the system.

(a) The NICS will retain NICS Index records that indicate that receipt of a firearm by the individuals to whom the records pertain would violate Federal or state law. The NICS will retain such records indefinitely, unless they are canceled by the originating agency. In cases where a firearms disability is not permanent, e.g., a disqualifying restraining order, the NICS will automatically purge the pertinent record when it is no longer disqualifying. Unless otherwise removed, records contained in the NCIC and III files that are accessed during a background check will remain in those files in accordance with established policy.

(b) The FBI will maintain an automated NICS Audit Log of all incoming and outgoing transactions that pass through the system.

(1) Contents. The NICS Audit Log will record the following information: Type of transaction (inquiry or response), line number, time, date of inquiry, header, message key, ORI or FFL identifier, and inquiry/response data (including the name and other identifying information about the prospective transferee and the NTN).

(i) NICS Audit Log records relating to denied transactions will be retained for 10 years, after which time they will be transferred to a Federal Records Center for storage;

(ii) NICS Audit Log records relating to transactions in an open status, except the NTN and date, will be destroyed after not more than 90 days from the date of inquiry; and

(iii) In cases of NICS Audit Log records relating to allowed transactions, all identifying information submitted by or on behalf of the transferee will be destroyed within 24 hours after the FFL receives communication of the determination that the transfer may proceed. All other information, except the NTN and date, will be destroyed after not more than 90 days from the date of inquiry.

(2) Use of information in the NICS Audit Log. The NICS Audit Log will be used to analyze system performance, assist users in resolving operational problems, support the appeals process, or support audits of the use and performance of the system. Searches may be conducted on the Audit Log by time frame, i.e., by day or month, or by a particular state or agency. Information in the NICS Audit Log pertaining to allowed transactions may be accessed directly only by the FBI and only for the purpose of conducting audits of the use and performance of the NICS, except that:

(i) Information in the NICS Audit Log, including information not yet destroyed under §5.9(b)(1)(iii), that indicates, either on its face or in conjunction with other information, a violation or potential violation of law or regulation, may be shared with appropriate authorities responsible for investigating, prosecuting, and/or enforcing such law or regulation; and

(ii) The NTNs and dates for allowed transactions may be shared with ATF in Individual FFL Audit Logs as specified in §25.9(b)(4).

(3) Limitation on use. The NICS, including the NICS Audit Log, may not be used by any Department, agency, officer, or employee of the United States to establish any system for the registration of firearms, firearm owners, or firearm transactions or dispositions, except with respect to persons prohibited from receiving a firearm by 18 U.S.C. 922(g) or (n) or by state law. The NICS Audit Log will be monitored and reviewed on a regular basis to detect any possible misuse of NICS data.

(4) Creation and Use of Individual FFL Audit Logs. Upon written request from ATF containing the name and license number of the FFL and the proposed date of inspection of the named FFL by ATF, the FBI may extract information from the NICS Audit Log and create an Individual FFL Audit Log for transactions originating at the named FFL for a limited period of time. An Individual FFL Audit Log shall contain all information on denied transactions, and, with respect to all other transactions, only non–identifying information from the transaction. In no instance shall an Individual FFL Audit Log contain more than 60 days worth of allowed or open transaction records originating at the FFL. The FBI will provide POC states the means to provide to the FBI information that will allow the FBI to generate Individual FFL Audit Logs in connection with ATF inspections of FFLs in POC states. POC states that elect not to have the FBI generate Individual FFL Audit Logs for FFLs in their states must develop a means by which the POC will provide such Logs to ATF.

(c) The following records in the FBI–operated terminals of the NICS will be subject to the Brady Act's requirements for destruction:

(1) All inquiry and response messages (regardless of media) relating to a background check that results in an allowed transfer; and

(2) All information (regardless of media) contained in the NICS Audit Log relating to a background check that results in an allowed transfer.

(d) The following records of state and local law enforcement units serving as POCs will be subject to the Brady Act's requirements for destruction:

(1) All inquiry and response messages (regardless of media) relating to the initiation and result of a check of the NICS that allows a transfer that are not part of a record system created and maintained pursuant to independent state law regarding firearms transactions; and

(2) All other records relating to the person or the transfer created as a result of a NICS check that are not part of a record system created and maintained pursuant to independent state law regarding firearms transactions.

§ 25.10 Correction of erroneous system information.

(a) An individual may request the reason for the denial from the agency that conducted the check of the NICS (the **"denying agency,"** which will be either the FBI or the state or local law enforcement agency serving as a POC). The FFL will provide to the denied individual the name and address of the denying agency and the unique transaction number (NTN or STN) associated with the NICS background check. The request for the reason for the denial must be made in writing to the denying agency. (POCs at their discretion may waive the requirement for a written request.)

(b) The denying agency will respond to the individual with the reasons for the denial within five business days of its receipt of the individual's request. The response should indicate whether additional information or documents are required to support an appeal, such as fingerprints in appeals involving questions of identity (i.e., a claim that the record in question does not pertain to the individual who was denied).

(c) If the individual wishes to challenge the accuracy of the record upon which the denial is based, or if the individual wishes to assert that his or her rights to possess a firearm have been restored, he or she may make application first to the denying agency, i.e., either the FBI or the POC. If the denying agency is unable to resolve the appeal, the denying agency will so notify the individual and shall provide the name and address of the agency that originated the document containing the information upon which the denial was based. The individual may then apply for correction of the record directly to the agency from which it originated. If the record is corrected as a result of

the appeal to the originating agency, the individual may so notify the denying agency, which will, in turn, verify the record correction with the originating agency (assuming the originating agency has not already notified the denying agency of the correction) and take all necessary steps to correct the record in the NICS.

(d) As an alternative to the above procedure where a POC was the denying agency, the individual may elect to direct his or her challenge to the accuracy of the record, in writing, to the FBI, NICS Operations Center, Criminal Justice Information Services Division, 1000 Custer Hollow Road, Module C–3, Clarksburg, West Virginia 26306–0147. Upon receipt of the information, the FBI will investigate the matter by contacting the POC that denied the transaction or the data source. The FBI will request the POC or the data source to verify that the record in question pertains to the individual who was denied, or to verify or correct the challenged record. The FBI will consider the information it receives from the individual and the response it receives from the POC or the data source. If the record is corrected as a result of the challenge, the FBI shall so notify the individual, correct the erroneous information in the NICS, and give notice of the error to any Federal department or agency or any state that was the source of such erroneous records.

(e) Upon receipt of notice of the correction of a contested record from the originating agency, the FBI or the agency that contributed the record shall correct the data in the NICS and the denying agency shall provide a written confirmation of the correction of the erroneous data to the individual for presentation to the FFL. If the appeal of a contested record is successful and thirty (30) days or less have transpired since the initial check, and there are no other disqualifying records upon which the denial was based, the NICS will communicate a **"Proceed"** response to the FFL. If the appeal is successful and more than thirty (30) days have transpired since the initial check, the FFL must recheck the NICS before allowing the sale to continue. In cases where multiple disqualifying records are the basis for the denial, the individual must pursue a correction for each record.

(f) An individual may also contest the accuracy or validity of a disqualifying record by bringing an action against the state or political subdivision responsible for providing the contested information, or responsible for denying the transfer, or against the United States, as the case may be, for an order directing that the contested information be corrected or that the firearm transfer be approved.

(g) An individual may provide written consent to the FBI to maintain information about himself or herself in a Voluntary Appeal File to be established by the FBI and checked by the NICS for the purpose of preventing the future erroneous denial or extended delay by the NICS of a firearm transfer. Such file shall be used only by the NICS for this purpose. The FBI shall remove all information in the Voluntary Appeal File pertaining to an individual upon receipt of a written request by that individual. However, the FBI may retain such information contained in the Voluntary Appeal File as long as needed to pursue cases of identified misuse of the system. If the FBI finds a disqualifying record on the individual after his or her entry into the Voluntary Appeal File, the FBI may remove the individual's information from the file.

§25.11 Prohibited activities and penalties.

(a) State or local agencies, FFLs, or individuals violating this subpart A shall be subject to a fine not to exceed $10,000 and subject to cancellation of NICS inquiry privileges.

(b) Misuse or unauthorized access includes, but is not limited to, the following:

(1) State or local agencies', FFLs', or individuals' purposefully furnishing incorrect information to the system to obtain a **"Proceed"** response, thereby allowing a firearm transfer;

(2) State or local agencies', FFLs', or individuals' purposefully using the system to perform a check for unauthorized purposes; and

(3) Any unauthorized person's accessing the NICS.

FIREARMS LAW ADMINISTERED BY THE POSTAL SERVICE

TITLE 18, UNITED STATES CODE, CHAPTER 83

§1715. Firearms as nonmailable; regulations

Pistols, revolvers, and other firearms capable of being concealed on the person are nonmailable and shall not be deposited in or carried by the mails or delivered by any officer or employee of the Postal Service. Such articles may be conveyed in the mails, under such regulations as the Postal Service shall prescribe, for use in connection with their official duty, to officers of the Army, Navy, Air Force, Coast Guard, Marine Corps, or Organized Reserve Corps; to officers of the National Guard or Militia of a State, Territory, Commonwealth, Possession, or District; to officers of the United States or of a State, Territory, Commonwealth, Possession, or District whose official duty is to serve warrants of arrest or commitments; to employees of the Postal Service; to officers and employees of enforcement agencies of the United States; and to watchmen engaged in guarding the property of the United States, a State, Territory, Commonwealth, Possession, or District. Such articles also may be conveyed in the mails to manufacturers of firearms or bona fide dealers therein in customary trade shipments, including such articles for repairs or replacement of parts, from one to the other, under such regulations as the Postal Service shall prescribe.

Whoever knowingly deposits for mailing or delivery, or knowingly causes to be delivered by mail according to the direction thereon, or at any place to which it is directed to be delivered by the person to whom it is addressed, any pistol, revolver, or firearm declared nonmailable by this section, shall be fined under this title or imprisoned not more than two years, or both.

RULINGS, PROCEDURES, AND INDUSTRY CIRCULARS

TABLE OF CONTENTS

Editor's Note:

The **Rulings, Procedures, and Industry Circulars** set out herein cannot address all subjects under all possible circumstances. When there is doubt, please contact your nearest ATF office.

Editorial Note: On July 1, 1972, ATF became its own Bureau under the Department of the Treasury (37 FR 11696). As a result, all references to Title 26 CFR Part 178 and 179 were re-designated under Title 27 of the CFR beginning April 15, 1975 (40 FR 16835).

Editorial Note: Pursuant to the Homeland Security Act of 2002 (Public Law No. 107-296), ATF transferred from the Department of the Treasury to the Department of Justice, effective January 24, 2003. At that time, Title 27 CFR Part 47, 178, and 179 were re-designated as 27 CFR Part 447, 478, and 479 (68 FR 3744).

"Rev. Rul." or **"RR"** is a revenue ruling published by the Alcohol, Tobacco and Firearms Division of the Internal Revenue Service, the predecessor of the Bureau of Alcohol, Tobacco, Firearms and Explosives. **"ATF Rul."** means ATF Ruling; **"ATFP"** means ATF Procedure; **"I.C."** means Industry Circular; **"ATFB"** means ATF Bulletin; **"ATFQB"** means ATF Quarterly Bulletin; **"ATF C.B."** means ATF Cumulative Bulletin; and **"C.B."** means Internal Revenue Service Cumulative Bulletin.

RULINGS

26 CFR 178.41: General (Also 178.42, 178.50)

Firearms or ammunition may not be sold at gun shows by a licensed dealer, but orders may be taken under specified conditions; Revenue Ruling 66-265 superseded.

Rev. Rul. 69-59

Advice has been requested whether a person who is licensed under 18 U.S.C. Chapter 44 (which superseded the Federal Firearms Act (15 U.S.C. Chapter 18)) or who is continuing operations under a license issued to him under the Federal Firearms Act as a manufacturer, importer or dealer in firearms or ammunition may sell firearms or ammunition at a gun show held on premises other than those covered by his outstanding license.

Under 18 U.S.C. 923(a), **"a separate fee"** is required to be paid for each place at which business as a licensee is to be conducted. Further, each applicant for a license is required to have in a State **"premises from which he conducts business"** (18 U.S.C. 923(d)(1)(E)) and to specify such premises in the license application. In addition, records are required to be maintained at the business premises covered by the license (18 U.S.C. 923(g)).

Therefore, a person holding a valid license may engage in the business covered by the license only at the specific business premises for which his license has been obtained. Thus, a licensee may

not sell firearms or ammunition at a gun show held on premises other than those covered by his license. He may, however, have a booth or table at such a gun show at which he displays his wares and takes orders for them, provided that the sale and delivery of the firearms or ammunition are to be lawfully effected from his licensed business premises only and his records properly reflect such transactions.

There are no provisions in the law for the issuance of temporary licenses to cover sales at gun shows, and licenses will be issued only for premises where the applicant regularly intends to engage in the business to be covered by the license.

This ruling does not apply to the activities of licensed collectors with respect to the receipt or disposition of curios and relics by such collectors. For provisions relating to transactions by licensed collectors of curios and relics, see 26 CFR 178.50 of the regulations.

Rev. Rul. 66-265, 1966-2, C.B. 559, is hereby superseded.

Editor's Note:

In 1986, the GCA was amended to allow licensees to sell firearms at gun shows in the State in which their licensed premises are located, and was further amended in 1997 to allow licensees to sell curio or relic firearms to other licensees at any location. Moreover, the interstate controls no longer apply to ammunition sales. However, the ruling is still appli-

cable to licensees' off-premises sales not addressed by these amendments.

26 CFR 178.114: IMPORTATION BY MEMBERS OF THE U.S. ARMED FORCES

Rev. Rul. 69-309

This ruling was revoked by T.D. ATF-426.

26 CFR 178.41: GENERAL (Also 178.42, 178.45, 178.49)

Licensed firearms dealers operating at multiple locations may establish a common expiration date for all licenses.

ATF Rul. 73-9

Advice has been requested whether a common expiration date may be established for all licenses issued in one region to dealers in firearms or ammunition operating several premises for which licenses are required for dealing in firearms (other than destructive devices) or ammunition except that for destructive devices.

Firearms or ammunition dealers, such as chain-store proprietors, have pointed out that the furnishing of applications for renewal of licenses (for each location at which an activity requiring a license is conducted) is creating severe administrative problems where the licenses involved have various expiration dates occurring throughout the calendar year.

In regard to licensing of dealers, 18 U.S.C. 923 provides in pertinent part, that no person shall engage in business as a firearms or ammunition dealer until he has filed an application with, and received a license to do so from, the Secretary (of the Treasury or his delegate). Furthermore, each such applicant who is not a dealer in destructive devices or a pawnbroker shall pay a fee of $10 per year.

Authority to implement the above requirements by such rules and regulations as the Secretary may deem necessary is found at 18 U.S.C. 926.

In regard to renewal of licenses for dealers in firearms or ammunition, regulations in 26 CFR 178.45, written under the authority of 18 U.S.C. 926, require that if a licensee intends to continue in the business of dealing in firearms or ammunition during any portion of the ensuing year he shall execute and file prior to the date of expiration of his license an application for license renewal, Form 8 (Firearms) (Part 3), accompanied by the required fee. (Such application should be filed with the Director of the Internal Revenue Service Center for the internal revenue region in which the business or activity is operated). As to the duration of a license, regulations in 26 CFR 178.49 state that a license shall not be issued for a period of less than one year and that the license shall entitle the dealer to engage in the specific business or activity for the period stated on the license unless sooner terminated.

Held, licensed firearms or ammunition dealers operating more than one location for which a license is required by 26 CFR 178.41 may, upon approval of a regional director, Bureau of Alcohol, Tobacco and Firearms, establish a common expiration date for all licenses issued to their several locations. Dealers wishing to establish such a date for all licenses issued to them may make application in writing to the regional director of the region in which the businesses or activities are operated. The application should set out the requested common expiration date and should list all licensed premises in the region covered by the application. The regional director will advise the dealer whether the request may be approved and, if approved, will provide the necessary instructions and renewal applications. It is pointed out that approval of a request will probably require that two renewal applications be submitted for each premises, i.e., one to cover the period from the currently required renewal date to the requested common expiration date

and another to cover the period from the common expiration date to the expiration date of the following twelve-month period. Also, it is pointed out that although the first renewal license will be cancelled on and after the requested common expiration date, the regulations do not provide for refund of any part of the $10 license fee paid for such license.

Editor's Note:

The fees for a dealer (01), including pawnbroker (02), in firearms other than destructive devices is now $200 for a 3 year period. Request to establish a common expiration date for all licenses issued to a dealer operating at several locations should be directed to the Chief, Federal Firearms Licensing Center, at 244 Needy Road, Martinsburg, West Virginia 25405. Approval of a request may result in lost time on some license expirations. All licenses included in the request for a common expiration date will be changed to accommodate the earliest expiration date of the combination of licenses. Additional renewal fees and/or renewals are not needed. Ninety (90) days prior to the common expiration dates for all licenses a renewal form will be generated by the Federal Firearms Licensing Center to the licensee.

[73 ATF C.B. 102]

26 CFR 178.11: MEANING OF TERMS

(Also 178.23, 178.44)

Because of the nature of operations conducted by a gunsmith, he shall not be required to have business premises open to the general public or to have regular business hours.

ATF Rul. 73-13

Advice has been requested whether a gunsmith must establish business premises open to the general public and regular business hours in the same manner as dealers engaged in the business of selling firearms or ammunition at wholesale or retail in order to be eligible for a license under the provisions of Chapter 44, Title 18, United States Code and regulations issued pursuant thereof.

It has been pointed out that generally, the clientele of a gunsmith who is not engaged in the business of selling firearms or ammunition at wholesale or retail are only those persons seeking to have firearms repaired or altered as opposed to

those seeking to purchase firearms or ammunition. Furthermore, a gunsmith often does not have regular business hours but conducts business during hours which are best suited to himself or his clientele. Thus, the number of hours which such a gunsmith is open to serve his clientele may be fewer than those of a dealer engaged in the business of buying and selling firearms or ammunition.

The term **"dealer"** is defined in 18 U.S.C. 921(a)(11) to include **"any person engaged in the business of repairing firearms or of making or fitting special barrels, stocks, or trigger mechanisms to firearms."**

One of the requirements under 18 U.S.C. 923(d)(1) for approval of a dealer's license is that the applicant has in a State premises from which he conducts business subject to a license or from which he intends to conduct such business within a reasonable period of time. The term **"business premises"** is defined n 26 CFR 178.11 to mean: **"The property on which firearms or ammunition importing, manufacturing or dealing business is or will be conducted. A private dwelling, no part of which is open to the public, shall not be recognized as coming within the meaning of the term."** The type of business premises, as well as the business hours, is required by 26 CFR 178.44 to be included on the application for a firearms dealer's license.

It is also provided in 18 U.S.C. 923(g) that the premises of a licensee may be entered during business hours for the purpose of inspecting or examining records or documents required to be kept as well as any firearms or ammunition kept or stored at such premises.

Held, because of the nature of operations conducted by a gunsmith, any applicant for a license who intends to engage solely in this type of business and so specifies on his application will not be required to maintain regular business hours. Further, if the business is conducted from a private dwelling, a separate portion should be designated as the business premises, which need not be open to all segments of the public but only accessible to the clientele that the business is set up to serve. However, the licensed premises of the gunsmith are subject to the inspection requirements of 18 U.S.C. 923(g) and 26 CFR 178.23 and the gunsmith must maintain the required records as specified in 26 CFR 178.121 et seq.

It is further held that, since under the law a gunsmith is a licensed firearms dealer, if he engages in the business of buying and selling firearms, he must record his transactions on Form 4473 (Firearms Transaction Record) for each sale, and maintain the firearms acquisition and disposition records required of all licensed dealers. However, if a gunsmith engages in the business of buying and selling firearms during the term of his current license, he may be required to submit a new Form 7 (Firearms) at the time of renewal in accordance with 26 CFR 178.45 and meet the requirements of an applicant engaging in the business of buying and selling firearms, such as having business premises open to the general public and having regular business hours.

[Amplified by ATF Rul. 77-1]

Editor's Note:

Gunsmiths are no longer required to hold a separate license. These activities are now covered under a type 01 license.

27 CFR 178.11: MEANING OF TERMS

(Also 178.23, 178.44, 178.99, 178.124)

Because of the nature of operations conducted by a consultant or expert, he shall not be required to have business premises open to the general public or to have regular business hours.

ATF Rul. 73-19

The purpose of this ruling is to provide guidelines under which bona fide firearms consultants or experts may obtain licenses as firearms dealers and thereby be permitted to receive firearms in interstate commerce from nonlicensed individuals for testing or examination purposes.

Revenue Ruling 69-248, C.B. 1969-1, 360 (Internal Revenue) permits firearms licensees to ship, transport, or deliver firearms and ammunition in interstate commerce to their nonlicensed employees, agents, or representatives for business purposes. As was clarified in Industry Circular 72-23 the ruling also permits firearms licensees to similarly transfer firearms and ammunition to nonlicensed professional writers, consultants, and evaluators for research or evaluation.

The Bureau of Alcohol, Tobacco and Firearms has received inquiries from firearms consultants, evaluators, and examiners concerning the receipt, and subsequent return, in interstate commerce of firearms from nonlicensed individuals for testing and examination for the purpose of furnishing expert testimony about those firearms.

18 U.S.C. 922(a)(2)(A) permits an individual to ship (and have returned to him) in interstate commerce a firearm to a firearms licensee for repair or customizing. Furthermore, the definition of a firearms dealer in 18 U.S.C. 921 and 26 CFR 178.11 is sufficiently broad that it can be interpreted to include a qualified firearms consultant or expert who is engaged in the business of testing or examining firearms. In view of these provisions, the Bureau has determined that firearms consultants or experts may be licensed as firearms dealers in order that they may receive firearms from nonlicensed individuals for testing and examination.

Because of the nature of operations conducted by a firearms consultant or expert, any licensed dealer who engages solely in this type of business will not be required to maintain regular business hours. If the business is conducted from a private residence, a separate portion of the dwelling should be designated as **"business premises."** Such premises need not be open to all segments of the public but only accessible to the clientele that the business is set up to serve. However, the licensed premises of the firearms consultant-expert shall be subject to inspection under the authority of 18 U.S.C. 923(g) and 26 CFR 178.23.

A licensed firearms consultant or expert shall maintain records of receipt and delivery of firearms, as is required by 26 CFR 178, Subpart H, except that the licensee need not prepare Forms 4473, Firearms Transaction Record, reflecting the firearms examined. However, shipments and deliveries of firearms shall not be made in care of persons ineligible to receive them under Section 922(h), Title 18, U.S.C., or under Title VII of the Omnibus Crime Control and Safe Streets Act of 1968 (18 U.S.C. Appendix, Sections 1201-1203).

A firearms consultant or expert who desires to obtain a license as a dealer in firearms shall file Form 7 (Firearms), Application for License Under 18 U.S.C. Chapter 44, Firearms, in the manner prescribed by 26 CFR 178.44. The application shall include a statement that the applicant is engaged in business as a bona fide firearms consultant or expert and, where the applicant intends to perform testing or examination services for one or more persons on a continuing basis, the statement shall include the name, address, and nature of business of such persons. A license as a dealer in firearms will be issued only after the Regional Director is satisfied that the applicant is a bona fide consultant or expert and is otherwise qualified under the law.

Since a licensed firearms consultant or expert is a firearms dealer if he engages in the business of buying and selling firearms, he must record his transactions on Form 4473 (Firearms Transaction Record) for each sale, and maintain the firearms acquisition and disposition records required of all licensed dealers. If a firearms consultant or expert engages in the business of buying and selling firearms during the term of his current license, he may be required to submit a new Form 7 (Firearms) at the time of renewal in accordance with 26 CFR 178.45 and meet the requirements of an applicant engaging in the business of buying and selling firearms, such as having business premises open to the general public and having regular business hours.

For ruling respective to licensing of gunsmiths, see ATF Ruling 73-13, 1973-8 ATF Bulletin 9.

Editor's Note:

ATF Ruling 73-19 was modified by ATF Ruling 2010-1. Industry Circular 72-23 is no longer in effect (see ATF Ruling 2010-1).

ATF underwent a restructuring in 1998 and replaced its Regional Directors with a Director of Industry Operations (DIO) in each field division. Unless otherwise noted, the DIO is the approving official for action formerly taken by the regional directors.

[This ATF ruling does not apply to firearms within the purview of the National Firearms Act (26 U.S.C. Chapter 53)]

[73 ATF C.B. 93]

26 CFR 179.104: REGISTRATION OF FIREARMS BY CERTAIN GOVERNMENTAL ENTITIES

When NFA firearms are registered on Form 10 by governmental entities, subsequent transfers of such firearms shall be made only to other governmental entities.

ATF Rul. 74-8

Advice has been requested whether the Bureau will approve transfer of National Firearms Act weapons by a state or political subdivision (Police department) to a special occupational taxpayer where such firearms were registered in the National Firearms Registration and Transfer Record pursuant to 26 CFR 179.104.

Section 179.104 of the regulations provides that any state, any political subdivision thereof, or any official police organization of such a government entity engaged in criminal investigations, which acquires for official use a firearm not registered to it, such as by abandonment or by forfeiture, will register such firearm with the Director by filing Form 10 (Firearms), Registration of Firearms Acquired by Certain Governmental Entities, and that such registration shall become a part of the National Firearms Registration and Transfer Record.

The purpose of the above regulation was to permit the limited registration of firearms by certain governmental entities for official use only. Section 179.104 may not be used as a vehicle to register otherwise unregisterable firearms for the purpose of introducing such firearms into ordinary commercial channels. Accordingly when registration of firearms by governmental entities is approved on Form 10, the form will be marked **"official use only."** The Bureau will approve subsequent transfers of such firearms only to other governmental entities for official use. Otherwise such firearms must be destroyed or abandoned to the Bureau.

[74 ATF C.B. 67]

26 CFR 178.114: IMPORTATION BY MEMBERS OF THE U.S. ARMED FORCES

A member of the U.S. Armed Forces who is a resident of any State or territory which requires that a permit or other authorization be issued prior to possessing or owning a handgun shall submit evidence of compliance with State law before an application to import a handgun may be approved.

ATF Rul. 74-13

State and local authorities have called to the attention of the Bureau of Alcohol, Tobacco and Firearms instances in which members of the United States Armed Forces have transported, shipped, received, or imported handguns into the United States to their place of residence pursuant to 18 U.S.C. 925(a)(4) without obtaining the required permit or other authorization required by their state of residence to acquire, possess, or own (as opposed to a license to purchase) handguns.

Title 18 U.S.C. 925(a)(4) provides that when established to the satisfaction of the Secretary of the Treasury to be consistent with the provisions of 18 U.S.C., Chapter 44 and other applicable Federal and State laws and published ordinances, the Secretary may authorize the transportation, shipment, receipt, or importation into the United States to the place of residence of any member of the United States Armed Forces who is on active duty outside the United States (or has been on active duty outside the United States within the 60-day period immediately preceding the transportation, shipment, receipt or importation), of any firearm or ammunition which is (a) determined by the Secretary to be generally recognized as particularly suitable for sporting purposes, or determined by the Department of Defense to be a type of firearm normally classified as a war souvenir; and (b) intended for the personal use of such member.

26 CFR 178.114(a) provides that an application for a permit to import a firearm or ammunition into the United States to the place of residence of any military member of the United States Armed Forces on active duty outside the United States shall include a certification by the applicant that the transportation, receipt or possession of the firearm or ammunition to be imported would not constitute a violation of any State law or local ordinance at the place of the applicant's residence.

In order to assure that the transportation, shipment, receipt or importation of handguns under 18 U.S.C. 925(a)(4) is not in violation of applicable State laws, it is held that, any member of the United States Armed Forces who is a resident of any State or territory which requires that a permit or authorization be obtained prior to acquiring, possessing or owning a handgun, must submit a copy of the license, permit, certificate of registration, or firearm identification card, as applicable, as required by his State in order for an application to import a handgun under 18 U.S.C. 925(a)(4) to be approved.

[74 ATF C.B. 60]

26 CFR 178.124: FIREARMS TRANSACTION RECORDS

(Also 178.123, 178.125, 178.147)

Form 4473 shall not be required to record disposition of a like replacement firearm when such firearm is delivered by a licensee to the person from whom the malfunctioning or damaged firearm was received, provided such disposition is recorded in the licensee's permanent records.

ATF Rul. 74-20

The Bureau of Alcohol, Tobacco and Firearms has been requested to state its position in regard to the requirement for the execution of a Firearms Transaction Record, Form 4473, when a defective, damaged or otherwise malfunctioning firearm is replaced by a federal licensee as an alternative to the repair and return to the purchaser of the defective firearm.

The proviso under 26 CFR 178.124(a) states that Form 4473 shall not be required to record the disposition made of a firearm that is delivered to a licensee for the sole purpose of repair or customizing when such firearm is returned to the person from whom received. No specific mention is made in that statement in regard to a **"replacement"** that may be furnished to the customer in lieu of repairing and returning the damaged or malfunctioning merchandise. However, 26 CFR 178.147 states **". . . notwithstanding any other provisions of this part, the . . . licensed manufacturer . . . may return in interstate commerce . . . a replacement firearm of the same kind and type."** In view of the above, 26 CFR 178.147 may be so interpreted as to allow the disposition of such firearms without the execution of Form 4473 to record the transaction.

It is held, therefore, that a firearms transaction record, Form 4473, shall not be required to record the disposition of a replacement firearm of the same kind and type where such a firearm is delivered by a licensee to the person from whom the malfunctioning or damaged firearm was received.

It should be noted, however, that the licensee is required by 26 CFR 178.125 to maintain in his permanent records the disposition of such a replacement firearm.

[74 ATF C.B. 61]

26 CFR 178.11: MEANING OF TERMS

(Also 179.11)

A small caliber weapon ostensibly designed to expel only tear gas, similar substances, or pyrotechnic signals, which may readily be converted to expel a projectile by means of an explosive, classified as a firearm.

ATF Rul. 75-7

The Bureau has re-examined its position with respect to the applicability of Titles I and II of the Gun Control Act of 1968 (Chapter 44 of Title 18 U.S.C. and Chapter 53 of Title 26 U.S.C. (National Firearms Act)), to small caliber weapons (commonly known as **"penguns"**) ostensibly designed to expel only tear gas, similar substances or pyrotechnic signals by the action of an explosive.

Revenue Ruling 56-29, C.B. 1956-1, 552 (Internal Revenue) held that a tear gas gun designed to expel only a gas or other mist rather than a shot or a projectile was not a **"firearm"** as that term was previously defined in the repealed Federal Firearms Act and in the National Firearms Act. Revenue Ruling 56-29 also held that if such a device was capable of firing other than the shells or cartridges designed for use therewith, such as fixed metallic cartridges or shotgun shells, it would be a **"firearm"** within the purview of the National and/or Federal Firearms Acts, depending upon the individual characteristics of the device.

The Bureau has had occasion to re-examine such weapons for the purpose of determining their status under the Gun Control Act of 1968. Such weapons are readily concealable and ostensibly designed to expel only tear gas, similar substances or pyrotechnic signals by the action of an explosive, normally by means of a firing mechanism designed to accept a plastic or extruded aluminum helix-type disposable cartridge.

Unlike the definition of **"firearm"** in the repealed Federal Firearms Act, the term **"firearm"** as used in 18 U.S.C. 921(a)(3) includes **"any weapon (including a starter gun) which will or is designed to or may readily be converted to expel a projectile by the action of an explosive."** Tests performed on these weapons have established that they may readily be converted to expel a projectile by the action of an explosive, normally by means of a minor alteration of the expanded Helix cartridge and/or the simple attachment of a barrel/chamber to the firing mechanism.

Held, a small caliber weapon ostensibly designed to expel only tear gas, similar substances or pyrotechnic signals by the action of an explosive, which may readily be converted to expel a projectile by means of an explosive, constitutes, a **"firearm"** within the purview of 18 U.S.C. 921(a)(3)(A).

Such weapons manufactured within the United States on or after June 1, 1975, will be subject to all of the provisions of Chapter 44 and 26 CFR Part 178. Such weapons manufactured before June 1, 1975, will not be treated as subject to the provisions of Chapter 44 and 26 CFR Part 178 in order to allow persons manufacturing and dealing in such weapons to comply with the provisions of Chapter 44 and 26 CFR Part 178.

Since such weapons are not generally recognized as particularly suitable for or readily adaptable to sporting purposes (18 U.S.C. 925(d)(3)), the importation of such weapons is prohibited unless such importation comes within one of the statutory exceptions provided in 18 U.S.C. 925. The importation of such weapons pursuant to an unexpired permit will not be affected by this Ruling.

The Bureau has long held that such weapons when actually converted to fire other than the gas or pyrotechnic cartridges originally designed for use therewith are **"firearms"** under Chapter 44 and the National Firearms Act, depending upon the individual characteristics of the weapon. See Revenue Ruling 56-29, supra, and Revenue Ruling 56-597, C.B. 1956-2, 931 (Internal Revenue). These determinations with respect to the converted weapon were not altered by the amended definitions of the term "firearm" now found in 18 U.S.C. 921(a)(3) and the term **"any other weapon"** in 26 U.S.C. 5845(e). Accordingly, any such weapon which is capable of being concealed on the person which has originally been designed or converted to discharge a shot through the energy of an explosive will remain subject to the provisions of the National Firearms Act as an **"any other weapon"** (26 U.S.C. 5845(e)).

Revenue Ruling 56-29, C.B. 1956-1, 552 (Internal Revenue), is hereby revoked.

[75 ATF C.B. 55]

27 CFR 178.94: SALES OR DELIVERIES BETWEEN LICENSEES

A firearms licensee may continue operations until his renewal application for a license is finally acted upon.

ATF Rul. 75-27

Under 5 U.S.C. 558, when a licensee has made timely and sufficient application for a renewal in accordance with agency rules, a license with reference to an activity of a continuing nature does not expire until the application has been finally determined by the agency. In accordance with section 558, a firearms licensee who timely applies for renewal of his license is authorized to continue his firearms operations as authorized by his license until his renewal application is finally acted upon. As provided by 27 CFR 178.94, a transferor licensee is authorized to continue to make shipments to a licensee for not more than 45 days following the expiration date of the transferee's license.

Held, a transferor licensee may continue to make firearms and ammunition shipments to a licensee who has timely applied for renewal of his license but has not had his application acted upon within 45 days after the expiration of his license. The transferor licensee shall, however, in cases where the 45-day period has passed, obtain appropriate evidence that the transferee's license renewal application is still pending in the office of the (Compliance), Bureau of Alcohol, Tobacco and Firearms. Such evidence should consist of a letter from the Regional Director (Compliance), to the transferee licensee stating that his renewal application has been timely received and that action thereon is currently pending.

Editor's Note:

Renewal applications are processed by the Federal Firearms Licensing Center in Martinsburg, West Virginia. A request for a Letter of Authorization (LOA) can be obtained from the Chief, Federal Firearms Licensing Center, stating that the transferee's renewal application has been timely received and that action is currently pending. Upon issuance of an LOA, the FFL EzCheck system will show the date the LOA was issued and the date it will expire.

[75 ATF C.B. 60]

27 CFR 178.92: IDENTIFICATION OF FIREARMS

Importers may adopt serial numbers placed on certain firearms by foreign manufacturers.

ATF Rul. 75-28

This ruling is superseded by ATF Rul. 2013-3.

27 CFR 178.11: MEANING OF TERMS

(Also 179.11)

A hand-held device designed to expel by means of an explosive two electrical contacts (barbs) connected by two wires attached to a high voltage source in the device classified as a firearm.

ATF Rul. 76-6

The Bureau has been asked to determine the applicability of Titles I and II of the Gun Control Act of 1968 (Chapter 44 of Title 18 U.S.C., and Chapter 53 of Title 26 U.S.C. (National Firearms Act)) to a device known as the Taser, a hand-held device designed to expel by means of an explosive two electrical contacts (barbs) connected by two wires attached to a high voltage source in the device. Upon contact with an individual, a high voltage electrical charge is carried to the barbs by the wires which temporarily immobilizes the victim.

The term **"firearm"** as used in 18 U.S.C. 921(a)(3)(A) includes **"any weapon (including a starter gun) which will be or is designed to or may readily be converted to expel a projectile by the action of an explosive."** The Bureau has determined that the device is a weapon and notwithstanding the fact that the barbs and wires remain attached to the hand-held device after expulsion, these items are projectiles within the meaning of the statute. Since the projectiles are expelled by the action of an explosive, the weapon is a firearm under 18 U.S.C. 921(a)(3)(A).

With respect to the National Firearms Act, the term **"any other weapon"** in 26 U.S.C. 5845(e) generally means a weapon or device capable of being concealed on the person from which a shot can be discharged through the energy of an explosive. Such term does not include a pistol or a revolver having a rifled bore, or rifled bores, or weapons designed, made, or intended to be fired from the shoulder and not capable of firing fixed ammunition. Since the Taser meets the statutory definition, it is an **"any other weapon"** (26 U.S.C. 5845(e)).

Held, a hand-held device designed to expel by means of an explosive two electrical contacts (barbs) connected by two wires attached to a high voltage source in the device is a **"firearm"** within the purview of 18 U.S.C. 921(a)(3)(A). It is also an **"any other weapon"** under the National Firearms Act (26 U.S.C. 5845(e)).

In order to allow persons manufacturing and dealing in such weapons to comply with the provisions of Chapter 44 and 27 C.F.R. Part 178, this ruling will be applicable to such weapons manufactured within the United States on or after May 1, 1976. Such weapons manufactured before May 1, 1976, will not be treated as subject to the provisions of Chapter 44 and 27 C.F.R. Part 178. With respect to the **"any other weapon"** classification under the National Firearms Act, pursuant to 26 U.S.C. 7805(b), this ruling will not be applied to such weapons manufactured before May 1, 1976. Accordingly, such weapons manufactured on or after May 1, 1976, will be subject to all the provisions of the National Firearms Act and 27 CFR Part 179.

(Amplified by ATFR 80-20) [76 ATF C.B. 96]

27 CFR 178.124: FIREARMS TRANSACTION RECORD

(Also 178.125, 178.126a)

Certain reporting and recordkeeping requirements of pawnbrokers are explained.

ATF Rul. 76-15

The Bureau of Alcohol, Tobacco and Firearms has been asked to comment on the proper use of Form 4473, Firearms Transaction Record, the entering of information into the firearms acquisition and disposition record, and the reporting of multiple sales of pistols and revolvers by a pawnbroker.

Some confusion apparently exists among pawnbrokers as to whether Form 4473 should be executed at the time a firearm is accepted as a pledge by a pawnbroker and whether another Form 4473 should be executed at the time the firearm is redeemed by the pledgor. The question has also arisen whether a pawnbroker, upon receiving a firearm as a pledge, should enter such receipt into his permanent acquisition and disposition record. Furthermore, some pawnbrokers are unsure whether the provisions of section 178.126a relating to reporting multiple sales or other dispositions of pistols and revolvers apply in connection with the redemption or sale of pawned firearms.

18 U.S.C. 923(g) provides in part that each licensee shall maintain such records of shipment, receipt, sale, or other disposition of firearms and ammunition in such manner as the Secretary may by regulation prescribe. 18 U.S.C. 922(b)(5) provides that it shall by unlawful for any firearms licensee to sell or deliver any firearm or ammunition to any person unless the licensee notes in his records the name, age, and place of residence of such person. 27 CFR 178.124(a) provides in part that a licensed dealer shall not sell or otherwise dispose of, temporarily or permanently, any firearm to any person, other than another licensee, unless he records the transaction on a firearms transaction record, Form 4473. Pursuant to 27 CFR 178.125(e), each licensed dealer is required to enter into a permanent record each receipt and disposition of firearms. The regulations in 27 CFR 178.126a provide that each licensee shall report multiple sales or other disposition of pistols and revolvers whenever he disposes of, at one time or during any five consecutive business days, two or more pistols or revolvers or any combination of such weapons totaling two or more to any unlicensed person.

The regulations do not require that a pawnbroker execute Form 4473 when a firearm is pledged for a loan. However, he must record the receipt thereof in his permanent acquisition and disposition record as required by 27 CFR 178.125(e). At the time a firearm is redeemed by a nonlicensee pledgor, Form 4473 must be executed and the appropriate entry made in the permanent acquisition and disposition record. Although a redemption is not considered a sale, it is a disposition for purposes of 27 CFR 178.124(a), 178.125(e), and 178.126a. See *Huddleston v. United States*, 415 U.S. 814 (1974).

However, no report of multiple sales and other dispositions is required to be filed with ATF when the handguns are returned to the person from whom received.

Held, Form 4473, Firearms Transaction Record, need not be executed when a pawnbroker accepts a firearm as a pledge for a loan. However, if a non-licensee pledgor redeems the firearm or if disposition of the firearm is made to any other nonlicensee, Form 4473 must be executed.

Held further, pawnbrokers must enter into their permanent acquisition and disposition record the receipt of a firearm as a pledge for a loan and any disposition, including redemption, of such firearm.

Held further, pawnbrokers must submit reports of multiple sales and other dispositions of pistols and revolvers as required by 27 CFR 178.126(a) when the person receiving them is not the person who pawned the firearms.

[76 ATF C.B. 100]

27 CFR 179.11: MEANING OF TERMS

Mere possession of a license and a special tax stamp as a dealer in firearms does not qualify a person to receive firearms transfer-tax-free.

ATF Rul. 76-22

The Bureau has had occasion to consider the status of a person who represented himself as a dealer; registered and paid the special (occupational) tax, imposed under 26 U.S.C. 5801, as a **"dealer"** in firearms as defined in section 5845(a) of the National Firearms Act (Chapter 53, Title 26, U.S.C.), made application for and was duly issued a license as a **"dealer"** under Title I of the Gun Control Act of 1968 (Chapter 44, Title 18, U.S.C.); and it was subsequently established that the person procured the special tax stamp and the license to facilitate the purchase of firearms for his personal collection.

Held, the mere possession of a license and a special (occupational) tax stamp as a dealer in firearms does not qualify a person to receive firearms transfer-tax-free. Any person holding a license and a special tax stamp as a dealer in firearms and not actually engaged within the United States in the business of selling NFA firearms may not lawfully receive NFA firearms without the transfer tax having been paid by the transferor. Where it is, therefore, determined that the proposed transferee on a Form 3, Application for Tax-exempt Transfer of Firearm and Registration to Special (Occupational) Taxpayer, is not actually engaged in the business of deal-

ing in NFA firearms, such application will be denied. In addition, if such person receives NFA firearms without the transfer tax having been paid, such firearms may be subject to seizure for forfeiture as having been unlawfully transferred without payment of the transfer tax.

Rev. Rul. 58-432, 1958-2 C.B. 875 (Internal Revenue), is superseded.

[76 ATF C.B. 103]

27 CFR 178.121: GENERAL (Also 178.147)

Recordkeeping requirements for firearms from which parts are salvaged for use in repairing firearms are clarified.

ATF Rul. 76-25

The Bureau of Alcohol, Tobacco and Firearms has been asked to clarify the recordkeeping requirements for Title I firearms from which parts are salvaged for use in repairing firearms and the recordkeeping requirements for frames or receivers of firearms exchanged between customers and licensees.

Section 921(a)(3) of Title 18, United States Code, and the regulations at 27 CFR 178.11 define the term **"firearm"** to include any weapon which will, or is designed to, or may readily be converted to, expel a projectile by the action of an explosive, and the frame or receiver of any such weapon.

The regulations in 27 CFR 178.122, 178.123 and 178.125 require each licensed importer, licensed manufacturer, and licensed dealer, respectively, to maintain such records of acquisition (including by manufacture) or disposition, whether temporary or permanent, of firearms as therein prescribed.

Held, a licensee who purchases a damaged firearm for the purpose of salvaging parts therefrom shall enter receipt of the firearm in his firearms acquisition and disposition record. If the frame or receiver of the firearm is damaged to the extent that it cannot be repaired, or if the licensee does not desire to repair the frame or receiver, he may destroy it and show the disposition of the firearm in his records as having been destroyed. Before a firearm may be considered destroyed, it must be cut, severed or mangled in such manner as to render the firearm completely inoperative and such that it cannot be restored to an operative condition.

Where the repair of a customer's firearm results in an exchange of a frame or receiver, an entry shall be made in the licensee's records to show the transfer of such replacement part, as it is a **"firearm"** as defined in 18 U.S.C. 921(a)(3). Further, as held in ATF Ruling 74-20, 1974 ATF C.B. 61, a Form 4473, Firearms Transaction Record, shall not be required to record the disposition of a replacement firearm of the same kind and type where the firearm is delivered by the licensee to the person from whom the malfunctioning or damaged firearm was received. The frame or receiver received from the customer shall be entered as an acquisition, and if destroyed, it shall be entered in the disposition record as destroyed.

With regard to National Firearms Act firearms as defined in 26 U.S.C. 5845(a), in addition to the above record-keeping requirements, the registration and transfer procedures of 27 CFR Part 179 must be complied with.

[76 ATF C.B. 99]

27 CFR 178.121: GENERAL (RECORDS)

The recordkeeping requirements for licensed gunsmiths are clarified. ATF Rul. 73-13 amplified.

ATF Rul. 77-1

The Bureau of Alcohol, Tobacco and Firearms has received several requests for a clarification as to the authorized activities of and recordkeeping requirements for licensed gunsmiths.

It has specifically come to our attention that several gunsmiths desire to do on-the-spot repairs at organized shooting events. It has also come to our attention that several gunsmiths are uncertain as to what transactions are exempt from the recordkeeping requirements.

ATF Ruling 73-13, 1973 ATF C.B. 92, held that a licensed gunsmith must maintain the required records as specified in 27 CFR 178.121 et seq., and if a gunsmith engages in the business of buying and selling firearms, he must record these transactions on a Form 4473 (Firearms Transaction Record) for each sale. However, as provided in section 178.124(a), a Form 4473 is not required to record the disposition made of a firearm delivered to a gunsmith for repair or customizing when the firearm is returned to the person from whom received.

The Bureau recognizes the necessity for having on-the-spot repairs made to firearms at skeet, trap, target, and similar organized events. It is, therefore, held that licensed gunsmiths may make immediate on-the-spot repairs to firearms at skeet, trap, target, and similar organized shooting events.

Held further, a licensed gunsmith must enter into his bound acquisition and disposition record, required to be maintained by 27 CFR 178.125(e), each receipt and disposition of firearms, except that a firearm need not be entered in the bound acquisition and disposition record if the firearm is brought in for adjustment or repair and the owner waits while it is being adjusted or repaired or if the gunsmith returns the firearm to the owner during the same business day it is brought in. If the firearm is retained from one business day to another or longer, it must be recorded in the bound acquisition and disposition record.

Held further, a licensed gunsmith is not required to prepare a Form 4473 (Firearms Transaction Record) where a firearm is delivered to him for the sole purpose of customizing, adjustment, or repair and the firearm is returned to the person from whom received. However, if a licensed gunsmith engages in the business of selling firearms, he must record these transactions on a Form 4473 for each sale in addition to maintaining the bound firearms acquisition and disposition record required by 27 CFR 178.125(e).

ATF Rul. 73-13, 1973 ATF C.B. 92, is hereby amplified.

[77 ATF C.B. 185]

27 CFR 178.124: FIREARMS TRANS-ACTION RECORD

Means of identification furnished by a nonlicensee purchasing a firearm.

ATF Rul. 79-7

This ruling is superseded by ATF Rul. 2001-5.

27 CFR 178.112: IMPORTATION BY A LICENSED IMPORTER

APPLICATIONS TO IMPORT SURPLUS MILITARY FIREARMS OR NONSPORTING FIREARMS OR AMMUNITION FOR INDIVIDUAL LAW ENFORCEMENT OFFICERS FOR OFFICIAL USE MUST BE ACCOMPANIED BY THE AGENCY'S PURCHASE ORDER.

ATF Rul. 80-8

The Bureau of Alcohol, Tobacco and Firearms has received several inquiries from firearms importers and dealers, law enforcement agencies, and the public requesting clarification of the statutes, regulations and procedures regarding the importation of firearms for law enforcement agencies.

Importation of surplus military firearms or firearms not particularly suitable for or readily adaptable to sporting purposes is generally prohibited by section 925(d)(3) of Title 18, United States Code. However, section 925(a)(1) provides that this prohibition does not apply to the importation of firearms or ammunition sold or shipped to, or issued for the use of the United States or any department or agency thereof or any State or any department, agency, or political subdivision thereof.

Pursuant to section 925(a)(1), the Bureau has previously allowed the importation of surplus military firearms and nonsporting firearms for individual law enforcement officers for official use. In approving such importation applications, the Bureau required federal firearms licensees to obtain from the agency employing the officer a certificate of the chief law enforcement officer stating that the firearm or ammunition is for use in the performance of official duties.

However, once these firearms are imported for the individual officer for **"official use,"** there is no prohibition in the law against the officer's resale or retention of the firearms for personal use. The purpose of section 925(a)(1) is to permit importation of firearms for the exclusive use of government agencies. The statute was not intended and may not be used as a vehicle by which unimportable firearms can be introduced into ordinary commercial channels in the United States.

Held, a licensee's application to import surplus military firearms or nonsporting firearms or ammunition for law enforcement officers will not be approved unless accompanied by a purchase order from a department or agency of the United States or any department, agency or political subdivision of any State. The term **"State"** includes the District of Columbia, the Commonwealth of Puerto Rico, and the possessions of the United States.

[80 ATF C.B. 20]

27 CFR 178.11: MEANING OF TERMS

(Also 27 CFR 179.11)

The Bureau has been asked to classify a device known as the Taser Model TF76 and Model TF76A. Generally, the Taser can be described as a hand-held device designed to expel, by means of an explosive, two electrical contacts (barbs) which are connected by two wires to a high voltage source in the device.

ATF Rul. 80-20.

The Bureau has determined in ATF Rul. 76-6 that the Taser Model TF1 was a firearm as that term is defined in Title 18, United States Code (U.S.C.), Section 921(a)(3) and that the Model TF1 also met the **"any other weapon"** definition found in the National Firearms Act (NFA), Title 26, U.S.C., Section 5845(e). This ruling was limited in its application to Taser Models TF1 produced on or after May 1, 1976. The Taser Models TF76 and TF76A were subsequently developed and differ from the Taser Model TF1 in that these models each have a hand grip bent at an angle to the bore and the bore of each is rifled.

The changes in the design of the Taser Models TF76 and TF76A bring them within the exclusion found in the **"any other weapon"** definition of the NFA for pistols and revolvers having a rifled bore, or rifled bores.

Held, the Taser Models TF76 and TF76A are not subject to the provisions of the NFA. However, they are firearms as defined in Title 18, U.S.C., Section 921(a)(3) and are subject to the provisions of Title 18, U.S.C., Chapter 44 and Title 27, Code of Federal Regulations, Part 178.

ATF Rul. 76-6, ATF C.B. 1976, 96, is hereby amplified.

[ATFB 1980-4 24]

27 CFR 178.11: MEANING OF TERMS

An out-of-State college student may establish residence in a State by residing and maintaining a home in a college dormitory or in a location off-campus during the school term.

ATF Rul. 80-21

The Bureau has been asked to determine the State of residence of out-of-State college students for purposes of the Gun Control Act of 1968. **"State of residence"** is defined by regulation in 27 C.F.R. 178.11 as the State in which an individual regularly resides or maintains a home. The regulation also provides an example of an individual who maintains a home in State X and a home in State Y. The individual regularly resides in State X except for the summer months and in State Y for the summer months of the year. The regulation states that during the time the individual actually resides in State X he is a resident of State X, and during the time he actually resides in State Y he is a resident of State Y.

Applying the above example to out-of-State college students it is held, that during the time the students actually reside in a college dormitory or at an off-campus location they are considered residents of the State where the dormitory or off-campus home is located. During the time out-of-State college students actually reside in their home State they are considered residents of their home State.

[ATFB 1980-4 25]

27 CFR 178.111: GENERAL

Nonresident U.S. citizens returning to the United States and nonresident aliens lawfully immigrating to the United States may obtain a permit to import firearms acquired outside of the United States, provided such firearms may be lawfully imported.

ATF Rul. 81-3

The Bureau of Alcohol, Tobacco and Firearms (ATF) has received several inquiries regarding the proper procedures to be used by U.S. citizens residing overseas who are returning to the United States, and by persons lawfully immigrating to the United States, who desire to import personally owned firearms acquired outside of the United States.

A permit issued by ATF is, with certain exceptions, required to import or bring into the United States any firearm acquired outside of the United States. No permit is required to return a firearm to the United States provided that the person bringing in the firearm can establish to the satisfaction of the U.S. Customs Service officials at the port of entry that the firearm was previously taken out of the United States by such person. A firearm may be imported for resale only by a Federally licensed importer. However, a personally owned firearm may be imported through a Federally licensed importer, manufacturer, or dealer.

Section 922(a)(3) of Title 18, United States Code, makes it unlawful, with certain exceptions, for a person to bring into his State of residence a firearm which he acquired outside that State. An unlicensed resident of a State must, therefore, arrange for the importation of the firearm through a Federal firearms licensee.

The definition of **"State of residence"** in 27 CFR 178.11 provides that the State in which an individual regularly resides or maintains a home is the State of residence of that person. U.S. citizens who reside outside of the United States are not residents of a State while so residing. A person lawfully immigrating to the United States is not a resident of a State unless he is residing and has resided in a State for a period of at least 90 days. Therefore, such persons are not precluded by section 922(a)(3) from importing into the United States any firearms acquired outside of the United States that may be lawfully imported. The firearms must accompany such persons since once a person is in the United States and has acquired residence in a State he may import a firearm only by arranging for the importation through a Federal firearms licensee.

As applicable to this ruling, 18 U.S.C. 925(d) provides that firearms are importable if they are generally recognized as particularly suitable for, or readily adaptable to, sporting purposes, excluding National Firearms Act (NFA) firearms and surplus military firearms.

Held, a nonresident U.S. citizen returning to the United States after having resided outside of the United States, or a nonresident alien lawfully immigrating to the United States, may apply for a permit from ATF to import for personal use, not for resale, firearms acquired outside of the United States without having to utilize the services of a Federal firearms licensee. The application on ATF Form 6 Part I (7570.3A), Application and Permit for Importation of Firearms, Ammunition and Implements of War, should include a statement, on the application form or on an attached sheet, that:

(1) the applicant is a nonresident U.S. citizen who is returning to the United States from a residence outside of the United States or, in the case of an alien, is lawfully immigrating to the United States from a residence outside of the United States, and

(2) the firearms are being imported for personal use and not for resale.

[ATFB 1981-3 77]

27 CFR 179.11: MEANING OF TERMS

The AR15 auto sear is a machinegun as defined by 26 U.S.C. 5845(b).

ATF Rul. 81-4

The Bureau of Alcohol, Tobacco and Firearms has examined an auto sear known by various trade names including **"AR15 Auto Sear," "Drop In Auto Sear,"** and **"Auto Sear II,"** which consists of a sear mounting body, sear, return spring, and pivot pin. The Bureau finds that the single addition of this auto sear to certain AR15 type semiautomatic rifles, manufactured with M16 internal components already installed, will convert such rifles into machineguns. The National Firearms Act, 26 U.S.C. 5845(b), defines **"machinegun"** to include any combination of parts designed and intended for use in converting a weapon to shoot automatically more than one shot, without manual reloading, by a single function of the trigger.

Held: The auto sear known by various trade names including **"AR15 Auto Sear," "Drop In Auto Sear,"** and **"Auto Sear II,"** is a combination of parts designed and intended for use in converting a weapon to shoot automatically more than one shot, without manual reloading, by a single function of the trigger. Consequently, the auto sear is a machinegun as defined by 26 U.S.C. 5845(b).

With respect to the machinegun classification of the auto sear under the National Firearms Act, pursuant to 26 U.S.C. 7805(b), this ruling will not be applied to auto sears manufactured before November 1, 1981. Accordingly, auto sears manufactured on or after November 1, 1981, will be subject to all the provisions of the National Firearms Act and 27 CFR Part 179.

[ATFQB 1981-3 78]

Regardless of the date of manufacture of a drop in auto sear (i.e., before or after November 1, 1981) the possession or transfer of an unregistered drop in auto sear (a machinegun as defined) is prohibited by the National Firearms Act (NFA), 26 U.S.C. § 5861, and the Gun Control Act, 18 U.S.C. § 922(o). The last paragraph of ATF Ruling 81-4 only exempts the making, transfer, and special (occupational) taxes imposed by the NFA with respect to the making, manufacture, or transfer of drop in auto sears prior to November 1, 1981. See 26 U.S.C. §§ 5801, 5811, 5821, 7805(b)(8).

27 CFR 179.11: MEANING OF TERMS

The KG-9 pistol is a machinegun as defined in the National Firearms Act.

ATF Rul. 82-2

The Bureau of Alcohol, Tobacco and Firearms has examined a firearm identified as the KG-9 pistol. The KG-9 is a 9 millimeter caliber, semiautomatic firearm which is blowback operated and which fires from the open bolt position with the bolt incorporating a fixed firing pin. In addition, a component part of the weapon is a disconnector which prevents more than one shot being fired with a single function of the trigger.

The disconnector is designed in the KG-9 pistol in such a way that a simple modification to it, such as cutting, filing, or grinding, allows the pistol to operate automatically. Thus, this simple modification to the disconnector together with the configuration of the above design features (blowback operation, firing from the open bolt position, and fixed firing pin)in the KG-9 permits the firearm to shoot automatically more than one shot, without manual reloading, by a single function of the trigger. The above combination of design features as employed in the KG-9 is normally not found in the typical sporting firearm.

The National Firearms Act, 26 U.S.C. § 5845(b), defines a machinegun to include any weapon which shoots, is designed to shoot, or can be readily restored to shoot, automatically more than one shot, without manual reloading, by a single function of the trigger.

The **"shoots automatically"** definition covers weapons that will function automatically. The **"readily restorable"** definition defines weapons which pre-viously could shoot automatically but will not in their present condition. The **"designed"** definition includes those weapons which have not previously functioned as machineguns but possess design features which facilitate full automatic fire by simple modification or elimination of existing component parts.

Held: The KG-9 pistol is designed to shoot automatically more than one shot, without function of the trigger. Consequently, the KG-9 pistol is a machinegun as defined in section 5845(b) of the Act.

With respect to the machinegun classification of the KG-9 pistol under the National Firearms Act, pursuant to 26 U.S.C. § 7805(b), this ruling will not be applied to KG-9 pistols manufactured before January 19, 1982. Accordingly, KG-9 pistols manufactured on or after January 19, 1982, will be subject to all the provisions of the National Firearms Act and 27 CFR Part 179.

[ATFB 1982-1 18]

27 CFR 179.11: MEANING OF TERMS

The SM10 and SM11A1 pistols and SAC carbines are machineguns as defined in the National Firearms Act.

ATF Rul. 82-8

The Bureau of Alcohol, Tobacco and Firearms has reexamined firearms identified as SM10 pistols, SM11A1 pistols, and SAC carbines. The SM10 is a 9 millimeter or .45ACP caliber, semiautomatic firearm; the SM11A1 is a .380ACP caliber, semiautomatic firearm; and the SAC carbine is a 9 millimeter or .45ACP caliber, semiautomatic firearm. The weapons are blowback operated, fire from the open bolt position with the bolt incorporating a fixed firing pin, and the barrels of the pistols are threaded to accept a silencer. In addition, component parts of the weapons are a disconnector and a trip which prevent more than one shot being fired with a single function of the trigger.

The disconnector and trip are designed in the SM10 and SM11A1 pistols and in the SAC carbine (firearms) in such a way that a simple modification to them, such as cutting, filing, or grinding, allows the firearms to operate automatically. Thus, this simple modification to the disconnector or trip, together with the configuration of the above design features (blowback operating, firing from the open bolt position, and fixed firing pin) in the SM10 and SM11A1 pistols and in the SAC carbine, permits the firearms to shoot automatically, more than one shot, without manual reloading, by a single function of the trigger. The above combination of design features as employed in the SM10 and SM11A1 pistols and the SAC carbine are normally not found in typical sporting firearms.

The National Firearms Act, 26 U.S.C. § 5845(b), defines a machinegun to include any weapon which shoots, is designed to shoot, or can be readily restored to shoot, automatically more than one shot, without manual reloading, by a single function of the trigger.

The **"shoots automatically"** definition covers weapons that will function automatically. The **"readily restorable"** definition defines weapons which previously could shoot automatically but will not in their present condition. The **"designed"** definition includes those weapons which have not previously functioned as machineguns but possess design features which facilitate full automatic fire by a simple modification or elimination of existing component parts.

Held: The SM10 and SM11A1 pistols and the SAC carbine are designed to shoot automatically more than one shot, without manual reloading, by a single function of the trigger. Consequently, the SM10 and SM11A1 pistols and SAC carbines are machineguns as defined in Section 5845(b) of the Act.

With respect to the machinegun classification of the SM10 and SM11A1 pistols and SAC carbines, under the National Firearms Act, pursuant to 26 U.S.C. § 7805(b), this ruling will not be applied to SM10 and SM11A1 pistols and SAC carbines manufactured or assembled before June 21, 1982. Accordingly, SM10 and SM11A1 pistols and SAC carbines, manufactured or assembled on or after June 21, 1982, will be subject to all the provisions of the National Firearms Act and 27 CFR Part 179.

[ATFB 1982-2 49]

27 CFR 179.11: MEANING OF TERMS

The YAC STEN MK II carbine is a machinegun as defined in the National Firearms Act.

ATF Rul. 83-5

The Bureau of Alcohol, Tobacco and Firearms has examined a firearm identified as the YAC STEN MK II carbine. The YAC STEN MK II carbine is a 9 millimeter caliber firearm which has identical design characteristics to the original selective fire STEN submachinegun designed by Reginald Vernon Shepherd and Harold John Turpin. The weapon is blowback operated and fires from the open bolt position with the bolt incorporating a fixed firing pin. In addition, a component part of the weapon is a trip lever (disconnector) which has been modified to prevent more than one shot being fired with a single function of the trigger.

The trip lever (disconnector) is designed in such a way that a simple modification to it, such as bending, breaking or cutting, allows the weapon to operate automatically. Thus, this simple modification to the trip lever (disconnector), together with STEN submachinegun design features and components in the YAC STEN MK II carbine, permits the firearm to shoot automatically, more than one shot, without manual reloading by a single function of the trigger. The above combination of machinegun design features as employed in the YAC STEN MK II carbine are not normally found in the typical sporting firearm.

The National Firearms Act, 26 U.S.C. 5845(b), defines a machinegun to include any weapon which shoots, is designed to shoot, or can be readily restored to shoot, automatically more than one shot, without manual reloading, by a single function of the trigger.

The **"shoots automatically"** definition covers weapons that will function automatically. The **"readily restorable"** definition defines weapons which previously could shoot automatically but will not in their present condition. The **"designed"** definition includes weapons which have not previously functioned as machineguns but possess specific machinegun design features which facilitate automatic fire by simple alteration or elimination of existing component parts.

Held: The YAC STEN MK II carbine is designed to shoot automatically more than one shot, without manual reloading, by a single function of the trigger. Consequently, the STEN MK II semiautomatic carbine is a machinegun as defined in Section 5845(b) of the Act.

[ATFB 1983-3 35]

27 CFR 179.111: IMPORTATION PROCEDURE

A National Firearms Act (NFA) firearm may not be imported for use as a sample for sales to law enforcement agencies if the firearm is a curio or relic unless it is established that the firearm is particularly suitable for use as a law enforcement weapon.

ATF Rul. 85-2

The Bureau of Alcohol, Tobacco and Firearms has approved a number of applications to import National Firearms Act (NFA) firearms for the use of registered importers to generate orders for such firearms from law enforcement agencies.

A review of the characteristics of the NFA firearms approved for importation as sales samples indicates that some of the firearms are not being imported for the purposes contemplated by the statute. Some of the NFA firearms imported are, in fact, curios or relics and are more suitable for use as collector's items than law enforcement weapons.

Importations of NFA firearms are permitted by 26 U.S.C. 5844, which provides in pertinent part:

"No firearms shall be imported or brought into the United States or any territory under its control or jurisdiction unless the importer establishes, under regulations as may be prescribed by the Secretary, that the firearm to be imported or brought in is:

(1) being imported or brought in for the use of the United States or any department, independent establishment, or agency thereof or any State or possession or any political subdivision thereof; or

(2) ***

(3) being imported or brought in solely for ... use as a sample by a registered importer or registered dealer;

except that, the Secretary may permit the conditional importation or bringing in of a firearm for examination and testing in connection with classifying the firearm."

The sole purpose of the statute permitting the importation of NFA firearms as sales samples is to permit registered importers to generate orders for firearms from government entities, primarily law enforcement agencies, on the basis of the sample.

The implementing regulation, 27 CFR Section 179.111, provides that the person importing or bringing a firearm into the United States or any territory under its control or jurisdiction has the burden of proof to affirmatively establish that the firearm is being imported or brought in for one of the authorized purposes. In addition, a detailed explanation of why the importation falls within one of the authorized purposes must be attached to the application to import. The mere statement that an NFA firearm is being imported as a sales sample for demonstration to law enforcement agencies does not meet the required burden of proof and is not a detailed explanation of why the importation falls within the import standards.

Held, an application to import a National Firearms Act firearm as a sample in connection with sales of such firearms to law enforcement agencies will not be approved if the firearm is determined to be a curio or relic unless it is established by specific information that the firearm is particularly suitable for use as a law enforcement weapon. For example, the importer must provide detailed information as to why a sales sample of a particular weapon is suitable for law enforcement purposes and the expected customers who would require a demonstration of the weapon. Information as to the availability of firearms to fill subsequent orders would help meet the burden of establishing use as a sales sample. Also, letters from law enforcement agencies expressing a need for a particular model or interest in seeing a demonstration of a particular firearm would be relevant.

[ATFB 85-2 62]

Editor's Note:

The importation of machineguns for use as sales samples must also meet the requirements of 27 CFR 179.105(d).

27 CFR 178.118: IMPORTATION OF CERTAIN FIREARMS CLASSIFIED AS CURIOS OR RELICS

(Also 178.11 and 178.26)

Surplus military firearms frames or receivers alone not specifically classified as curios or relics by ATF will be denied importation.

ATF Rul. 85-10

The Bureau of Alcohol, Tobacco and Firearms has been asked whether applications will be approved for permits to import frames or receivers of surplus military firearms classified as curios or relics for purposes of 18 U.S.C. Chapter 44. Section 233 of the Trade and Tariff Act of 1984, 98 Stat. 2991, amended 18 U.S.C. § 925 to allow licensed importers to import firearms **"listed"** by the Secretary as curios or relics, excluding handguns not generally recognized as particularly suitable for or readily adaptable to sporting purposes. The amendment had the effect of allowing the importation of surplus military curio or relic firearms that were previously prohibited from importation by 18 U.S.C. § 925(d)(3).

Congressional intent was expressed by Senator Robert Dole in 130 CONG. REC. S2234 (daily ed. Mar. 2, 1984), as follows:

First. This provision is aimed at allowing collectors to import fine works of art and other valuable weapons.

Second. This provision would allow the importation of certain military surplus firearms that are classified as curios and relics by regulations of the Secretary of the Treasury.

Third. In order for an individual or firm to import a curio or relic it must first be put on a list by petitioning the Secretary of the Treasury. The Secretary must find the firearm's primary value is that of being a collector's item.

Fourth. The only reason a person would purchase these firearms is because of their peculiar collector's status. And, in fact, they must be special firearms and classified as such in order to import.

This language clearly shows that Congress intended to permit the importation of surplus military firearms of special interest and value to collectors and recognized by ATF as meeting the curio or relic definition in 27 C.F.R. § 178.11. The regulation defines **"curios or relics"** as firearms of **"special interest to collectors by reason of some quality other than is ordinarily associated with firearms intended for sporting use or as offensive or defensive weapons."** The regulation further defines curios or relics to include **"firearms which derive a substantial part of their monetary value from the fact that they are novel, rare, bizarre, or because of their association with some historical figure,**

period or event."

In classifying firearms as curios or relics under this regulation, ATF has recognized only assembled firearms as curios or relics. Moreover, ATF's classification of surplus military firearms as curios or relics has extended only to those firearms in their original military configuration. Frames or receivers of curios or relics and surplus military firearms not in their original military configuration were not generally recognized as curios or relics by ATF since they were not of special interest or value as collector's items. More specifically, they did not meet the definition of curio or relic in section 178.11 as firearms of special interest to collectors by reason of a quality other than is ordinarily associated with sporting firearms or offensive or defensive weapons. Furthermore, they did not ordinarily have monetary value as novel, rare, or bizarre firearms; nor were they generally considered curios or relics because of their association with some historical figure, period or event.

It is clear from the legislative history that Congress did not intend for frames or receivers alone of surplus military firearms, or any other surplus military firearms not in their original military configuration, to be importable under section 925(e). It is also clear that only those firearms classified by ATF as curios or relics were intended to be approved by ATF for importation.

Held, to be importable under 18 U.S.C. 925(e), surplus military firearms must be classified as curios or relics by ATF. Applications by licensed importers to import frames or receivers alone of surplus military curio or relic firearms will not be approved under section 925(e). Surplus military firearms will not be classified as curios or relics, and applications for permits to import such firearms approved, unless they are assembled in their original military configuration.

[ATFB 85-3 46]26 U.S.C. § 5845(f)(2): DESTRUCTIVE DEVICE

(Nonsporting shotgun having a bore of more than one-half inch in diameter)

The USAS-12 shotgun has a bore of more than one-half inch in diameter and is not generally recognized as particularly suitable for sporting purposes. Therefore, it is classified as a destructive device for purposes of the National Firearms Act, 26 U.S.C.

Chapter 53.

ATF Rul. 94-1

The Bureau of Alcohol, Tobacco and Firearms (ATF) has examined a firearm identified as the USAS-12 shotgun to determine whether it is a destructive device as that term is used in the National Firearms Act (NFA), 26 U.S.C. Chapter 53.

The USAS-12 is a 12-gauge, gas operated, autoloading semiautomatic shotgun which is chambered for 12-gauge 2 3/4-inch ammunition. It has an 18 1/4-inch barrel, is approximately 38 inches long, and weighs 12.4 pounds unloaded and approximately 15 pounds with a loaded magazine, depending on the capacity of the magazine. The USAS-12 is equipped with a 12-round detachable box magazine, but a 28-round detachable drum magazine is also available. The shotgun is approximately 11 inches deep with a box magazine. There is an integral carrying handle on top of the receiver, which houses a rifle-type aperture rear and adjustable post-type front sight. The USAS-12 has a separate combat-style pistol grip located on the bottom of the receiver, forward of the buttstock. An optional telescopic sight may be attached to the carrying handle. The barrel is located below the operating mechanism in such fashion that the barrel is in a straight line with the center of the buttstock.

Section 5845(f), Title 26, U.S.C., classifies certain weapons as **"destructive devices"** which are subject to the registration and tax provisions of the NFA. Section 5845(f)(2) provides as follows:

(f) Destructive device. – The term **"destructive device"** means * * *

(2) any type of weapon by whatever name known which will, or which may be readily converted to, expel a projectile by the action of an explosive or other propellant, the barrel or barrels of which have a bore of more than one-half inch in diameter, except a shotgun or shotgun shell which the Secretary or his delegate finds is generally recognized as particularly suitable for sporting purposes;

A **"sporting purposes"** test which is almost identical to that in section 5845(f) (2) appears in 18 U.S.C. § 925(d)(3). This provision of the Gun Control Act of 1968 (GCA) provides that the Secretary shall authorize a firearm to be imported into the United States if the firearm is **"generally recognized as particularly suitable for or readily adaptable to sporting purposes."** With the exception of the **"readily adaptable"** language, this provision is identical to the sporting shotgun exception to the destructive device definition. The definition of **"destructive device"** in the GCA (18 U.S.C. § 921(a)(4)) is identical to that in the NFA.

In determining whether shotguns with a bore of more than one-half inch in diameter are **"generally recognized as particularly suitable for sporting purposes"** and thus are not destructive devices under the NFA, we believe it is appropriate to use the same criteria used for evaluating shotguns under the **"sporting purposes"** test of section 925(d)(3). Congress used virtually identical language in describing the weapons subject to the two statutory schemes, and the language was added to the GCA and NFA at the same time.

In connection with the determination of importability, ATF determined that the USAS-12 shotgun was not eligible for importation under the sporting purposes test in section 925(d)(3). In reaching this determination, ATF evaluated the weight, size, bulk, designed magazine capacity, configuration, and other characteristics of the USAS-12. It was determined that the weight of the USAS-12, 12.4 pounds, made it much heavier than traditional 12-gauge sporting shotguns, which made it awkward to carry for extended periods, as in hunting, and cumbersome to fire at multiple small moving targets, as in skeet and trap shooting. The width of the USAS-12 with drum magazine, approximately 6 inches, and the depth with box magazine, in excess of 11 inches, far exceeded that of traditional sporting shotguns, which do not exceed 3 inches in width or 4 inches in depth. The large size and bulk of the USAS-12 made it extremely difficult to maneuver quickly enough to engage moving targets as is necessary in hunting, skeet, and trap shooting. The detachable box magazine with 12-cartridge capacity and the detachable drum magazine with 28-cartridge capacity were of a larger capacity than traditional repeating sporting shotguns, which generally contain tubular magazines with a capacity of 3-5 cartridges. Additionally, detachable magazines permit more rapid reloading than do tubular magazines. Finally, the combat-style pistol grip, the barrel-to-buttstock configuration, the bayonet lug, and the overall appearance and general shape of the weapon

were radically different from traditional sporting shotguns and strikingly similar to shotguns designed specifically for or modified for combat and law enforcement use.

Section 7805(b), Title 26, U.S.C., provides that the Secretary may prescribe the extent, if any, to which any ruling relating to the internal revenue laws shall be applied without retroactive effect. Accordingly, all rulings issued under the Internal Revenue Code are applied retroactively unless they specifically provide otherwise. Pursuant to section 7805(b), the Director, as the delegate of the Secretary, may prescribe the extent to which any ruling will apply without retroactive effect.

Held: The USAS-12 is a shotgun with a bore of more than one-half inch in diameter which is not particularly suitable for sporting purposes. The weight, size, bulk, designed magazine capacity, configuration, and other factors indicate that the USAS-12 is a semiautomatic version of a military-type assault shotgun. Accordingly, the USAS-12 is a destructive device as that term is used in 26 U.S.C. § 5845(f)(2). Pursuant to section 7805(b), this ruling is applied prospectively effective March 1, 1994, with respect to the making, transfer, and special (occupational) taxes imposed by the NFA. All other provisions of the NFA apply retroactively effective March 1, 1994.

[ATFB 1993-94-1 21]

26 U.S.C. § 5845(f)(2): DESTRUCTIVE DEVICE

(Nonsporting shotgun having a bore of more than one-half inch in diameter)

The Striker-12/Streetsweeper shotgun has a bore of more than one-half inch in diameter and is not generally recognized as particularly suitable for sporting purposes. Therefore, it is classified as a destructive for purposes of the National Firearms Act, 26 U.S.C. Chapter 53.

ATF Rul. 94-2

The Bureau of Alcohol, Tobacco and Firearms (ATF) has examined a firearm identified as the Striker-12/Streetsweeper shotgun to determine whether it is a destructive device as that term is used in the National Firearms Act (NFA), 26 U.S.C. Chapter 53.

The Striker-12 and Streetsweep-

er shotguns are virtually identical 12-gauge shotguns with a spring-driven revolving magazine. The magazine has a 12-round capacity. The shotgun has a fixed stock or folding shoulder stock and may be fired with the folding stock collapsed. The shotgun with an 18-inch barrel is 37 inches in length with the stock extended, and 26.5 inches in length with the stock folded. The shotgun is 5.7 inches in width and weighs 9.24 pounds unloaded. The Striker/Streetsweeper has two pistol grips, one in the center of the firearm below the buttstock, and one on the forearm. The Striker/Streetsweeper was designed and developed in South Africa as a military, security, and antiterrorist weapon. Various types of 12-gauge cartridges can be fired from the shotgun, and a rapid indexing procedure allows various types of ammunition to be loaded into the cylinder and selected for firing. All 12 rounds can be fired from the shotgun in 3 seconds or less.

Section 5845(f), Title 26, U.S.C., classifies certain weapons as **"destructive devices"** which are subject to the registration and tax provisions of the NFA. Section 5845(f)(2) provides as follows:

(f) Destructive device. – The term "destructive device" means * * *

(2) any type of weapon by whatever name known which will, or which may be readily converted to, expel a projectile by the action of an explosive or other propellant, the barrel or barrels of which have a bore of more than one-half inch in diameter, except a shotgun or shotgun shell which the Secretary or his delegate finds is generally recognized as particularly suitable for sporting purposes; ..."

A **"sporting purposes"** test which is almost identical to that in section 5845(f)(2) appears in 18 U.S.C. § 925(d)(3). This provision of the Gun Control Act of 1968 (GCA) provides that the Secretary shall authorize a firearm to be imported into the United States if the firearm is **"generally recognized as particularly suitable for or readily adaptable to sporting purposes."** With the exception of the readily adaptable language, this provision is identical to the sporting shotgun exception to the destructive devices definition. The definition of **"destructive device"** in the GCA (18 U.S.C. § 921(a)(4)) is identical to that in the NFA.

In determining whether shotguns

with a bore of more than one-half inch in diameter are **"generally recognized as particularly suitable for sporting purposes"** and thus are not destructive devices under the NFA, we believe it is appropriate to use the same criteria used for evaluating shotguns under the **"sporting purposes"** test of section 925(d)(3). Congress used virtually identical language in describing the weapons subject to the two statutory schemes, and the language was added to the GCA and NFA at the same time.

In 1984, ATF ruled that the Striker-12 was not eligible for importation under section 925(d)(3) since it is not particularly suitable for sporting purposes. In making this determination, the 1984 letter-ruling notes that the Striker was being used in a number of **"combat"** shooting events. In a letter dated June 30, 1986, ATF again denied importation to the Striker-12, on the basis that it did not meet the **"sporting purposes"** test of section 925(d)(3). This letter states that, **"We believe the weapon to have been specifically designed for military and law enforcement uses."**

In evaluating the physical characteristics of the Striker 12/Streetsweeper, ATF concludes that the weight, bulk, designed magazine capacity, configuration, and other features indicate that it was designed primarily for military and law enforcement use and is not particularly suitable for sporting purposes.

The weight of the Striker-12/Streetsweeper, 9.24 pounds unloaded, is on the high end for traditional 12-gauge sporting shotguns, which generally weigh between 7 and 10 pounds. Thus, the weight of the Striker-12/Streetsweeper makes it awkward to carry for extended periods, as in hunting, and cumbersome to fire at multiple small moving targets, as in skeet and trap shooting. The width of the Striker-12/Streetsweeper, 5.7 inches, far exceeds that of traditional sporting shotguns, which do not exceed three inches in width or four inches in depth. The large size and bulk of the Striker-12/Streetsweeper make it extremely difficult to maneuver quickly enough to engage moving targets as is necessary in hunting, skeet, and trap shooting. The spring driven revolving magazine with 12-cartridge capacity is a much larger capacity than traditional repeating sporting shotguns, which generally contain tubular magazines with a capacity of 3-5 cartridges. The folding shoulder stock and the two pistol grips are not typical of sporting-type shotguns. Finally, the overall appearance

and general shape of the weapon are radically different from traditional sporting shotguns and strikingly similar to shotguns designed specifically for or modified for combat and law enforcement use.

Section 7805(b), Title 26, U.S.C., provides that the Secretary may prescribe the extent, if any, to which any ruling relating to the internal revenue laws shall be applied without retroactive effect. Accordingly, all rulings issued under the Internal Revenue Code are applied retroactively unless they specifically provide otherwise. Pursuant to section 7805(b), the Director, as the delegate of the Secretary, may prescribe the extent to which any ruling will apply without retroactive effect.

Held: The Striker-12/Streetsweeper is a shotgun with a bore of more than one-half inch in diameter which is not particularly suitable for sporting purposes. The weight, size, bulk, designed magazine capacity, configuration, and other factors indicate that the Striker-12/Streetsweeper is a military-type shotgun, as opposed to a shotgun particularly suitable for sporting purposes. Accordingly, the Striker-12/Streetsweeper is a destructive device as that term is used in 26 U.S.C. § 5845(f)(2). Pursuant to section 7805(b), this ruling is applied prospectively effective March 1, 1994, with respect to the making, transfer, and special (occupational) taxes imposed by the NFA. All other provisions of the NFA apply retroactively effective March 1, 1994.

[ATFB 1993-1994-1 23]

18 U.S.C. § 921(a)(4): DESTRUCTIVE DEVICE

26 U.S.C. § 5845(f)(2): DESTRUCTIVE DEVICE

(Firearm having a bore of more than one-half inch in diameter)

37/38 mm gas/flare guns possessed with cartridges containing wood pellets, rubber pellets or balls, or bean bags are classified as destructive devices for purposes of the Gun Control Act, 18 U.S.C. Chapter 44, and the National Firearms Act, 26 U.S.C. Chapter 53.

ATF Rul. 95-3

The Bureau of Alcohol, Tobacco and Firearms (ATF) has examined various 37/38 mm gas/flare guns in combination with certain types of ammunition to determine whether these are destructive

devices as defined in the Gun Control Act (GCA), 18 U.S.C. Chapter 44, and the National Firearms Act (NFA), 26 U.S.C. Chapter 53.

Section 5845(f), Title 26, United States Code, classifies certain weapons as destructive devices" which are subject to the registration and tax provisions of the National Firearms Act (NFA). Section 5845(f)(2) provides as follows:

(f) Destructive device. — The term **"destructive device"** means * * *

(2) any type of weapon by whatever name known which will, or which may be readily converted to, expel a projectile by the action of an explosive or other propellant, the barrel or barrels of which have a bore of more than one-half inch in diameter, except a shotgun or shotgun shell which the Secretary or his delegate finds is generally recognized as particularly suitable for sporting purposes."

Section 5845(f)(3) excludes from the term **"destructive device"** any device which is neither designed or redesigned for use as a weapon and any device, although originally designed for use as a weapon, which is redesigned for use as a signaling, pyrotechnic, line throwing, safety, or similar device.

The definition of **"destructive device"** in the GCA (18 U.S.C. § 921(a)(4)) is identical to that in the NFA.

ATF has previously held that devices designed for expelling tear gas or pyrotechnic signals are not weapons and are exempt from the destructive device definition. However, ammunition designed to be used against individuals is available for these 37/38 mm devices. This **"anti-personnel"** ammunition consists of cartridges containing wood pellets, rubber pellets or balls, and bean bags.

When a gas/flare gun is possessed with **"anti-personnel"** type ammunition, it clearly becomes an instrument of offensive or defensive combat and is capable of use as a weapon. Since these gas/flare guns have a bore diameter of greater than one-half inch, fire a projectile by the means of an explosive, and, when possessed with **"anti-personnel"** ammunition, are capable of use as weapons, the combination of the gas/flare gun and **"anti-personnel"** ammunition is a destructive device as defined in the GCA and NFA. As a result, registration as a destructive device is required. Any person possessing a gas/flare gun with which **"anti-personnel"** ammunition will be used must register the making of a destructive device prior to the acquisition of any **"anti-personnel"** ammunition. In addition, the gas/flare guns are classified as firearms as defined by the GCA when possessed with **"anti-personnel"** type ammunition.

Each gas/flare gun possessed with anti-personnel ammunition will be required to be identified as required by law and regulations (27 CFR §§ 178.92 and 179.102), including a serial number. Any person manufacturing the gas/flare gun and the **"anti-personnel"** ammunition must, if selling them in combination, have the appropriate Federal firearms license as a manufacturer of destructive devices and must have paid the special (occupational) tax as a manufacturer of National Firearms Act firearms. Any person importing the gas/flare gun and the **"anti-personnel"** ammunition must, if importing them in combination, have the appropriate Federal firearms license as an importer of destructive devices and must have paid the special (occupational) tax as an importer of National Firearms Act firearms.

Further, the **"anti-personnel"** ammunition to be used in the gas/flare launchers is ammunition for destructive devices for purposes of the GCA. Any person manufacturing the **"anti-personnel"** ammunition must have the appropriate Federal firearms license as a manufacturer of ammunition for destructive devices. Any person importing the **"anti-personnel"** ammunition must have the appropriate Federal firearms license as an importer of ammunition for destructive devices.

Held: 37/38 mm gas/flare guns possessed with **"anti-personnel"** ammunition, consisting of cartridges containing wood pellets, rubber pellets or balls, or bean bags, are destructive devices as that term is used in 18 U.S.C. § 921(a)(4) and 26 U.S.C. 5845(f)(2).

[ATFB 95-3 28]

18 U.S.C. § 923 (a): ENGAGING IN THE BUSINESS OF DEALING IN FIREARMS (Auctioneers)

Auctioneers who regularly conduct consignment-type auctions of firearms, for example, held every 1-2 months, on behalf of firearms owners where the auctioneer takes possession of the firearms pursuant to a consignment contract with the owner of the firearms giving the auctioneer authority to sell the firearms and providing for a commission to be paid by the owner upon sale of the firearms are required to obtain a license as a dealer in firearms.

ATF Rul. 96-2

An association of auctioneers has asked the Bureau of Alcohol, Tobacco and Firearms (ATF) for a ruling concerning the auctions conducted by their members and whether the sale of firearms at such auctions requires a Federal firearms license as a dealer in firearms.

The auctioneers' association stated that their members generally conduct two types of auctions: estate-type auctions and consignment auctions. In estate-type auctions, articles to be auctioned, including firearms, are sold by the executor of the estate of an individual. In these cases the firearms belong to and are possessed by the executor. The auctioneer acts as an agent of the executor and assists the executor in finding buyers for the firearms. The firearms are possessed by the estate and their sale to third parties is controlled by the estate. The auctioneer is paid a commission on the sale of each firearm by the estate at the conclusion of the auction.

The association states that, in consignment-type auctions, an auctioneer may take possession of firearms in advance of the auction. The firearms are inventoried, evaluated, and tagged for identification. The firearms belong to individuals or businesses who have entered into a consignment agreement with the auctioneer giving the auctioneer authority to sell the firearms. The agreement states that the auctioneer has the exclusive right to sell the items listed on the contract at a location, time, and date to be selected by the auctioneer. The agreement also provides for the payment of a commission by the owner to the auctioneer. The consignment-type auctions generally involve accepting firearms for auction from more than one owner. Also, these auctions are held on a regular basis, for example, every 1-2 months.

Section 923(a), Title 18, U.S.C., provides that no person shall engage in the business of dealing in firearms until he has filed an application and received a license to do so. Section 922(a)(1), Title 18, U.S.C., provides that it is unlawful for any person, other than a licensee, to engage in the business of dealing in firearms. Licensees generally may not conduct business away from their licensed premises.

The term **"dealer"** is defined at 18 U.S.C. § 921(a)(11)(A) to include any person engaged in the business of selling firearms at wholesale or retail. The term **"engaged in the business"** as applied to a dealer in firearms means a person who devotes time, attention, and labor to dealing in firearms as a regular course of trade or business with the principal objective of livelihood and profit through the repetitive purchase and resale of firearms. A dealer can be **"engaged in the business"** without taking title to the firearms that are sold. However, the term does not include a person who makes occasional sales, exchanges, or purchases of firearms for the enhancement of a personal collection or for a hobby, or who sells all or part of his personal collection of firearms. 18 U.S.C. § 921(a)(21)(C).

In the case of estate-type auctions, the auctioneer acts as an agent of the executor and assists the executor in finding buyers for the estate's firearms. The firearms are possessed by the estate, and the sales of firearms are made by the estate. In these cases, the auctioneer does not meet the definition of **"engaging in the business"** as a dealer in firearms and would not require a license. An auctioneer engaged in estate-type auctions, whether licensed or not, may perform this function, including delivery of the firearms, away from the business premises.

In the case of consignment-type auctions held on a regular basis, for example, every 1-2 months, where persons consign their firearms to the auctioneer for sale pursuant to an agreement as described above, the auctioneer would be **"engaging in the business"** and would require a license. The auctioneer would be disposing of firearms as a regular course of trade or business within the definition of a **"dealer"** under § 921(a)(11)(A) and must comply with the licensing requirements of the law.

As previously stated, licensed auctioneers generally must engage in the business from their licensed premises. However, an auctioneer may conduct an auction at a location other than his licensed premises by displaying the firearms at the auction site, agreeing to the terms of sale of the firearms, then returning the firearms to the licensed premises for delivery to the purchaser.

Held: Persons who conduct estate-type auctions at which the auctioneer assists the estate in selling the estate's firearms, and the firearms are possessed and transferred by the estate, do not require a Federal firearms license.

Held further: Persons who regularly conduct consignment-type auctions, for example, held every 1-2 months, where the auctioneer takes possession of the firearms pursuant to a consignment contract giving the auctioneer the exclusive right and authority to sell the firearms at a location, time and date to be selected by the auctioneer and providing for a commission to be paid upon sale are required to obtain a license as a dealer in firearms pursuant to 18 U.S.C. § 923(a).

[ATFB 96-2 101]

18 U.S.C. 921(a)(4): DESTRUCTIVE DEVICE

26 U.S.C. 5845(f)(2): DESTRUCTIVE DEVICE

(Nonsporting shotgun having a bore of more than one-half inch in diameter)

The registration period for the USAS-12, Striker-12, and Streetsweeper shotguns closed on May 1, 2001, pursuant to ATF Rul. 2001-1.

ATF Rul. 2001-1

Pursuant to ATF Rulings 94-1 (ATF Q.B. 1994-1, 22) and 94-2 (ATF Q.B. 1994-1, 24), the Bureau of Alcohol, Tobacco and Firearms (ATF) classified the USAS-12, Striker 12, and Streetsweeper shotguns as destructive devices under the National Firearms Act (NFA), 26 U.S.C. Chapter 53. The NFA requires that certain **"firearms"** be registered and imposes taxes on their making and transfer. The term **"firearm"** is defined in section 5845 to include **"destructive devices."** The term **"destructive device"** is defined in section 5845(f)(2) as follows:

[T]he term 'destructive device' means ... (2) any type of weapon by whatever name known which will, or which may be readily converted to, expel a projectile by the action of an explosive or other propellant, the barrel or barrels of which have a bore of more than one-half inch in diameter, except a shotgun or shotgun shell which the Secretary finds is generally recognized as particularly suitable for sporting purposes;

The USAS-12, Striker 12, and Streetsweeper shotguns were classified as destructive devices pursuant to section 5845(f) because they are shotguns with a bore of more than one-half inch in diameter which are not generally recognized as particularly suitable for sporting purposes.

Pursuant to 26 U.S.C. 7805(b), ATF Ruls. 94-1 and 94-2 were issued prospectively with respect to the making, transfer, and special (occupational) taxes imposed by the NFA. Thus, although the classification of the three shotguns as NFA weapons was retroactive, the prospective application of the tax provisions allowed registration without payment of tax. ATF has contacted all purchasers of record of the shotguns to advise them of the classification of the weapons as destructive devices and that the weapons must be registered. ATF has registered approximately 8,200 of these weapons to date.

Held, the registration period for the USAS-12, Striker-12, and Streetsweeper shotguns will close on May 1, 2001. No further registrations will be accepted after that date. Persons in possession of unregistered NFA firearms are subject to all applicable penalties under 26 U.S.C. Chapter 53.

Date signed: February 2, 2001.

27 CFR 47.45: IMPORTATION OF SURPLUS MILITARY CURIO OR RELIC FIREARMS

Importers of surplus military curio or relic firearms must submit originals of all appropriate statements supporting the Form 6 application.

ATF Rul. 2001-3

This ruling is superseded by ATF Rul. 2010-9.

18 U.S.C. 922(t)(1)(C): IDENTIFICATION OF TRANSFEREE

27 CFR 178.124: FIREARMS TRANSACTION RECORD

Licensees may accept a combination of valid government-issued documents to satisfy the identification document requirements of the Brady Act. The required valid government-issued photo identification document bearing the name, photograph, and date of birth of the transferee may be supplemented by another valid, government-issued document showing the transferee's residence address. A member of the Armed Forces on active duty is a resident of the State in which his or her permanent duty station is located, and may satisfy the identification document requirement by presenting his or her military identification card along with official orders showing that his or her permanent duty station is within the State where the licensed premises are located.

ATF Rul. 2001-5

The Bureau of Alcohol, Tobacco and Firearms (ATF) has received numerous inquiries from Federal firearms licensees (FFLs) regarding the acceptance of identification documents that do not show the purchaser's current residence address. FFLs have asked whether they may accept other documents, such as tax bills or vehicle registration documents, to establish the current residence address of the purchaser.

It has been ATF's longstanding position that licensees may accept a combination of documents to establish the identity of a firearm purchaser. ATF Rul. 79-7, ATFQB 79-1, 26, interpreted a licensee's obligation to obtain satisfactory identification from a purchaser in the manner customarily used in commercial transactions, pursuant to the existing regulations under the Gun Control Act of 1968 (GCA). The ruling held that satisfactory identification of a firearms purchaser must include the purchaser's name, age or date of birth, place of residence, and signature. The ruling also held that while a particular document may not be sufficient to meet the statutory requirement for identifying the purchaser, any combination of documents that together disclosed the required information would be acceptable.

ATF Rul. 79-7 has been superseded by an amendment to the GCA. The Brady Handgun Violence Prevention Act (Brady Act), which took effect in 1994, mandated the use of photo identification documents for transfers subject to the Act. Under the permanent provisions of the Brady Act, which went into effect on November 30, 1998, a licensed importer, manufacturer, or dealer is generally required to initiate a background check through the National Instant Criminal Background Check System (NICS) prior to transferring a firearm to an unlicensed individual.

The Brady Act requires a licensee to identify the nonlicensed transferee by examining a valid government-issued identification document that contains the photograph of the holder. See 18 U.S.C. 922(t)(1)(C). This requirement applies to all over-the-counter transfers, even where the transferee holds a permit that qualifies as an exception to the requirement for a NICS check at the time of transfer. 27 CFR 178.124(c)(3)(i).

The Brady Act incorporates the definition of an **"identification document"** provided by 18 U.S.C. 1028(d)(2), which is set forth in relevant part as follows:

[A] document made or issued by or under the authority of the United States Government, a State, political subdivision of a State, a foreign government, political subdivision of a foreign government, an international governmental or an international quasi-governmental organization which, when completed with information concerning a particular individual, is of a type intended or commonly accepted for the purpose of identification of individuals.

ATF regulations further require that the identification document must contain the name, residence address, date of birth, and photograph of the holder. 27 CFR 178.11.

ATF has received questions from licensees regarding purchasers who present a State-issued driver's license or other identification document that shows either an out-of-date residence address or a mailing address (such as a post office box) in lieu of a residence address. ATF has advised that these identification documents, standing alone, would not satisfy the requirements of the regulations implementing the Brady Act.

It is ATF's position that a combination of documents may be used to satisfy the Brady Act's requirement for an identification document. The prospective transferee must present at least one valid document that meets the statutory definition of an identification document; i.e., it must bear the transferee's name and photograph, it must have been issued by a governmental entity, and it must be of a type intended or commonly accepted for identification purposes. ATF recognizes, however, that some valid government-issued identification documents do not include the bearer's current residence address. Such an identification document may be supplemented with another valid government-issued document that contains the necessary information.

Thus, for example, a licensee may accept a valid driver's license that accurately reflects the purchaser's name, date of birth, and photograph, along with a vehicle registration issued by the State indicating the transferee's current address. Licensees should note that if the law of the State that issued the driver's license provides that the driver's license is invalid due to any reason (i.e., the license is expired or is no longer valid due to an unreported change of address), then the driver's license may not be used for identification purposes under the Brady Act. If a licensee has reasonable cause to question the validity of an identification document, he or she should not proceed with the transfer until those questions can be resolved.

The licensee must record on the Form 4473 the type of identification document(s) presented by the transferee, including any document number. Examples of documents that may be accepted to supplement information on a driver's license or other identification document include a vehicle registration, a recreation identification card, a fishing or hunting license, a voter identification card, or a tax bill. However, the document in question must be valid and must have been issued by a government agency.

ATF has also received questions from licensees as to how to comply with the identification document requirement in the case of purchasers who are in the military. Some active duty military personnel may not have driver's licenses from the State in which they are stationed. The only identification document carried by some active duty military personnel is a military identification card that bears the holder's name, date of birth, and photograph, but does not reflect the holder's residence address.

Section 921(b) of the GCA provides that a member of the Armed Forces on active duty is a resident of the State in which his permanent duty station is located. The purchaser's official orders showing that his or her permanent duty station is within the State where the licensed premises are located suffice to establish the purchaser's residence for GCA purposes. In combination with a military identification card, such orders will satisfy the Brady Act's requirement for an identification document, even though the purchaser may actually reside in a home that is not located on the military base.

Licensees should note that for purposes of the GCA, military personnel may in some cases have two States of residence. For example, a member of the Armed Forces whose permanent duty station is Fort Benning, Georgia, may actually reside in a home in Alabama. For GCA purposes, that individual is a resident of Georgia when he or she is in Georgia and a resident of Alabama when he or she is in Alabama. If such an individual wishes to purchase a firearm in Alabama, he or she must of course comply with the identification document requirement in the same way as any other Alabama resident.

Held: the Brady Act and the implementing ATF regulations require licensed importers, manufacturers, and dealers to examine a valid government-issued identification document that bears the name, residence address, date of birth, and photograph of the holder prior to making an over-the-counter transfer to any unlicensed transferee. Licensees may accept a combination of valid, government-issued documents to satisfy the identification document requirements of the Brady Act. A government-issued photo identification document bearing the name, photograph, and date of birth of the transferee may be supplemented by another valid, government-issued document showing the transferee's current residence address.

Held further, a purchaser who is a member of the Armed Forces on active duty is a resident of the State in which his or her permanent duty station is located, and may satisfy the identification document requirement by presenting his or her military identification card along with official orders showing that his or her permanent duty station is located within the State where the licensed premises are located.

ATF Ruling 79-7, ATFQB 79-1, 26, is hereby superseded.

Date signed: December 31, 2001

Editor's Note:

"Identification document" currently is defined in 18 U.S.C. 1028(d)(3).

27 CFR 179.105: TRANSFER AND POSSESSION OF MACHINEGUNS

Applications to transfer two (2) machineguns of a particular model to a Federal firearms licensee as sales

samples will be approved if documentation shows necessity for demonstration to government agencies.

ATF Rul. 2002-5

The Bureau of Alcohol, Tobacco and Firearms (ATF) has received inquiries from dealers in machineguns concerning the justification necessary to obtain more than one machinegun of a particular model as dealer sales samples. Specifically, the inquiries are from machinegun dealers who demonstrate machineguns to large police departments and Special Weapons and Tactics (SWAT) teams, which requires the firing of thousands of rounds of ammunition during a single demonstration. Section 922(o) of Title 18, United States Code, makes it unlawful for any person to transfer or possess a machinegun, except a transfer to or by or under the authority of the United States or any department or agency thereof or a State or a department, agency, or political subdivision of; or any lawful transfer or lawful possession of a machinegun lawfully possessed before May 19, 1986.

The regulations in 27 CFR 179.105(d) provide that applications to register and transfer a machinegun manufactured or imported on or after May 19, 1986, to dealers registered under the National Firearms Act (NFA), 26 U.S.C. Chapter 53, will be approved if three conditions are met. The conditions required to be established include (1) a showing of the expected government customers who would require a demonstration of the weapon; (2) information as to the availability of the machinegun to fill subsequent orders; and (3) letters from government entities expressing a need for a particular model or interest in seeing a demonstration of a particular weapon. The regulation further provides that applications to transfer more than one machinegun of a particular model must also establish the dealer's need for the quantity of samples sought to be transferred.

The dealer sales sample regulation in section 179.105(d) is a narrow exception to the general prohibition on possession of post-1986 machineguns imposed by section 922(o). It requires that dealers submit letters of interest from law enforcement agencies to ensure that dealers possess post-1986 machineguns only for the purposes permitted by law, i.e., for sale or potential sale to government agencies.

Qualified dealers in machineguns often demonstrate weapons to all officers of the department, requiring the machinegun to fire thousands of rounds of ammunition during a single demonstration. In the case of new model machineguns, a department may wish to have thousands of rounds fired from the weapon before they are fully satisfied of its reliability. ATF is aware that after firing hundreds of rounds a machinegun often gets too hot to safely handle, resulting in the dealer's inability to demonstrate the weapon until it cools. In addition, it is not uncommon for machineguns to jam or misfeed ammunition after a large quantity of ammunition has been fired. Accordingly, dealers who demonstrate machineguns to departments with a large number of officers have asked that ATF approve the transfer of two (2) machineguns of each model as dealer sales samples.

The purpose of the dealer sales sample provision is to permit properly qualified dealers to demonstrate and sell machineguns to law enforcement agencies. Neither the law nor the implementing regulations were intended to impose unnecessary obstacles to police departments and other law enforcement agencies in obtaining the weapons they need to carry out their duties. Accordingly, if a dealer can provide documentation that the dealer needs to demonstrate a particular model of machinegun to an entire police department or SWAT team, ATF will approve the transfer of two (2) machineguns of that model to the dealer as sales samples.

This ruling should not be interpreted to imply that under no circumstances may a Federal firearms licensee (FFL) receive more than two (2) machineguns as sales samples. Consistent with past practice, an FFL who can show a bona fide reason as to why they need more than two (2) machineguns, may be able to receive more than two (2) if the request is accompanied by specific documentation.

Held: applications to transfer two (2) machineguns of a particular model to a Federal firearms licensee as sales samples will be approved if the dealer provides documentation that the dealer needs to demonstrate the machinegun to all the officers of a police department or the department's SWAT team or special operations team. An FFL who offers other bona fide reasons for their need for two (2) or more machineguns may get more than two (2) with specific documentation.

Date signed: September 6, 2002

18 U.S.C. 923(i): LICENSING

26 U.S.C. 5842: IDENTIFICATION OF FIREARMS

27 CFR 179.102: IDENTIFICATION OF FIREARMS

27 CFR 178.92: IDENTIFICATION OF FIREARMS, ARMOR PIERCING AMMUNITION, AND LARGE CAPACITY AMMUNITION FEEDING DEVICES

In accordance with 27 CFR 178.92 and 27 CFR 179.102, identification of firearms, armor piercing ammunition, and large capacity ammunition feeding devices, the terms "conspicuously" and "legibly" as used therein mean, respectively, that the markings are wholly unobstructed from plain view and that the markings contain exclusively Roman letters and Arabic numerals.

ATF Rul. 2002-6

The Bureau of Alcohol, Tobacco and Firearms (ATF) has been asked by State and local law enforcement officials to trace firearms that are marked, in part, with non-Roman letters, and/or non-Arabic numbers. Specifically, ATF received a request to trace a Makarov type pistol made in Bulgaria. The original manufacturer marking was ИМ 18 355. Because the importer did not stamp the firearm with a unique identifier that could be recognized by either ATF or a State or local law enforcement official, and because the marking contained a Cyrillic character, the firearm was not properly recorded, resulting in a failed trace of the weapon.

Because markings with non-Roman characters or non-Arabic numbers are not easily recorded or transmitted through ordinary means by importers, dealers or distributors, many firearm traces have proved unsuccessful. In some cases, an importer attempts to translate portions of the markings into Roman letters and Arabic numbers and re-marks the weapon with **"translated"** symbols. For example, an imported SKS rifle was marked with the serial number ДМ7639И. The importer translated the marking as LM7639i, but rather than re-stamp the entire number merely added the letters **"L"** and **"i"** below the original markings. This practice often results in failed traces because those required to record the markings (importers, dealers, or distributors) may record only the translated portions or both sets of markings.

Moreover, law enforcement recovering a firearm with such markings may submit a trace request lacking some portion of the markings, further impeding efforts to successfully trace the firearm.

In addition, ATF has found that some traces have failed because the required markings on the firearms barrel were wholly or partially obstructed from plain view by a flash suppressor or bayonet mount, resulting in the Federal Firearms Licensee creating an inaccurate record. ATF has been unable to trace hundreds of firearms as a result of nonstandard or obscured markings.

As a result of these practices, some licensed importers may not be in compliance with the marking requirements set forth in 27 CFR 178.92 and 27 CFR 179.102 because they have marked using non-Roman letters (such as Greek or Russian letters, Δ or Д) or non-Arabic numbers (e.g., XXV).

The above regulations require markings that legibly identify each item or package and require that such markings are conspicuous. ATF has consistently taken the position that **"legibly"** marked means using exclusively Roman letters (A, a, B, b, C, c, and so forth) and Arabic numerals (1, 2, 3, 4, 5, 6, and so forth), and **"conspicuous"** means that all required markings must be placed in such a manner as to be wholly unobstructed from plain view. These regulations apply to licensed manufacturers and licensed importers relative to firearms, armor piercing ammunition, and large capacity ammunition feeding devices, and to makers of National Firearms Act firearms.

Firearms, armor piercing ammunition, and large capacity ammunition feeding devices which contain required markings or labels using non-Roman letters (such as Greek or Russian letters, Δ or Д) or non-Arabic numbers (e.g., XXV), must be completely remarked or relabeled with a new serial number or other required markings that satisfy the legibility requirements described above. It is not sufficient to simply add an additional Roman letter or Arabic numeral to a nonconforming marking; a new and unique marking using Roman letters and Arabic numerals is required. Where feasible, the new markings should be placed directly above the non-compliant markings.

Similarly, firearms and large capacity ammunition feeding devices which contain required markings obstructed in whole or in part from plain view must be remarked with required markings that satisfy the conspicuousness requirements described above. For example, required markings may not be placed on a portion of the barrel where the markings would be wholly or partially obstructed from view by another part of the firearm, such as a flash suppressor or bayonet mount.

In certain unavoidable circumstances owing mainly to firearms of unusual design or other limiting factor(s) which would limit the ability of the manufacturer or importer to comply with the above legibility and conspicuousness requirements, alternate means of identification may be authorized as described in 27 CFR 178.92(a)(3)(i), (ii), or (iii) and 27 CFR 178.92(c)(3)(ii).

Held, a Makarov type pistol imported from Bulgaria utilizing Cyrillic letters or non-Arabic numbers is not marked in accordance with 27 CFR 178.92 and 27 CFR 179.102.

Held further, an imported firearm with any part of the required marking partially or wholly obstructed from plain view is not marked in accordance with section 27 CFR 178.92 and 27 CFR 179.102.

Date signed: November 5, 2002

26 USC 5844, 18 USC 922(o), 22 USC 2778

IMPORTATION OF BROWNING M1919 TYPE RECEIVERS FOR UNRESTRICTED COMMERCIAL SALE

An ATF-approved method of destruction for the Browning M1919 type machinegun will result in the severed portions of the receiver being importable for unrestricted commercial sale.

ATF Rul. 2003–1

The Bureau of Alcohol, Tobacco and Firearms (ATF) has received inquiries about modifications necessary to the receiver of a Browning M1919 type machinegun to make it importable under 26 U.S.C. 5844 and 18 U.S.C. 922(o) for unrestricted commercial sale.

The Browning M1919 is a machinegun as defined in 26 U.S.C. 5845(b). The receiver of a Browning M1919 is also a machinegun as defined. Various manufacturers made Browning M1919 style machineguns in caliber .30-06 and 7.62x51mm (.308). The M1919 is a recoil- operated, belt-fed machinegun designed to be fired from a mount.

Section 5844 of Title 26, United States Code, makes it unlawful to import any firearm into the United States, unless the firearm to be imported or brought in is: (1) being imported for use by the United States or any department, independent establishment, or agency thereof or any State or possession or any political subdivision thereof; or (2) the firearm is being imported for scientific or research purposes; or (3) it is being imported solely for testing or use as a model by a registered manufacturer or solely for use as a sample by a registered importer or dealer. Additionally, the Secretary may permit the conditional importation of a firearm for examination and testing in connection with classifying the firearm.

Section 922(o) of Title 18, United States Code, makes it unlawful for any person to transfer or possess a machinegun, except a transfer to or by the United States or any department or agency thereof or a State or a department, agency, or political subdivision thereof; or any lawful transfer or lawful possession of a machinegun lawfully possessed before May 19, 1986.

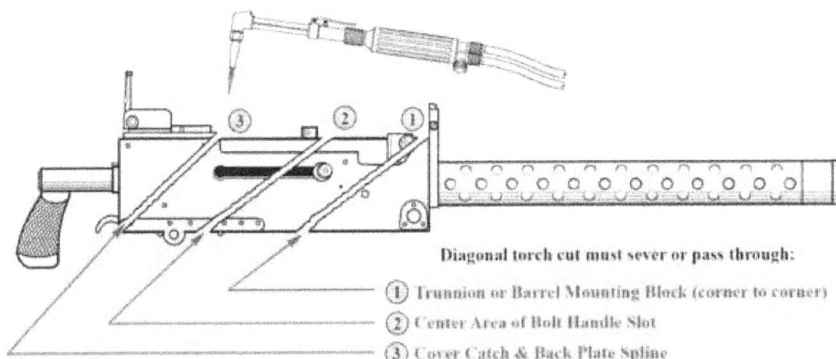

Browning M1919 Type Firearm

Diagonal torch cut must sever or pass through:
① Trunnion or Barrel Mounting Block (corner to corner)
② Center Area of Bolt Handle Slot
③ Cover Catch & Back Plate Spline

A review of the statutes above indicates that machineguns and machinegun receivers cannot be lawfully imported for unrestricted commercial sale. Accordingly, machinegun receivers may be imported for commercial sale only if they are destroyed in a manner that will prevent their function and future use as a firearm. The resulting severed receiver portions would not be subject to the provisions of 26 U.S.C. 5844 or 18 U.S.C. 922(o); however, these articles would be subject to the provisions of the Arms Export Control Act, 18 U.S.C. 925, 22 U.S.C. 2778, and implementing regulation at 27 CFR Part 47. It is important to note that these machinegun receivers must be destroyed and cannot be imported whether they are serviceable or unserviceable.

An ATF-approved method of destruction for a Browning M1919 type machinegun receiver requires three diagonal torch cuts that sever or pass through the following areas: (1) the trunnion or barrel mounting block (corner to corner), (2) the center area of the bolt handle slot, and (3) the cover catch and back plate spline. All cutting must be done with a cutting torch having a tip of sufficient size to displace at least ¼ inch of material at each location. Each cut must completely sever the receiver in the designated areas and must be done with a diagonal torch cut. Using a band-saw or a cut-off wheel to destroy the receiver does not ensure destruction of the weapon.

This method of destruction is illustrated in the diagram below.

Alternative methods of destruction may also be acceptable. These alternative methods must be equivalent in degree to the approved method of destruction. Receivers that are not sufficiently modified cannot be approved for importation. To ensure compliance with the law, it is recommended that the importer submit in writing the alternative method of destruction to the ATF Firearms Technology Branch (FTB) for review and approval prior to importation.

Held, an ATF-approved method of destruction for a Browning M1919 type machinegun receiver will result in the severed portions of the receiver being importable for unrestricted commercial sale. The severed articles would not be subject to the provisions of 26 U.S.C. 5844 or 18 U.S.C. 922(o), but would continue to be subject to the provisions of the Arms Export Control Act,

22 U.S.C. 2778. Alternative methods of destruction may also be acceptable. It is recommended that such methods be reviewed and approved by the ATF Firearms Technology Branch prior to the weapon's importation.

Date signed: January 24, 2003

26 USC 5844, 18 USC 922(o), 22 USC 2778

IMPORTATION OF FN FAL TYPE RECEIVERS FOR UNRESTRICTED COMMERCIAL SALE

An ATF-approved method of destruction for the FN FAL type machinegun will result in the severed portions of the receiver being importable for unrestricted commercial sale.

ATF Rul. 2003–2

The Bureau of Alcohol, Tobacco and Firearms (ATF) has received inquiries about modifications necessary to the receiver of an FN FAL type machinegun to make it importable under 26 U.S.C. 5844 and 18 U.S.C. 922(o) for unrestricted commercial sale.

The FN FAL is a machinegun as defined in 26 U.S.C. 5845(b). The receiver of an FAL is also a machinegun as defined. Various manufacturers made FAL style machineguns in caliber 7.62x51mm (.308). The FAL is a gas-operated, shoulder-fired, magazine-fed, selective-fire machinegun.

Section 5844 of Title 26, United States Code, makes it unlawful to import any firearm into the United States, unless the firearm to be imported or brought in is: (1) being imported for use by the United States or any department, independent establishment, or agency thereof or any State or possession or

any political subdivision thereof; or (2) the firearm is being imported for scientific or research purposes; or (3) it is being imported solely for testing or use as a model by a registered manufacturer or solely for use as a sample by a registered importer or dealer. Additionally, the Secretary may permit the conditional importation of a firearm for examination and testing in connection with classifying the firearm.

Section 922(o) of Title 18, United States Code, makes it unlawful for any person to transfer or possess a machinegun, except a transfer to or by the United States or any department or agency thereof or a State or a department, agency, or political subdivision thereof; or any lawful transfer or lawful possession of a machinegun lawfully possessed before May 19, 1986.

A review of the statutes above indicates that machineguns and machinegun receivers cannot be lawfully imported for unrestricted commercial sale. Accordingly, machinegun receivers may be imported for commercial sale only if they are destroyed in a manner that will prevent their function and future use as a firearm. The resulting severed receiver portions would not be subject to the provisions of 26 U.S.C. 5844 or 18 U.S.C. 922(o); however, these articles would be subject to the provisions of the Arms Export Control Act, 18 U.S.C.925, 22 U.S.C. 2778, and implementing regulation at 27 CFR Part 47. It is important to note that these machinegun receivers must be destroyed and cannot be imported whether they are serviceable or unserviceable.

An ATF-approved method of destruction for an FN FAL type machinegun receiver requires three diagonal torch cuts that sever or pass through the following

FN FAL Type Firearm

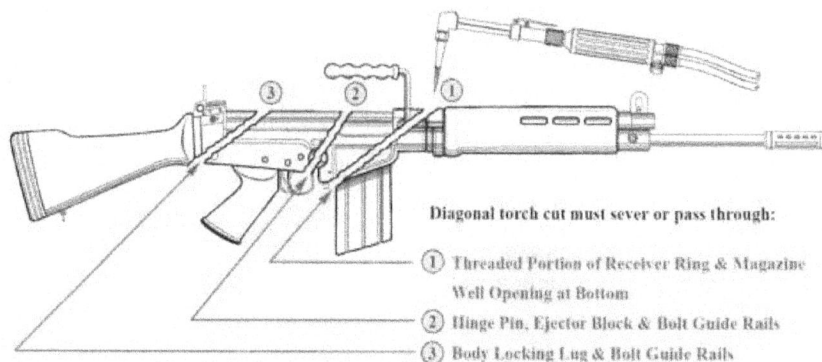

Diagonal torch cut must sever or pass through:

① Threaded Portion of Receiver Ring & Magazine Well Opening at Bottom

② Hinge Pin, Ejector Block & Bolt Guide Rails

③ Body Locking Lug & Bolt Guide Rails

areas: (1) the threaded portion of the receiver ring and magazine well opening at bottom, (2) the hinge pin, ejector block and bolt guide rails, and (3) the body locking lug and bolt guide rails. All cutting must be done with a cutting torch having a tip of sufficient size to displace at least ¼ inch of material at each location. Each cut must completely sever the receiver in the designated areas and must be done with a diagonal torch cut. Using a band- saw or a cut-off wheel to destroy the receiver does not ensure destruction of the weapon.

This method of destruction is illustrated in the diagram below.

Alternative methods of destruction may also be acceptable. These alternative methods must be equivalent in degree to the approved method of destruction. Receivers that are not sufficiently modified cannot be approved for importation. To ensure compliance with the law, it is recommended that the importer submit in writing the alternative method of destruction to the ATF Firearms Technology Branch (FTB) for review and approval prior to importation.

Held, an ATF-approved method of destruction for an FN FAL type machinegun receiver will result in the severed portions of the receiver being importable for unrestricted commercial sale. The severed articles would not be subject to the provisions of 26

U.S.C. 5844 or 18 U.S.C. 922(o), but would continue to be subject to the provisions of the Arms Export Control Act, 22 U.S.C. 2778. Alternative methods of destruction may also be acceptable. It is recommended that such methods be reviewed and approved by the ATF Firearms Technology Branch prior to the weapon's importation.

Date signed: January 24, 2003.

26 USC 5844, 18 USC 922(o), 22 USC 2778

IMPORTATION OF HECKLER & KOCH G3 TYPE RECEIVERS FOR UNRESTRICTED COMMERCIAL SALE

An ATF-approved method of destruction for the Heckler & Koch G3 type machinegun will result in the severed portions of the receiver being importable for unrestricted commercial sale.

ATF Rul. 2003–3

The Bureau of Alcohol, Tobacco and Firearms (ATF) has received inquiries about modifications necessary to the receiver of a Heckler and Koch G3 type machinegun to make it importable under 26 U.S.C. 5844 and 18 U.S.C. 922(o) for unrestricted commercial sale.

The G3 is a machinegun as defined in 26 U.S.C. 5845(b). The receiver of a G3 is also a machinegun as defined. Various manufacturers made G3 style machineguns in caliber

7.62x51mm (.308). The G3 is a delayed blowback, shoulder-fired, magazine-fed, selective-fire machinegun.

Section 5844 of Title 26, United States Code, makes it unlawful to import any firearm into the United States, unless the firearm to be imported or brought in is: (1) being imported for use by the United States or any department, independent establishment, or agency thereof or any State or possession or any political subdivision thereof; or (2) the firearm is being imported for scientific or research purposes; or (3) it is being imported solely for testing or use as a model by a registered manufacturer or solely for use as a sample by a registered importer or dealer. Additionally, the Secretary may permit the conditional importation of a firearm for examination and testing in connection with classifying the firearm.

Section 922(o) of Title 18, United States Code, makes it unlawful for any person to transfer or possess a machinegun, except a transfer to or by the United States or any department or agency thereof or a State or a department, agency, or political subdivision thereof; or any lawful transfer or lawful possession of a machinegun lawfully possessed before May 19, 1986.

A review of the statutes above indicates that machineguns and machinegun receivers cannot be lawfully imported for unrestricted commercial sale. Accordingly, machinegun receivers may be imported for commercial sale only if they are destroyed in a manner that will prevent their function and future use as a firearm. The resulting severed receiver portions would not be subject to the provisions of 26 U.S.C. 5844 or 18 U.S.C. 922(o); however, these articles would be subject to the provisions of the Arms Export Control Act, 18 U.S.C. 925, 22 U.S.C. 2778, and implementing regulation at 27 CFR Part 47. It is important to note that these machinegun receivers must be destroyed and cannot be imported whether they are serviceable or unserviceable.

An ATF-approved method of destruction for a Heckler and Koch G3 type machinegun receiver requires four diagonal torch cuts that sever or pass through the following areas: (1) the chamber area, (2) the grip assembly locking pin hole, (3) the ejection port, and (4) the buttstock locking pin hole. All cutting must be done with a cutting torch having a tip of sufficient size to displace at least ¼ inch of material at each location. Each cut must completely sever the receiver in the designated areas and must be done with a diagonal torch cut. Using a bandsaw or a cut-off wheel to destroy the receiver does not ensure destruction of the weapon.

This method of destruction is illustrated in the diagram below.

Alternative methods of destruction may also be acceptable. These alternative methods must be equivalent in degree to the approved method of destruction. Receivers that are not sufficiently modified cannot be approved for importation. To ensure compliance with the law, it is recommended that the importer submit in writing the alternative method

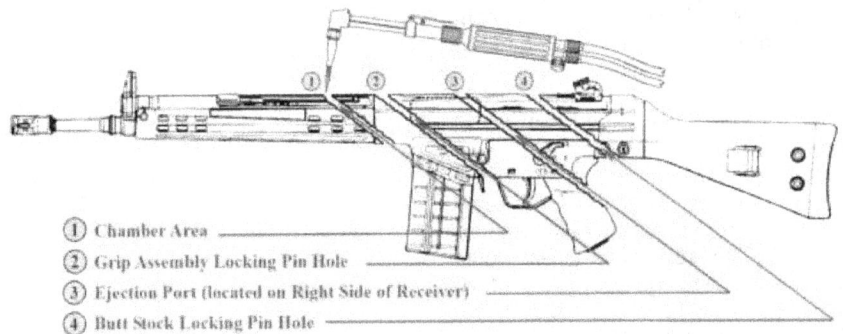

Heckler & Koch G3 Type Firearm

① Chamber Area
② Grip Assembly Locking Pin Hole
③ Ejection Port (located on Right Side of Receiver)
④ Butt Stock Locking Pin Hole

of destruction to the ATF Firearms Technology Branch (FTB) for review and approval prior to importation.

Held, an ATF-approved method of destruction for a Heckler and Koch G3 type machinegun receiver will result in the severed portions of the receiver being importable for unrestricted commercial sale. The severed articles would not be subject to the provisions of 26 U.S.C. 5844 or 18 U.S.C. 922(o), but would continue to be subject to the provisions of the Arms Export Control Act, 22 U.S.C. 2778. Alternative methods of destruction may also be acceptable. It is recommended that such methods be reviewed and approved by the ATF Firearms Technology Branch prior to the weapon's importation.

Date signed: January 24, 2003.

26 USC 5844, 18 USC 922(o), 22 USC 2778

IMPORTATION OF STEN TYPE RECEIVERS FOR UNRESTRICTED COMMERCIAL SALE

An ATF-approved method of destruction for the Sten type machinegun will result in the severed portions of the receiver being importable for unrestricted commercial sale.

ATF Rul. 2003–4

The Bureau of Alcohol, Tobacco and Firearms (ATF) has received inquiries about modifications necessary to the receiver of a Sten type machinegun to make it importable under 26 U.S.C. 5844 and 18 U.S.C. 922(o) for unrestricted commercial sale.

The Sten is a machinegun as defined in 26 U.S.C. 5845(b). The receiver of a Sten is also a machinegun as defined. Various manufacturers made Sten style machineguns in caliber 9x19mm (9mm Luger). The Sten is a blowback-operated, shoulder-fired, magazine-fed, selective-fire submachinegun.

Section 5844 of Title 26, United States Code, makes it unlawful to import any firearm into the United States, unless the firearm to be imported or brought in is: (1) being imported for use by the United States or any department, independent establishment, or agency thereof or any State or possession or any political subdivision thereof; or (2) the firearm is being imported for scientific or research purposes; or (3) it is being imported solely for testing or use as a model by a registered manufacturer or solely for use as a sample by a registered importer or dealer. Additionally, the Secretary may permit the conditional importation of a firearm for examination and testing in connection with classifying the firearm.

Section 922(o) of Title 18, United States Code, makes it unlawful for any person to transfer or possess a machinegun, except a transfer to or by the United States or any department or agency thereof or a State or a department, agency, or political subdivision thereof; or any lawful transfer or lawful possession of a machinegun lawfully possessed before May 19, 1986.

A review of the statutes above indicates that machineguns and machinegun receivers cannot be lawfully imported for unrestricted commercial sale. Accordingly, machinegun receivers may be imported for commercial sale only if they are destroyed in a manner that will prevent their function and future use as a firearm. The resulting severed receiver portions would not be subject to the provisions of 26 U.S.C. 5844 or 18 U.S.C. 922(o); however, these articles would be subject to the provisions of the Arms Export Control Act, 18 U.S.C. 925, 22 U.S.C. 2778, and implementing regulation at 27 CFR Part 47. It is important to note that these machinegun receivers must be destroyed and cannot be imported whether they are serviceable or unserviceable.

An ATF-approved method of destruction for a Sten type machinegun receiver requires three diagonal torch cuts that sever or pass through the following areas: (1) the threaded portion of the receiver/chamber area, (2) the return spring cap socket, and (3) the sear slot in the lower side of the receiver. All cutting must be done with a cutting torch having a tip of sufficient size to displace at least ¼ inch of material at each location. Each cut must completely sever the receiver in the designated areas and must be done with a diagonal torch cut. Using a bandsaw or a cut-off wheel to destroy the receiver does not ensure destruction of the weapon.

This method of destruction is illustrated in the diagram below.

Alternative methods of destruction may also be acceptable. These alternative methods must be equivalent in degree to the approved method of destruction. Receivers that are not sufficiently modified cannot be approved for importation. To ensure compliance with the law, it is recommended that the importer submit in writing the alternative method of destruction to the ATF Firearms Technology Branch (FTB) for review and approval prior to importation.

Held, an ATF-approved method of destruction for a Sten type machinegun receiver will result in the severed portions of the receiver being importable for unrestricted commercial sale. The severed articles would not be subject to the provisions of 26 U.S.C. 5844 or 18 U.S.C. 922(o), but would continue to be subject to the provisions of the Arms Export Control Act, 22 U.S.C. 2778. Alternative methods of destruction may also be acceptable. It is recommended that such methods be reviewed and approved by the ATF Firearms Technology Branch prior to the weapon's importation.

Date signed: January 24, 2003.

18 U.S.C. 925(d): EXCEPTIONS

22 U.S.C. 2778: IMPORTATION

26 U.S.C. 5844: IMPORTATION

27 CFR 478.111, 478.112, 478.113: IMPORTATION

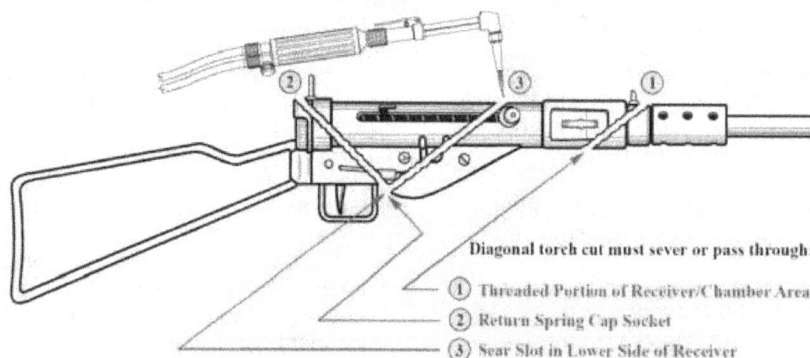

Sten Type Firearm

Diagonal torch cut must sever or pass through:
① Threaded Portion of Receiver/Chamber Area
② Return Spring Cap Socket
③ Sear Slot in Lower Side of Receiver

27 CFR 479.111, 479.112, 479.113: IMPORTATION

27 CFR 447.42: APPLICATION FOR PERMIT

Persons with a valid Federal Firearms license and/or registered as an importer of articles enumerated on the U.S. Munitions Import List seeking to import firearms, ammunition and implements of war may submit the ATF Form 6, Application and Permit for Importation of Firearms, Ammunition and Implements of War, electronically using the eForm 6 online electronic filing system, provided such persons have met certain registration requirements.

ATF Rul. 2003-6

This ruling is superseded by ATF Rul. 2013-1

18 U.S.C. 922(b)(3): TRANSFER OF FIREARMS

27 CFR 478.11, 478.124(c)(3)(ii): DEFINITIONS, TRANSFER OF FIREARMS

A Federal firearms licensee (FFL) may not lawfully transfer a firearm to a nonimmigrant alien who has not resided in a State continuously for at least 90 days immediately prior to the FFL conducting a National Instant Criminal Background Check System (NICS) check.

ATF Rul. 2004-1

This ruling is superseded by Interim Final Rule **"Residency Requirements for Aliens Acquiring Firearms (2011R–23P)"**, effective July 9, 2012 (77 FR 33630).

26 U.S.C. 5812, 5841, 5844, 5861, 5872

27 CFR 479.11, 479.26, 479.105,

479.111, 479.112, 479.114 – 479.119: IMPORTATION OF FIREARMS SUBJECT TO THE NATIONAL FIREARMS ACT

18 U.S.C. 921(a)(3), 922(l), 922(o), 923(e), 924(d), 925(d)(3)

27 CFR 478.11, 478.22, 478.111 – 478.113: IMPORTATION OF MACHINEGUNS, DESTRUCTIVE DEVICES, SHORT-BARREL SHOTGUNS,

SHORT-BARREL RIFLES, FIREARMS SILENCERS, AND OTHER FIREARMS SUBJECT TO THE NATIONAL FIREARMS ACT

22 U.S.C. 2778

27 CFR 447.11, 447.21: TEMPORARY IMPORTATION OF DEFENSE ARTICLES

The Bureau of Alcohol, Tobacco, Firearms and Explosives has approved an alternate method or procedure for importers to use when temporarily importing firearms subject to the National Firearms Act, the Gun Control Act and the Arms Export Control Act for inspection, testing, calibration, repair, or incorporation into another defense article.

ATF Rul. 2004-2

The Bureau of Alcohol, Tobacco, Firearms and Explosives (ATF) has received numerous inquiries from importers who wish to temporarily import firearms subject to the Gun Control Act of 1968 (GCA), 18 U.S.C. Chapter 44, and the National Firearms Act (NFA), 26 U.S.C. Chapter 53, for inspection, testing, calibration, repair, or incorporation into another defense article. Importers advise ATF that they generally obtain a temporary import license, DSP-61, from the Department of State authorizing the importation or comply with one of the regulatory exemptions from licensing in 22 CFR 123.4. They ask whether such a license or exemption is sufficient to satisfy the requirements of the GCA and NFA.

Statutory Background

1. The National Firearms Act

The NFA imposes restrictions on certain firearms, including registration requirements, transfer approval requirements, and import restrictions. 26 U.S.C. 5812, 5841, 5844. The term **"firearm"** is defined in 26 U.S.C. 5845(a) to include machineguns, short-barrel shotguns, short-barrel rifles, silencers, destructive devices, and **"any other weapons."** Section 5844 of the NFA provides that no firearm may be imported into the United States unless the importer establishes that the firearm to be imported is—

(1) Being imported or brought in for the use of the United States or any department, independent establishment, or agency thereof or any State or possession or any political

subdivision thereof; or

(2) Being imported or brought in for scientific or research purposes; or

(3) Being imported or brought in solely for testing or use as a model by a registered manufacturer or solely for use as a sample by a registered importer or registered dealer.

Regulations implementing the NFA in 27 CFR Part 479 require importers to obtain an ATF Form 6, Application and Permit for Importation of Firearms, Ammunition and Implements of War, prior to importing NFA firearms into the United States. 27 CFR 479.111. In addition, the regulations require importers to register the firearms they import by filing with the Director an accurate notice on Form 2, Notice of Firearms Manufactured or Imported, executed under the penalties of perjury, showing the importation of a firearm. 27 CFR 479.112. When an NFA firearm is to be exported from the United States, the exporter must file with the Director an application on Form 9, Application and Permit for Exportation of Firearms, to obtain authorization to export the firearm. 27 CFR 479.114-119.

Regulations in 27 CFR Part 479 indicate that NFA firearms may be imported for scientific or research purposes or for testing or use as a model by a registered manufacturer or as a sample by a registered importer or registered dealer. 27 CFR 479.111(a). However, section 479.105(c), implementing section 922(o) of the GCA, clarifies that machineguns manufactured on or after May 19, 1986, may be imported only with a purchase order for transfer to a governmental entity, or as a dealer's sales sample pursuant to section 479.105(d).

The regulations in Part 479 give the Director the authority to approve an alternate method or procedure in lieu of a method or procedure specifically prescribed in the regulations when it is found that:

(1) Good cause is shown for the use of the alternate method or procedure;

(2) The alternate method or procedure is within the purpose of, and consistent with the effect intended by, the specifically prescribed method or procedure and that the alternate method or procedure is substantially equivalent to that specifically prescribed method or procedure; and

(3) The alternate method or procedure will not be contrary to any provision of law and will not result in an increase in cost to the Government or hinder the effective administration of the GCA or regulations issued thereunder.

27 CFR 479.26.

2. *The Gun Control Act*

Import provisions of the GCA, 18 U.S.C. 922(l) and 925(d)(3), generally prohibit the importation of firearms subject to the NFA, except for the use of governmental entities. 18 U.S.C. 925(a)(1). The term **"firearm"** is defined in section 921(a)(3) to include any weapon which will or is designed to or may be readily converted to expel a projectile by the action of an explosive; the frame or receiver of such weapon; any firearm silencer; and any destructive device. In addition, section 922(o) of the GCA prohibits the transfer or possession of a machinegun manufactured on or after May 19, 1986, except for the official use of governmental entities.

Regulations implementing the GCA in 27 CFR Part 478 require that persons importing firearms into the United States obtain an approved ATF Form 6, Application and Permit for Importation of Firearms, Ammunition and Implements of War, prior to bringing the firearms into the United States. 27 CFR 478.111-114. Regulations in Part 478 provide that the Director may approve an alternate method or procedure in lieu of a method or procedure specifically prescribed by the GCA and regulations when it is found that:

(1) Good cause is shown for the use of the alternate method or procedure;

(2) The alternate method or procedure is within the purpose of, and consistent with the effect intended by, the specifically prescribed method or procedure and that the alternate method or procedure is substantially equivalent to that specifically prescribed method or procedure; and

(3) The alternate method or procedure will not be contrary to any provision of law and will not result in an increase in cost to the Government or hinder the effective administration of the GCA or regulations issued thereunder.

27 CFR 478.22.

3. *The Arms Export Control Act*

The Arms Export Control Act (AECA), 22 U.S.C. 2778, gives the President the authority to control the export and import of defense articles and defense services in furtherance of world peace and the security and foreign policy of the United States. Authority to administer the permanent import provisions of the AECA was delegated to the Attorney General, while the authority to administer the export and temporary import provisions of the AECA was delegated to the Secretary of State. Executive Order 11958 of January 18, 1977, as amended by Executive Order 13333 of January 23, 2003, 3 CFR Executive Order 13284.

The term **"defense article"** is defined in 27 CFR 447.11 as any item designated in sections 447.21 or 447.22. Section 447.21, the U.S. Munitions Import List, includes a number of defense articles that are also subject to the GCA and NFA. Category I, **"Firearms,"** includes non-automatic and semiautomatic firearms to caliber .50 inclusive, combat shotguns, shotguns with barrels less than 18 inches in length, and firearms silencers and suppressors. All Category I firearms are subject to the GCA. **"Combat shotguns"** include the USAS-12 shotgun and the Striker-12/Streetsweeper shotgun, which have been classified as destructive devices under the GCA and NFA. In addition, all shotguns with barrels of less than 18 inches in length are subject to both the GCA and NFA. All rifles with barrels of less than 16 inches in length are subject to both the GCA and NFA, and silencers are subject to both the GCA and NFA.

Category II, **"Artillery Projectors,"** includes guns over caliber .50, howitzers, mortars, and recoilless rifles. Firearms over .50 caliber have a bore of more than one-half inch in diameter and are **"destructive devices"** as defined in the GCA and NFA.

Category IV, **"Launch Vehicles, Guided Missiles, Ballistic Missiles, Rockets, Torpedoes, Bombs and Mines,"** includes rockets, bombs, grenades, torpedoes, and land and naval mines. All these articles are **"destructive devices"** as defined in the GCA and NFA.

Regulations of the Department of State implementing the AECA generally require a temporary import license, DSP-61, for the temporary import and subsequent export of unclassified defense articles, unless otherwise exempted. 22 CFR 123.3.

Regulations in 22 CFR 123.4 provide an exemption from licensing if the item temporarily imported:

(1) Is serviced (e.g., inspection, testing, calibration or repair, including overhaul, reconditioning and one-to-one replacement of any defective items, parts or components, but excluding any modification, enhancement, upgrade or other form of alteration or improvement that changes the basic performance of the item), and is subsequently returned to the country from which it was imported. Shipment may be made by the U.S. importer or a foreign government representative of the country from which the goods were imported; or

(2) Is to be enhanced, upgraded or incorporated into another item which has already been authorized by the Office of Defense Trade Controls for permanent export; or

(3) Is imported for the purpose of exhibition, demonstration or marketing in the United States and is subsequently returned to the country from which it was imported; or

(4) Has been rejected for permanent import by the Department of the Treasury [after January 24, 2003, the Department of Justice] and is being returned to the country from which it was shipped; or

(5) Is approved for such import under the U.S. Foreign Military Sales (FMS) program pursuant to an executed U.S. Department of Defense Letter of Offer and Acceptance (LOA).

Willful violations of the AECA are punishable by imprisonment for not more than 10 years, a fine of not more than $1,000,000, or both. 22 U.S.C. 2778(c). Articles imported in violation of the AECA are also subject to seizure and forfeiture. 18 U.S.C. 545.

Discussion

A temporary import license authorizing the temporary importation and subsequent export of a defense article by the Department of State satisfies all legal requirements under the AECA. Importers may also comply with AECA requirements if the importation meets one of the exemptions in 22 CFR 123.4. However, if the defense article is subject to the GCA and NFA, the importer must also comply with the requirements of those statutes. Neither the GCA nor NFA make a distinction between temporary importation and permanent importation, as is the case under the AECA. Regulations implementing the GCA and NFA make it clear that an **"importation"** occurs when firearms are brought within the territory of the United States. 27 CFR 478.11 and 479.11. Accordingly, any bringing of firearms into the territory of the United States is subject to the import provisions of the GCA and NFA. Issuance of a temporary import license by the Department of State, or exemption from licensing under regulations in 22 CFR Part 123, will not excuse compliance with the GCA and NFA.

The statutes and regulations outlined above do not address the importation of machineguns manufactured after May 19, 1986, for scientific or research purposes or for testing, repair, or use as a model by a manufacturer or importer. Nor do the regulations address the importation of post-86 machineguns for repair, inspection, calibration, or incorporation into another defense article.

For other **"defense articles"** that are subject to the requirements of the GCA and NFA, such as silencers, destructive devices, and short-barrel weapons, ATF has the authority to approve the importation of such firearms for scientific or research purposes or for testing or use as a model or sample by a registered importer or registered dealer. However, such importations must comply with all applicable provisions of the NFA, including filing of a Form 2, Notice of Firearms Manufactured or Imported, to effect registration. If such articles are subsequently exported, a Form 9, Application and Permit for Permanent Exportation of Firearms, must also be approved prior to exportation.

As with post-86 machineguns, neither the law nor regulations specifically address the importation of firearms subject to the NFA for purposes of repair, inspection, calibration, or for incorporation into another defense article.

ATF recognizes that inspection, repair, calibration, incorporation into another defense article, and reconditioning of machineguns, destructive devices, and other NFA firearms is often necessary for National defense. These defense articles are frequently sold to allies of the United States for their legitimate defense needs. Accordingly, ATF believes it is appropriate to recognize an alternate method

that allows importers to temporarily import these firearms, subject to requirements to ensure the security of these articles while they are in the United States and accountability of the persons who import them.

Pursuant to 27 CFR 478.22 and 479.26, ATF hereby authorizes an alternate method or procedure for importers of defense articles to use for temporary importation of such articles for inspection, calibration, repair, or incorporation into another defense article when such articles are subject to the requirements of the NFA and GCA. The procedure requires that importers

(1) Be qualified under the GCA and NFA to import the type of firearms sought for importation;

(2) Obtain a temporary import license, DSP-61, from the Department of State in accordance with 22 CFR 123.3 OR qualify for a temporary import license exemption pursuant to 22 CFR 123.4;

(3) Within 15 days of the release of the firearms from Customs custody, file an ATF Form 2, Notice of Firearms Manufactured or Imported, showing the importation of the firearms. The DSP-61 must be attached to the Form 2. If the importation is subject to a licensing exemption under 22 CFR 123.4, the importer must submit with the ATF Form 2 a statement, under penalty of perjury, attesting to the exemption and stating that the article will be exported within four years of its importation into the United States;

(4) Maintain the defense articles in a secure place and manner to ensure that the articles are not diverted to criminal or terrorist use; and

(5) Export the articles within 4 years of importation into the United States.

Importers who follow the procedures outlined above will be in compliance with all the provisions of the GCA, NFA, and AECA administered and enforced by ATF. All other provisions of the law must be followed.

ATF finds that the procedure outlined above meets the legal requirements for an alternate method or procedure because there is good cause to authorize the importation of defense articles for repair, inspection, calibration, or incor-

poration into another defense article. Because such defense articles are often provided to allies of the United States, it is imperative that the original manufacturers have a lawful method of importing such articles for repair and routine maintenance. The alternate method or procedure is consistent with the effect intended by the procedure set forth in the GCA and NFA, because the firearms must be registered and stored securely. Finally, the alternate method is consistent with the requirements of the GCA and NFA and will not result in any additional costs to ATF or the Department of State.

"Transfers" of NFA Weapons After Importation

ATF recognizes that temporarily imported NFA firearms are sometimes **"transferred"** from the importer to a contractor within the United States for inspection, testing, calibration, repair, or incorporation into another defense article. ATF has approved a procedure for authorizing the transportation or delivery of temporarily imported NFA firearms to licensed contractors for repair or manipulation, as noted above.

Conveyance of an NFA weapon to a licensee for purposes of inspection, testing, calibration, repair, or incorporation into another defense article is generally not considered to be a **"transfer"** under 26 U.S.C. 5845(j). ATF has taken the position that temporary custody by a licensee is not a transfer for purposes of the NFA since no sale, lease, or other disposal is intended by the owner. However, in order to document the transaction as a temporary conveyance and make clear that an actual **"transfer"** of a firearm has not taken place, ATF strongly recommends that the importer submit a Form 5, Application for Tax Exempt Transfer and Registration of Firearm, for approval prior to conveying a firearm for repair or manipulation. In the alternative, the importer should convey the weapon with a letter to the contractor, stating: (1) the weapon is being temporarily conveyed for inspection, testing, calibration, repair, or incorporation into another defense article; and (2) the approximate time period the weapon is to be in the contractor's possession. The transferee must be properly licensed to engage in an NFA firearms business.

Held, pursuant to 27 CFR 478.22 and 479.26, the Bureau of Alcohol, Tobacco, Firearms and Explosives has approved an alternate method or procedure for importers to use when temporarily importing firearms subject to the Gun Control Act, National Firearms Act, and the Arms Export Control Act for inspection, testing, calibration, repair, or incorporation into another defense article. This procedure applies to all defense articles that are also subject to the NFA and GCA. The procedure requires that importers

(1) Be qualified under the GCA and NFA to import the type of firearms sought for importation;

(2) Obtain a temporary import license, DSP-61, from the Department of State in accordance with 22 CFR 123.3 or qualify for a temporary import license exemption pursuant to 22 CFR 123.4;

(3) Within 15 days of the release of the firearms from Customs custody, file an ATF Form 2, Notice of Firearms Manufactured or Imported, showing the importation of the firearms. The DSP-61 must be attached to the Form 2. If the importation is subject to a licensing exemption under 22 CFR 123.4, the importer must submit with the ATF Form 2 a statement, under penalty of perjury, attesting to the exemption and stating that the article will be exported within four years of its importation into the United States;

(4) Maintain the defense articles in a secure place and manner to ensure that the articles are not diverted to criminal or terrorist use; and

(5) Export the articles within 4 years of importation into the United States.

Held further, temporary conveyance of NFA weapons from the importer to a contractor within the United States for purposes of inspection, testing, calibration, repair, or incorporation into another defense article may be accomplished through advance approval of ATF Form 5, Application for Tax Exempt Transfer and Registration of Firearm, or with a letter from the importer to the contractor stating: (1) the weapon is being temporarily conveyed for inspection, testing, calibration, repair, or incorporation into another defense article; and (2)the approximate time period the weapon is to be in the contractor's possession. The transferee must be properly licensed to engage in an NFA firearms business.

Date signed: April 7, 2004

26 U.S.C. 5845(b): DEFINITIONS (MACHINEGUN)

27 CFR 179.11: MEANING OF TERMS

The 7.62mm Aircraft Machine Gun, identified in the U.S. military inventory as the "M-134" (Army), "GAU-2B/A" (Air Force), and "GAU-17/A" (Navy), is a machinegun as defined by 26 U.S.C. 5845(b). Rev. Rul. 55-528 modified.

ATF Rul. 2004-5

The Bureau of Alcohol, Tobacco, Firearms and Explosives (ATF) has examined the 7.62mm Aircraft Machine Gun, commonly referred to as a **"Minigun."** The Minigun is a 36-pound, six-barrel, electrically powered machinegun. It is in the U.S. military inventory and identified as the **"M-134"** (Army), **"GAU-2B/A"** (Air Force), and **"GAU-17/A"** (Navy). It is a lightweight and extremely reliable weapon, capable of discharging up to 6,000 rounds per minute. It has been used on helicopters, fixed-wing aircraft, and wheeled vehicles. It is highly adaptable, being used with pintle mounts, turrets, pods, and internal installations.

The Minigun has six barrels and bolts which are mounted on a rotor. The firing sequence begins with the manual operation of a trigger. On an aircraft, the trigger is commonly found on the control column, or joystick. Operation of the trigger causes an electric motor to turn the rotor. As the rotor turns, a stud on each bolt travels along an elliptical groove on the inside of the housing, which causes the bolts to move forward and rearward on tracks on the rotor. A triggering cam, or sear shoulder, trips the firing pin when the bolt has traveled forward through the full length of the bolt track. One complete revolution of the rotor discharges cartridges in all six barrels. The housing that surrounds the rotor, bolts and firing mechanism constitutes the frame or receiver of the firearm.

The National Firearms Act defines **"machinegun"** as **"any weapon which shoots, is designed to shoot, or can be readily restored to shoot, automatically more than one shot, without manual reloading, by a single function of the trigger."** 26 U.S.C. 5845(b). The term also includes **"the frame or receiver of any such weapon, any part designed and intended solely and exclusively, or combination of parts designed and intended, for use in converting a weapon into a machinegun, and any combination of**

parts from which a machinegun can be assembled if such parts are in the possession or under the control of the person." Id.; see 18 U.S.C.b921(a)(23); 27 CFR 478.11, 479.11.

ATF and its predecessor agency, the Internal Revenue Service (IRS), have historically held that the original, crank-operated Gatling Gun, and replicas thereof, are not automatic firearms or machineguns as defined. See Rev. Rul. 55-528, 1955-2 C.B.

482. The original Gatling Gun is a rapid-firing, hand-operated weapon. The rate of fire is regulated by the rapidity of the hand-cranking movement, manually controlled by the operator. It is not a **"machinegun"** as that term is defined in 26 U.S.C. 5845(b) because it is not a weapon that fires automatically.

The Minigun is not a Gatling Gun. It was not produced under the 1862 - 1893 patents of the original Gatling Gun. While using a basic design concept of the Gatling Gun, the Minigun does not incorporate any of Gatling's original components and its feed mechanisms are entirely different. Critically, the Minigun shoots more than one shot, without manual reloading, by a single function of the trigger, as prescribed by 26 U.S.C. 5845(b). See United States v. Fleischli, 305 F.3d 643, 655-656 (7th Cir. 2002). See also Staples v. United States, 511 U.S. 600, 603 (1994) (automatic refers to a weapon that **"once its trigger is depressed, the weapon will automatically continue to fire until its trigger is released or the ammunition is exhausted"**); GEORGE C. NONTE, JR., FIREARMS ENCYCLOPEDIA 13 (Harper & Rowe 1973) the term **"automatic"** is defined to include **"any firearm in which a single pull and continuous pressure upon the trigger (or other firing device) will produce rapid discharge of successive shots so long as ammunition remains in the magazine or feed device - in other words, a machinegun"**; WEBSTER'S II NEW RIVERSIDE - UNIVERSITY DICTIONARY (1988) (defining automatically as **"acting or operating in a manner essentially independent of external influence or control"**); JOHN QUICK, PH.D., DICTIONARY OF WEAPONS AND MILITARY TERMS 40 McGraw-Hill 1973) (defining automatic fire as **"continuous fire from an automatic gun, lasting until pressure on the trigger is released"**).

The term **"trigger"** is generally held to be the part of a firearm that is used to initiate the firing sequence. See United States v. Fleischli, 305 F.3d at 655-56 (and cases cited therein); see also ASSOCIATION OF FIREARMS AND TOOLMARK EXAMINERS (AFTE) GLOSSARY 185 (1st ed. 1980) ("**that part of a firearm mechanism which is moved manually to cause the firearm to discharge**"); WEBSTER'S II NEW RIVERSIDE - UNIVERSITY DICTIONARY (1988) ("**lever pressed by the finger in discharging a firearm**").

Held, the 7.62mm Minigun is designed to shoot automatically more than one shot, without manual reloading, by a single function of the trigger. Consequently, the 7.62mm Minigun is a machinegun as defined in section 5845(b) of the National Firearms Act. See United States v. Fleischli, 305 F.3d at 655-56. Similarly, the housing that surrounds the rotor is the frame or receiver of the Minigun, and thus is also a machinegun. Id.; see 18 U.S.C. 921(a)(23); 27 CFR 478.11, 479.11.

To the extent this ruling is inconsistent with Revenue Ruling 55-528 issued by the IRS, Revenue Ruling 55-528, 1955-2 C.B. 482, is hereby modified.

Date signed: August 18, 2004

18 U.S.C. 921(a)(3): DEFINITIONS (FIREARM)

18 U.S.C. 921(a)(24): DEFINITIONS (FIREARM MUFFLER AND FIREARM SILENCER)

26 U.S.C. 5845(a)(7): DEFINITIONS (FIREARM)

26 U.S.C. 5845(i): DEFINITIONS (MAKE)

Certain integral devices intended to diminish the report of paintball guns are not "firearm silencers" or "firearm mufflers" under the Gun Control Act of 1968 or the National Firearms Act.

ATF Rul. 2005-4

The Bureau of Alcohol, Tobacco, Firearms and Explosives (ATF) has received requests from manufacturers of paintball guns for evaluation and classification of integral devices intended to diminish the report of a paintball gun. Specifically, the manufacturers have asked whether the device would be considered a **"silencer"** as defined in the Gun Control Act of 1968 (GCA), 18 U.S.C. Chapter 44, and the National Firearms Act (NFA), 26 U.S.C. Chapter 53.

The sample submitted is a paintball gun with a ported device attached to the barrel. The paintball gun uses compressed air to expel a projectile. The paintball gun has a barrel with a smooth bore 12 1/4 inches long and 1 inch in diameter. The barrel is permanently welded to the paintball gun. The section of the barrel that the device is attached to has an internal diameter of .68 inches and is ported with 20 openings. Ten of the openings are rectangular in shape and are approximately .430 inches wide and 1 inch in length. The other 10 openings are oval in shape and approximately .25 inches wide and 1 inch in length. Two end caps, each with a diameter of 1.5 inches, are permanently welded to the barrel at either end of the ported section of the barrel. A section of plastic tube approximately 8 inches in length and 2 inches in diameter covers the ported section of the barrel. It is attached by sliding over the paintball gun barrel and is held in place by two rubber **"O"** rings that are affixed to metal bushings. These metal bushings are permanently welded to the barrel, one toward the rear and one at the muzzle end of the barrel. The plastic tube, however, can be removed.

To determine whether the ported barrel and outer sleeve would function as a silencer, it was removed by cutting with a hack saw. A threaded adaptor with tape wrapped around it was utilized so the device could be attached to a .22 caliber Ruger Mark II pistol. With the device attached, a sound meter test was performed, with five shots being fired with the device attached and five shots fired without the device attached. The testing indicated that the attachment of the device resulted in a 7.98-decibel sound reduction.

The GCA defines the term **"firearm"** as:

(A) any weapon (including a starter gun) which will or is designed to or may readily be converted to expel a projectile by the action of an explosive; (B) the frame or receiver of any such weapon; (C) any firearm muffler or firearm silencer; or (D) any destructive device. Such term does not include an antique firearm.

18 U.S.C. 921(a)(3)(C).

The definition of **"firearm silencer"** and **"firearm muffler"** in 18 U.S.C. 921(a)(24) provides as follows:

The terms **"firearm silencer"** and **"firearm muffler"** mean any device for silencing, muffling, or diminishing the report of a portable firearm, including any combination of parts, designed or redesigned, and intended for use in assembling or fabricating a firearm silencer or firearm muffler, and any part intended only for use in such assembly or fabrication.

The NFA defines the term **"firearm"** to include any silencer as defined in section 921 of the GCA. 26 U.S.C. 5845(a)(7).

The term **"make"** is defined in the NFA to include manufacturing, putting together,

altering, any combination of these, or otherwise producing a firearm. 26 U.S.C. 5845(i).

The paintball gun examined by ATF is not a **"firearm"** as defined, because it does not, is not designed to, and may not be readily converted to expel a projectile by the action of an explosive and does not utilize the frame or receiver of a firearm. The sole issue presented is whether the ported barrel and outer sleeve are a firearm muffler or firearm silencer as defined in the GCA and NFA.

The design characteristics of the ported barrel and outer sleeve are similar to those of conventional commercial silencers. The barrel is ported to allow the escape of gases from a fired round and the outer sleeve dampens or muffles the sound when a round is fired. Moreover, the sound meter test indicates a reduction of 7.98 decibels when the ported barrel and sleeve were attached to a .22 caliber pistol, which is consistent with the sound reduction resulting from the use of commercial silencers.

Noteworthy, the definition of **"firearm silencer"** and **"firearm muffler"** requires that the device be one for diminishing the report of a portable firearm. The device under consideration is permanently attached to and an integral part of a paintball gun, which is not a firearm as defined in the GCA or NFA. The device cannot be removed from the paintball gun without destroying the barrel and rendering the paintball gun unusable. Under these circumstances, the integral device is not a firearm muffler or firearm silencer.

However, once the device is cut from the paintball gun, it can be used to diminish the report of a firearm. As stated previously, the design characteristics of the device are consistent with those of commercial silencers, and testing indicates that the device functions to reduce the report of the firearm. Moreover, removal of the device from the paintball gun indicates some intention to utilize the device for something other than reducing the report of the paintball gun. Because the device will no longer be permanently attached to an unregulated item, and because of its silencer design characteristics, removal will result in the making of a silencer under the NFA and GCA. This is consistent with the definition of **"make"** in the NFA, as removal of the device results in production of a silencer.

Making of a silencer by an unlicensed person for personal use does not require any license or notification to ATF under the GCA. However, under the NFA the making of a silencer must be approved in advance by ATF. Unlicensed persons must file an ATF Form 1, Application to Make a Firearm, pay a $200 making tax, and comply with all other provisions of the law prior to making the firearm. Approval of the Form 1 results in registration of the silencer to the maker. Subsequent transfers of the registered silencer must be approved in advance by ATF. Possession of a silencer not registered to the possessor, and a making or transfer that does not comply with the NFA and regulations are punishable by imprisonment for up to 10 years, a fine of $250,000, or both.

Held, a device for an unregulated paintball gun, having a permanently affixed, integral ported barrel and other components, that functions to reduce the report of the paintball gun is not a **"firearm silencer"** or **"firearm muffler"** as defined, as the device is not one for diminishing the report of a portable firearm.

Held further, removal of the permanently affixed ported barrel and other components from a paintball gun is a **"making"** of a silencer under the GCA and NFA that requires advance approval from ATF.

Date approved: October 12, 2005

18 U.S.C. 922(o): Transfer or possession of machinegun

26 U.S.C. 5845(b): Definition of machinegun

18 U.S.C. 921(a)(23): Definition of machinegun

The definition of machinegun in the National Firearms Act and the Gun Control Act includes a part or parts that are designed and intended for use in converting a weapon into a machinegun. This language includes a device that, when activated by a single pull of the trigger, initiates an automatic firing cycle that continues until the finger is released or the ammunition supply is exhausted.

ATF Rul. 2006-2

The Bureau of Alcohol, Tobacco, Firearms and Explosives (ATF) has been asked by several members of the firearms industry to classify devices that are exclusively designed to increase the rate of fire of a semiautomatic firearm. These devices, when attached to a firearm, result in the firearm discharging more than one shot with a single function of the trigger. ATF has been asked whether these devices fall within the definition of machinegun under the National Firearms Act (NFA) and Gun Control Act of 1968 (GCA). As explained herein, these devices, once activated by a single pull of the trigger, initiate an automatic firing cycle which continues until either the finger is released or the ammunition supply is exhausted. Accordingly, these devices are properly classified as a part **"designed and intended solely and exclusively, or combination of parts designed and intended, for use in converting a weapon into a machinegun"** and therefore machineguns under the NFA and GCA.

The National Firearms Act (NFA), 26 U.S.C. Chapter 53, defines the term **"firearm"** to include a machinegun. Section 5845(b) of the NFA defines **"machinegun"** as **"any weapon which shoots, is designed to shoot, or can be readily restored to shoot, automatically more than one shot, without manual reloading, by a single function of the trigger. The term shall also include the frame or receiver of any such weapon, any part designed and intended solely and exclusively, or combination of parts designed and intended, for use in converting a weapon into a machinegun, and any combination of parts from which a machinegun can be assembled if such parts are in the possession or under the control of a person."** The Gun Control Act of 1968 (GCA), 18 U.S.C. Chapter 44, defines machinegun identically to the NFA. 18 U.S.C. 921(a)(23). Pursuant to 18 U.S.C. 922(o), machineguns manufactured on or after May 19, 1986, may only be transferred to or possessed by Federal, State, and local government agencies for official use.

ATF has examined several firearms accessory devices that are designed and intended to accelerate the rate of fire for semiautomatic firearms. One such device consists of the following components: two metal blocks; the first block replaces the original manufacturer's V-Block of a Ruger 10/22 rifle and has attached two rods approximately ¼ inch in diameter and approximately 6 inches in length; the second block, approximately 3 inches long, 1 ⅜ inches wide, and ¾ inch high, has been machined to allow the two guide rods of the first block to pass through. The second block supports the guide rods and attaches to the stock. Using ¼ inch rods, metal washers, rubber and metal bushings, two collars with set screws, one coiled spring, C-clamps, and a split ring, the two blocks are assembled together with the composite stock. As attached to the firearm, the device permits the entire firearm (receiver and all its firing components) to recoil a short distance within the stock when fired. A shooter pulls the trigger which causes the firearm to discharge. As the firearm moves rearward in the composite stock, the shooter's trigger finger contacts the stock. The trigger mechanically resets, and the device, which has a coiled spring located forward of the firearm receiver, is compressed. Energy from this spring subsequently drives the firearm forward into its normal firing position and, in turn, causes the trigger to contact the shooter's trigger finger. Provided the shooter maintains finger pressure against the stock, the weapon will fire repeatedly until the ammunition is exhausted or the finger is removed. The assembled device is advertised to fire approximately 650 rounds per minute. Live-fire testing of this device demonstrated that a single pull of the trigger initiates an automatic firing cycle which continues until the finger is released or the ammunition supply is exhausted.

As noted above, a part or parts designed and intended to convert a weapon into a machinegun, i.e., a weapon that will shoot automatically more than one shot, without manual reloading, by a single function of the trigger, is a machinegun under the NFA and GCA. ATF has determined that the device constitutes a machinegun under the NFA and GCA. This determination is consistent with the legislative history of the National Firearms Act in which the drafters equated **"single function of the trigger"** with **"single pull of the trigger."** See, e.g., National Firearms Act: Hearings Before the Comm. on Ways and Means, House of Representatives, Second Session on H.R. 9066, 73rd Cong., at 40 (1934). Accordingly, conversion parts that, when installed in a semiautomatic rifle, result in a weapon that shoots more than one shot, without manual reloading, by a single pull of the trigger, are a machinegun as defined in the National Firearms Act and the Gun Control Act.

Held, a device (consisting of a block replacing the original manufacturer's V-Block of a Ruger 10/22 rifle with two attached rods approximately ¼ inch in diameter and approximately 6 inches in length; a second block, approximately 3 inches long, 1 ⅜ inches wide, and ¾ inch high, machined to allow the two guide rods of the first block to pass through; the second block supporting the guide rods and attached to the stock; using ¼ inch rods; metal washers; rubber and metal bushings; two collars with set screws; one coiled spring; C-clamps; a split ring; the two blocks assembled together with the composite stock) that is designed to attach to a firearm and, when activated by a single pull of the trigger, initiates an automatic firing cycle that continues until either the finger is released or the ammunition supply is exhausted, is a machinegun under the National Firearms Act, 26 U.S.C. 5845(b), and the Gun Control Act, 18 U.S.C. 921(a)(23).

Held further, manufacture and distribution of any device described in this ruling must comply with all provisions of the NFA and the GCA, including 18 U.S.C. 922(o).

To the extent that previous ATF rulings are inconsistent with this determination, they are hereby overruled.

Date approved: December 13, 2006

18 U.S.C. 923(g): RECORDS REQUIRED TO BE MAINTAINED BY FEDERAL FIREARMS LICENSEES

27 CFR PART 478, SUBPART H: RECORDS

27 CFR 478.22: ALTERNATIVE METHODS OR PROCEDURES; EMERGENCY VARIATIONS FROM REQUIREMENTS

A Federal firearms licensee (FFL) who has an approved variance to use a computerized acquisition and disposition record may periodically download the required records to portable data storage devices instead of printing out such records, even if the original variance did not include this provision.

ATF Rul. 2007- 4

This ruling was superseded by ATF Rul. 2008-2.

18 U.S.C. 921(a)(3): DEFINITION OF FIREARM

18 U.S.C. 922(o): TRANSFER OR POSSESSION OF MACHINEGUN

18 U.S.C. 921(a)(23): DEFINITION OF MACHINEGUN

26 U.S.C. 5845(b): DEFINITION OF MACHINEGUN

27 CFR 478.11: DEFINITION OF FIREARM FRAME OR RECEIVER

The upper assembly of the Fabrique Nationale Herstal SA (FN) FNC rifle is classified as the receiver of the firearm for purposes of the Gun Control Act, 18 U.S.C. Chapter 44, and the National Firearms Act, 26 U.S.C. Chapter 53.

ATF Rul. 2008-1

The Bureau of Alcohol, Tobacco, Firearms and Explosives (ATF) has recently received inquiries concerning the installation of a registered sear into the FN FNC rifle. These sears, when installed, allow a semiautomatic FNC rifle to be converted into a machinegun. Prior to May 19, 1986, a number of these sears made for the FNC rifle were registered as machineguns in the National Firearms Registration and Transfer Record (NFRTR). These sears were required to be registered in the NFRTR as machineguns because they are parts designed and intended solely and exclusively for use in converting a weapon into a machinegun.

In order to install a registered FNC sear into a host FNC rifle, a hole must be drilled into the lower assembly and a portion of the solid area of the lower assembly between the magazine well and the compartment for the trigger mechanism must be milled out. If the lower assembly were to be classified as the receiver, any such modifications would create a new machinegun receiver, potentially in violation of 18 U.S.C. 922(o) (prohibiting the transfer or possession of a machinegun except for official Federal, State, or local government use).

The Gun Control Act of 1968 (GCA) at 18 U.S.C. 921(a)(23), and the National Firearms

Act (NFA) at 26 U.S.C. 5845(b), de-

fine the term **"machinegun"** as **"any weapon which shoots, is designed to shoot, or can be readily restored to shoot, automatically more than one shot, without manual reloading, by a single function of the trigger."** The term also includes **"the frame or receiver of any such weapon, any part designed and intended solely and exclusively, or combination of parts designed and intended, for use in converting a weapon into a machinegun, and any combination of parts from which a machinegun can be assembled if such parts are in the possession or under the control of a person."** (See also 27 CFR 478.11, 479.11). Title 27, Code of Federal Regulations, section 478.11 defines a **"firearm receiver"** as, **"[t]hat part of a firearm which provides housing for the hammer, bolt or breechblock, and firing mechanism, and which is usually threaded at its forward portion to receive the barrel."**

The FNC rifle consists of two major assemblies, the upper assembly and the lower assembly. The lower assembly houses the trigger, hammer, disconnector, safety/selector, and an automatic trip lever in the automatic version. It also incorporates a pistol grip and a magazine release. The upper assembly houses a barrel that is attached to the upper assembly by means of a barrel extension. It also houses the bolt carrier with gas piston affixed, gas tube and handguard, bolt, operating rod and spring. The two assemblies are mounted together with a front and rear takedown pin. Since 1981, ATF has classified the lower assembly as the receiver for purposes of the GCA and NFA.

ATF has reconsidered its classification of the lower assembly of the FNC rifle as the receiver. The upper assembly of the FNC rifle is more properly classified as the receiver. The upper assembly of the FNC rifle houses the bolt and provides a connection point for the barrel. Moreover, the upper assembly is classified as the receiver on similar types of firearms, to include other FN rifles, such as the FN FAL and FN SCAR. Reclassification of the upper assembly as the receiver will also allow the continued installation of a lawfully registered sear into an FNC rifle because no modification to the receiver, which is the upper assembly, is required to properly install the sear.

Held, the upper assembly of the FN FNC rifle is the receiver of the firearm.

Held further, in the event a licensed manufacturer in the United States manufactures a new firearm that is substantially similar to the FN FNC rifle, it must be marked with a model designation other than "FNC."

To the extent this ruling is inconsistent with any previous ATF classifications, they are hereby superseded.

Date approved: May 27, 2008

18 U.S.C. 923(g)(1)(A): RECORDS REQUIRED FOR FIREARMS LICENSEES

27 CFR PART 478, SUBPART H: RECORDS

27 CFR 478.22, 478.121, 478.122, 478.123, 478.125(e): LICENSEE ACQUISITION AND DISPOSITION RECORDS, ALTERNATE METHODS OR PROCEDURES

Under specified conditions, approval is granted to utilize computerized records as required records under 27 CFR 478.121, 478.122, 478.123, and 478.125(e), Subpart H.

ATF Rul. 2008-2

This ruling was superseded by ATF Rul. 2013-5.

18 U.S.C. 923(g)(1)(A): RECORDS REQUIRED

18 U.S.C. 922(b)(5): RECORDS REQUIRED

27 CFR 478.22: ALTERNATE METHODS OR PROCEDURES

27 CFR 478.121: RECORDS REQUIRED

27 CFR 478.124: FIREARMS TRANSACTION RECORD

27 CFR 479.26: ALTERNATE METHODS OR PROCEDURES

27 CFR 479.131: RECORDS REQUIRED

Pursuant to 27 CFR 478.22 and 479.26, ATF authorizes an alternate method or procedure from the provisions of 27 CFR 478.121, 478.124, and 479.131 that require licensees to complete ATF Form 4473 (5300.9), Firearms Transaction Record Part I–Over-the-Counter, by handwriting the information. ATF authorizes licensees to use an electronic version of Form 4473, as an alternate method or procedure, provided all of the requirements stated in this ruling are met.

ATF Rul. 2008-3

The Gun Control Act of 1968 (GCA), 18 U.S.C. Chapter 44, provides, in part, that each licensed importer, licensed manufacturer, and licensed dealer (licensees) must maintain records of sale or other disposition of firearms at their place of business for such period, and in such form, as the Attorney General may by regulations prescribe. With certain exceptions, a licensee may not sell or otherwise dispose, temporarily or permanently, of any firearm to any person, other than another licensee, unless the licensee records the transaction on a Firearms Transaction Record, ATF Form 4473 (5300.9). The answers on Form 4473 must be handwritten, and contain the information specified by Title 27, Code of Federal Regulations (CFR), section 478.124.

Persons with a valid Federal firearms license may seek approval to use an alternate electronic system of recordkeeping. Title 27, Code of Federal Regulations, sections

478.22 and 479.26, provide that the Director may approve an alternate method or procedure in lieu of a method or procedure specifically prescribed in the regulations when he finds that: (1) Good cause is shown for the use of the alternate method or procedure; (2) The alternate method or procedure is within the purpose of, and consistent with the effect intended by, the specifically prescribed method or procedure and that the alternate method or procedure is substantially equivalent to that specifically prescribed method or

procedure; and (3) The alternate method or procedure will not be contrary to any provision of law and will not result in an increase in cost to the Government or hinder the effective administration of Parts 478 or 479, Title 27, CFR.

The Bureau of Alcohol, Tobacco, Firearms and Explosives (ATF) recognizes that, provided certain conditions are met, a computer generated and printed version of Form 4473 may reduce recordkeeping errors. ATF finds that there is good cause to authorize a variance to the provisions of 27 CFR 478.121, 478.124, and 479.131 requiring that Form 4473 be handwritten by the transferee (buyer) and transferor (seller), except for the signatures, dates, and the National Instant Criminal Background Check System (NICS) information.

ATF finds that the alternate method set forth in this ruling is consistent with the provisions of 27 CFR 478.121, 478.124, and 479.131 because it will ensure that all required information is captured on the electronic Form 4473, and that the answers are confirmed in writing. Further, this alternate method is not contrary to any provision of law, should not increase costs to ATF, and should not hinder the effective administration of the regulations.

Under the terms, conditions, and in the order set forth in this ruling, ATF authorizes an alternate method or procedure to the ATF Form 4473 (5300.9) requirements of 27 CFR 478.121, 478.124, and 479.131, when licensees comply with the following requirements:

(1) The licensee will follow the Notices, Instructions, and Definitions governing the use of the OMB-approved version of the ATF Form 4473, and either: (a) acquires and uses his/her own electronic Form 4473 software; or (b) downloads from ATF's internet website and saves to his/her computer ATF's most current electronic version of Form 4473, known as ATF e-Form 4473.

(2) The electronic Form 4473 displays clearly, either on the same screen as the questions, or upon mouse-clicking a conspicuously displayed screen item, all of the Notices, Instructions, and Definitions contained on the current Office of Management and Budget (OMB) approved ATF Form 4473.

(3) The transferee (buyer) of the firearm(s) answers the questions for Section A of Form 4473 while physically present at the seller's premises. The questions presented on the computer screen must be legible, and contain the same wording as the current OMB-approved version of ATF Form 4473 (5300.9), Firearms Transaction Record Part I, Over-the-counter. Where Form 4473 requires **"checking"** or **"marking,"** the transferee may mouse-click in an appropriate box to answer the question on the electronic Form 4473. Where Form 4473 requires **"printing"** or **"handwriting"** (except for signatures and dates), the transferee will enter the information on the electronic form.

(4) The electronic Form 4473 must allow the transferee to review and amend his/her answers until he/she selects or mouse-clicks a statement on the screen that he/she **"agrees with and certifies"** the same transferee certification statement as the current OMB-approved ATF Form 4473 (5300.9). Once the statement has been selected or mouse-clicked, the transferee can no longer make revisions to his/her answers.

(5) The transferor (seller) of the firearm(s) enters the information into the licensee's computer for Section B (except the NICS background check information) and Section D (except for the signature and date) of the electronic Form 4473. The questions presented on the computer screen must be legible, and contain the same wording as the current OMB-approved version of ATF Form 4473 (5300.9).

(6) The electronic Form 4473 allows the transferor to review and amend his/her answers until he/she selects or mouse-clicks a statement on the screen that he/she **"agrees with and certifies"** the same transferor certification as the current OMB-approved ATF Form 4473 (5300.9). Once the statement has been selected or mouse-clicked, the transferor can no longer make revisions to his/her answers.

(7) The transferor (seller) prints all pages of ATF Form 4473 on 8 1/2" x 11" or the previously prescribed 8 1/2" x 33" paper, 20 pound, matte finish, and paper of the same color of paper as the current edition (the printer and paper to be provided by the licensee); the ink or toner quality used in printing, whether ink-jet or laser, must remain visible and legible for the duration that the form is required to be kept by law and regulation. The entire printed Form 4473 must be a verbatim/exact image of the current OMB-approved ATF Form 4473 (5300.9). Each response to a question must be legibly printed and aligned in the particular space or box allotted for that question on ATF Form 4473. If paper is unavailable, licensees should use the current hardcopy of the OMB-approved version of the form available from ATF's distribution center.

(8) The transferee (buyer) confirms his/her answers by handwriting (in ink) his/her signature and date on the printed form, certifying that his/her answers are true, correct, and complete, and that he/she has read, understood, and complied with the conditions, notices, definitions, and instructions for the form.

(9) The transferor (seller) contacts the NICS (or the State Point-of-Contact (POC)) for the background check, except where the buyer has a valid permit from the State where the transfer is to take place, or the transaction involves an ATF approved transfer of solely National Firearms Act (NFA) firearm(s), which qualify as exemptions to NICS. If applicable, the NICS/POC background check information is handwritten by the licensee in Section B of the printed form.

(10) The licensee/transferor reviews the entire form for completeness and accuracy, and confirms his/her answers by handwriting (in ink) his/her signature and date on the printed form, certifying that his/her answers are true, correct, and complete, and that he/she has read, understood, and complied with the conditions, notices, definitions, and instructions for the form.

(11) The licensee staples the printed pages together and retains the complete package, including all Notices, Instructions, and Definitions for ATF Form 4473, at his/her licensed premises in accordance with law and regulations.

The laws, regulations, policies, and procedures applicable to the handwritten version of

ATF Form 4473 apply to electronic versions of Form 4473. Licensees who fail to abide by the terms and conditions set forth in this ruling may be advised by ATF that their privilege of utilizing an electronic Form 4473 has been terminated. Errors on printed forms may only be corrected prior to completion of the transaction by marking through the error, entering the correct information, and initialing and dating each correction. Licensees are not required to use an electronic Form 4473, and may continue to use the OMB-approved version of the ATF Form 4473 available through ATF's Distribution Center. Licensees are reminded of their responsibility to ensure accuracy and completeness of all of their required records.

This ruling does not apply to a licensee or computer software that gathers information other than that set forth on the current OMB-approved ATF Form 4473 (5300.9), Part I. Licensees who wish to collect additional information must clearly distinguish information requested by the licensee from information required by law and ATF. In no case shall any licensee or software make any representation contrary to law or regulation, or any representation that ATF has required a collection of any additional information.

Held, pursuant to 27 CFR 478.22 and 479.26, ATF authorizes an alternate method or procedure from the provisions of 27 CFR 478.121, 478.124, and 479.131 that require licensees to complete ATF Form 4473 (5300.9), Firearms Transaction Record Part I, Over- the-counter, by handwriting the information. ATF authorizes licensees to use an electronic version of Form 4473, including ATF's e-Form 4473, as an alternate method or procedure, provided all of the requirements stated in this ruling are met.

Held further, if legal or administrative difficulties arise due to the download and/or use of any electronic Form 4473, or if ATF finds for any reason that use of an electronic Form 4473 fails to meet the requirements of this ruling or 27 CFR 478.22 or 479.26, licensees may be required to discontinue use of the electronic Form 4473 and handwrite the answers on the Form 4473 available through ATF's Distribution Center.

Date approved: November 19, 2008

18 U.S.C. 921(a): DEFINITION

18 U.S.C. 922(a)(I)(A): LICENSES REQUIRED

18 U.S.C. 923(a): LICENSES REQUIRED

27 CFR 478.11: DEFINITIONS

27 CFR 478.41(a): LICENSES REQUIRED

Any person who engages in an activity or process that primarily adds to or changes a firearm's appearance, by camouflaging a firearm by painting, dipping, or applying tape, or by engraving the external surface of a firearm, does not need to be licensed as a manufacturer under the Gun Control Act. Any person who is licensed as a dealer/gunsmith, and who camouflages or engraves firearms as described in this ruling does not need to be licensed as a manufacturer under the Gun Control Act. Any person who is engaged in the business of camouflaging or engraving firearms as described in this ruling must be licensed as a dealer, which includes a gunsmith, under the Gun Control Act.

ATF Rul. 2009-1

The Bureau of Alcohol, Tobacco, Firearms and Explosives (ATF) has received inquiries from Federally licensed manufacturers and dealers/gunsmiths seeking clarification as to whether camouflaging firearms, or cutting designs into firearms by engraving, constitute manufacturing activities that require a manufacturer's license.

Camouflaging refers to a patterned treatment using a variety of different colors that enables a firearm to blend into a particular outdoor environment. This typically involves painting, dipping, or applying a tape over the firearm's wood and/or metal parts.

Engraving firearms is a process in which a decorative pattern is placed on the external

metal of a firearm primarily for ornamental purposes. The engraving can be cut by hand or machine, or pressed into the metal. There are other engraving techniques that cut designs into firearms, such as checkering or scalloping.

The Gun Control Act of 1968 (GCA), Title 18, United States Code (U.S.C.), Chapter 44, provides, in part, that no person shall engage in the business of importing, manufacturing, or dealing in firearms until he has filed an application with and received a license to do so from the Attorney General. A **"firearm"** is defined by 18 U.S.C. 921(a)(3) to include any weapon (including a starter gun) which will or is designed to or may readily be converted to expel a projectile by the action of an explosive, and the frame or receiver of any such weapon. The term **"manufacturer"** is defined by 18 U.S.C. 921(a)(10) and 27 CFR 478.11 as any person engaged in the business of manufacturing firearms or ammunition for purposes of sale or distribution. The term **"dealer,"** which includes a gunsmith, is defined by 18 U.S.C. 921(a)(l1) and 27 CFR 478.11 to include any person engaged in the business of selling firearms at wholesale or retail, or repairing firearms or making or fitting special barrels, stocks, or trigger mechanisms to firearms.

In Revenue Ruling 55-342, ATF's predecessor agency interpreted the meaning of the terms **"manufacturer"** and **"dealer"** for the purpose of firearms licensing under the Federal Firearms Act, the precursor statute to the GCA. It was determined that a licensed dealer

could assemble firearms from component parts on an individual basis, but could not engage in the business of assembling firearms from component parts in quantity lots for purposes of sale or distribution without a manufacturer's license. Since then, ATF has similarly and consistently interpreted the term **"manufacturer"** under the GCA to mean any person who engages in the business of making firearms, by casting, assembly, alteration, or otherwise, for the purpose of sale or distribution.

Performing a cosmetic process or activity, such as camouflaging, that primarily adds to or changes the appearance or decoration of a firearm is not manufacturing. Unlike manufacturing processes that primarily enhance a firearm's durability, camouflaging is primarily cosmetic. Likewise, external engravings are cosmetic in nature and primarily affect only the appearance of a firearm.

Held, any person who engages in an activity or process that primarily adds to or changes a firearm's appearance by camouflaging the firearm by painting, dipping, or applying tape does not need to be licensed as a manufacturer under the Gun Control Act.

Held further, any person who engages in an activity or process that primarily adds to or changes a firearm's appearance by engraving the external surface of the firearm does not need to be licensed as a manufacturer under the Gun Control Act.

Held further, any person who is licensed as a dealer, which includes a gunsmith, and who camouflages or engraves firearms as described in this ruling does not need to be licensed as a manufacturer under the Gun Control Act.

Held further, any person who is engaged in the business of camouflaging or engraving firearms as described in this ruling must be licensed as a dealer, which includes a gunsmith, under the Gun Control Act.

Date approved: January 12, 2009

18 U.S.C. 921(a): DEFINITIONS

18 U.S.C. 922(a)(1)(A): LICENSES REQUIRED

18 U.S.C. 923(a): LICENSES REQUIRED

27 CFR 478.11: DEFINITIONS

27 CFR 478.41(a): LICENSES REQUIRED

Any person who installs "drop in" replacement parts in or on existing, fully assembled firearms does not manufacture a firearm, and does not need to be licensed as a manufacturer under the Gun Control Act. A "drop in" replacement part is one that can be installed in or on an existing, fully assembled firearm without drilling, cutting, or machining. A replacement part, whether factory original or otherwise, has the same design, function, substantially the same dimensions, and does not otherwise affect the manner in which the weapon expels a projectile by the action of an explosive. Any person who is licensed as a dealer, which includes a gunsmith, and who installs "drop in" replacement parts in or on existing, fully assembled firearms as described in this ruling does not need to be licensed as a manufacturer under the Gun Control Act. Any person who is engaged in the business of installing "drop in" replacement parts in or on existing, fully assembled firearms as described in this ruling must be licensed as a dealer, which includes a gunsmith, under the Gun Control Act.

ATF Rul. 2009-2

The Bureau of Alcohol, Tobacco, Firearms and Explosives (ATF) has received inquiries

from Federally licensed firearms manufacturers and dealers/gunsmiths seeking clarification as to whether installing **"drop in"** replacement parts in or on existing firearms constitutes a manufacturing activity that requires a manufacturer's license.

Persons may buy **"drop in"** replacement firearm parts to replace worn or broken original factory parts. Replacement parts, such as barrels, triggers, hammers, and sears have been designed so that they can be dropped in to replace existing parts on fully assembled firearms. A **"drop in"** replacement part is one that can be installed in or on an existing, fully assembled firearm (not solely a frame or receiver) without drilling, cutting, or machining.

The Gun Control Act of 1968 (GCA), Title 18, United States Code (U.S.C.), Chapter 44, provides, in part, that no person shall engage in the business of importing, manufacturing, or dealing in firearms until he has filed an application with and received a license to do so from the Attorney General. A **"firearm"** is defined by 18 U.S.C. 921(a)(3) to include any weapon (including a starter gun) which will or is designed to or may readily be converted to expel a projectile by the action of an explosive, and the frame or receiver of any such weapon. The term **"manufacturer"** is defined by 18 U.S.C. 921(a)(10) and 27 CFR 478.11 as any person engaged in the business of manufacturing firearms or ammunition for purposes of sale or distribution. The term **"dealer"** is defined by 18 U.S.C. 921(a)(11) and 27 CFR 478.11 to include any person engaged in the business of selling firearms at wholesale or retail, or repairing firearms or making or fitting special barrels, stocks, or trigger mechanisms to firearms.

In Revenue Ruling 55-342, ATF's predecessor agency interpreted the meaning of the terms **"manufacturer"** and **"dealer"** for the purpose of firearms licensing under the Federal Firearms Act, the precursor statute to the GCA. It was determined that a licensed dealer could assemble firearms from component parts on an individual basis, but could not engage in the business of assembling firearms from component parts in quantity lots for purposes of sale or distribution without a manufacturer's license. Since then, ATF has similarly and consistently interpreted the term **"manufacturer"** under the GCA to mean any person who engages in the business of making firearms, by casting, assembly, alteration, or otherwise, for the purpose of sale or distribution. Such persons must have a manufacturer's license under the GCA.

Installing **"drop in"** parts in or on existing, fully assembled firearms, whether factory original or otherwise, does not result in any alteration to the original firearms so long as they are replacement parts. A replacement part, whether factory original or otherwise, has the same design, function, substantially the same dimensions, and does not otherwise affect the manner in which the weapon expels a projectile by the action of an explosive.

Held, any person who installs **"drop in"** replacement parts in or on existing, fully assembled firearms does not need to be licensed as a manufacturer under the Gun Control Act.

Held further, a **"drop in"** replacement part is one that can be installed in or on an existing, fully assembled firearm without drilling, cutting, or machining. A replacement part, whether factory original or otherwise, has the same design, function, substantially the same dimensions, and does not otherwise affect the manner in which the weapon expels a projectile by the action of an explosive.

Held further, any person who is licensed as a dealer, which includes a gunsmith, and who installs **"drop in"** replacement parts in or on existing, fully assembled firearms as described in this ruling does not need to be licensed as a manufacturer under the Gun Control Act.

Held further, any person who is engaged in the business of installing **"drop in"** replacement parts in or on existing, fully assembled firearms as described in this ruling must be licensed as a dealer, which includes a gunsmith, under the Gun Control Act.

Date approved: January 12, 2009

18 U.S.C. 923(i): IDENTIFICATION OF FIREARMS

26 U.S.C. 5842(a): IDENTIFICATION OF FIREARMS

27 CFR 478.92(a): IDENTIFICATION OF FIREARMS

27 CFR 478.92(a)(4)(i): ALTERNATE MEANS OF IDENTIFICATION

27 CFR 479.102(a): IDENTIFICATION OF FIREARMS

27 CFR 479.102(c): ALTERNATE MEANS OF IDENTIFICATION

ATF authorizes licensed manufacturers who perform a manufacturing process on firearms for, or on behalf of, another licensed manufacturer not to place their serial numbers and other required identification markings on the firearms, provided such firearms already have been properly marked with a serial number and other identifying markings as required by 27 CFR 478.92(a) and 479.102(a) and that all of the other requirements stated in this ruling have been met.

ATF Rul. 2009-5

The Bureau of Alcohol, Tobacco, Firearms and Explosives (ATF) has received inquiries from Federally licensed manufacturers concerning the requirement that each manufacturer performing a manufacturing process on a firearm, including a frame or receiver, place their identifying markings on each firearm.

Many licensed manufacturers contract with other licensed manufacturers to perform various steps in the manufacturing process on firearms that already have a serial number and other required markings, and who then distribute those firearms to another licensed manufacturer, or into the wholesale or retail market. The manufacturers performing a manufacturing process for another manufacturer often ask ATF for approval not to place their marks of identification on the firearms. ATF has approved many of these **"non-marking variance"** requests after finding that certain conditions were met.

The Gun Control Act of 1968 (GCA), at Title 18, United States Code (U.S.C.), section 923(i), and the National Firearms Act (NFA), at Title 26, U.S.C., section 5842(a), require all licensed importers and manufacturers to identify each firearm imported or manufactured by means of a serial number engraved or cast on the frame or receiver of the weapon, in such manner as the Attorney General shall by regulations prescribe. Federal regulations at 27 CFR 478.92(a) and 479.102(a) prescribe the requirements for serialization and other marks of identification that must be placed on firearms.

Persons with a valid Federal importer or manufacturer license may seek approval to use an alternate means of identification of firearms. Sections 478.92(a)(4)(i) and 479.102(c) of Title 27, Code of Federal Regulations (CFR), provide that the Director may authorize other means of identification upon receipt of a letter application showing that such other identification is reasonable and will not hinder the effective administration of the firearms regulations.

The unique marks of identification of firearms serve several purposes. First, the marks are used by Federal firearms licensees to effectively track their firearms inventories and maintain all required records. Second, the marks enable law enforcement officers to trace specific firearms used in crimes from the manufacturer or importer to individual purchasers, and to identify particular firearms that have been lost or stolen. Further, marks help prove in certain criminal prosecutions that firearms used in a crime have travelled in interstate or foreign commerce.

Multiple identification markings may

be confusing to law enforcement and potentially hinder effective tracing of firearms used in crimes. Therefore, ATF finds that the other means to identify firearms specified in this ruling are reasonable and will not hinder the effective administration of the firearms regulations.

Licensed manufacturers who perform a manufacturing process on firearms for, or on behalf of, another licensed manufacturer need not place their serial numbers and other identification markings on firearms as required by 27 CFR 478.92(a) and 479.102(a), provided the following conditions have been met:

(1) The manufacturer is receiving the firearms, including frames or receivers, from another manufacturer.

(2) The manufacturer is performing a manufacturing process on the firearms as directed by another manufacturer before distributing those firearms to another manufacturer or into the wholesale or retail market.

(3) All manufacturers involved in the manufacturing process possess a valid Federal firearms manufacturer's license issued by ATF and are performing only the manufacturing processes authorized under that license.

(4) The firearms, including frames and receivers, are already properly marked with a serial number and all other identifying markings in accordance with 27 CFR 478.92(a) and, if applicable, 479.102(a).

(5) Prior to engaging in the manufacturing process, the manufacturer desiring not to mark must submit to ATF the following information:

(a) The manufacturer's name, address, and license number, and the name, address, and license number of the manufacturer for which the manufacturing process is being performed;

(b) A copy of the license held by each manufacturer;

(c) A description of the type of manufacturing process to be performed by the manufacturer desiring not to mark;

(d) The model(s), if assigned, of the firearms subject to the manufacturing process described;

(e) The serial numbers of the firearms in sequential order;

(f) The calibers or gauges of the firearms; and

(g) Any other information concerning the firearms manufacturer(s) or manufacturing process that ATF may require.

(6) The manufacturer desiring not to mark must maintain copies of its submission to ATF of the information required by this ruling with its permanent records of manufacture. The manufacturer availing itself of this ruling should retain proof of its submission to ATF (e.g., certified return receipt mail or tracking number). This proof of submission should show that it was sent to ATF's Firearms Technology Branch, or any other office that ATF may designate as the proper recipient of such information. Additionally, the manufacturer must allow ATF representatives to inspect such documents upon request at any time during business hours without a warrant.

(7) All manufacturers involved in the manufacturing process must maintain all records required by Federal law and regulation.

Once the manufacturer has submitted the necessary documentation to ATF pursuant to this ruling, and provided the manufacturer has complied with all other conditions set forth in this ruling, no **"non-marking variance"** approval from ATF is required, and the manufacturer may engage in the manufacturing process for, or on behalf of, another licensed manufacturer without placing its identifying markings on the firearms in accordance with 27 CFR 478.92(a) and 479.102(a).

Licensees are reminded of their responsibility to ensure accuracy and completeness of all required records. Additionally, licensees manufacturing firearms under the NFA and 27 CFR Part 479 are reminded that this ruling applies only to the relevant marking requirements for those firearms. Manufacturers still must abide by all other provisions relating to the manufacture, transfer, and possession of NFA firearms.

Held, pursuant to 27 CFR 478.92(a)(4)(i) and 479.102(c), ATF authorizes licensed manufacturers who perform a manufacturing process on firearms for, or on behalf of, another licensed manufacturer not to place their serial numbers and other required identification markings on the firearms, provided such firearms already have been properly marked with a serial number and other identifying markings as required by 27 CFR 478.92(a) and 479.102(a) and that all of the other requirements stated in this ruling have been met.

Held further, if ATF finds that a licensed manufacturer has failed to submit to ATF all information required by this ruling, ATF may require corrective or other action by the licensee. The licensee will be required to comply with the regulations contained at 27 CFR 478.92(a) and/or 27 CFR 479.102(a) unless and until such action is taken.

Held further, if ATF finds that a licensed manufacturer has failed to abide by the conditions of this ruling, or is engaged in any process that hinders the effective administration of the firearms laws or regulations, ATF may require that the licensed manufacturer comply with regulations contained at 27 CFR 478.92(a) and/or 27 CFR 479.102(a).

Date approved: November 9, 2009

18 U.S.C. 922(a)(2): INTERSTATE CONTROLS

18 U.S.C. 922(b)(3): INTERSTATE CONTROLS

18 U.S.C. 922(t): TRANSFER TO UNLICENSED PERSONS

18 U.S.C. 923(g)(1)(A): REQUIRED RECORDS

27 CFR 478.124: FIREARMS TRANSACTION RECORD

27 CFR 478.125: FIREARMS RECEIPT AND DISPOSITION RECORD

The temporary assignment of a firearm by an FFL to its unlicensed employees for bona fide business purposes, where the actual custody of the firearm is transferred for a limited period of time, and where title and control of the firearm remain with the FFL, is not a transfer for purposes of the Gun Control Act, and, accordingly, the FFL need not contact NICS for a background check, record a bound book disposition entry, nor complete an ATF Form 4473. The temporary assignment of a firearm by an FFL to its unlicensed agents, contractors, volunteers, or any other person who

is not an employee of the FFL, even for bona fide business purposes, is a transfer or disposition for purposes of the Gun Control Act, and, accordingly, the FFL must contact NICS for a background check, record a disposition entry, and complete an ATF Form 4473. Revenue Ruling 69-248 is superseded and ATF Ruling 73-19 is modified. Industry Circular 72-23 is no longer in effect.

ATF Rul. 2010-1

The Bureau of Alcohol, Tobacco, Firearms and Explosives (ATF) has received inquiries from firearms industry members asking whether the temporary assignment of a firearm by a Federal firearms licensee (FFL) to certain unlicensed persons constitutes a **"transfer"** under Federal law, thus requiring the FFL to contact the National Instant Criminal Background Check System (NICS) prior to assigning firearms to such persons.

FFLs use employees and sometimes the services of certain unlicensed persons, such as writers, evaluators, consultants, attorneys, engineers, and sales representatives, to carry out their business operations. In the course of their employment or contracted services, such persons may, for example, research, examine, photograph, and evaluate firearms on behalf of FFLs.

The Gun Control Act of 1968 (GCA), Title 18, United States Code (U.S.C.), Chapter 44, imposes Federal controls on FFLs who transfer firearms in interstate commerce. Section 922(a)(2) makes it unlawful for an FFL to ship or transport in interstate or foreign commerce any firearm to a person other than an FFL. Section 922(b)(3) makes it unlawful for an FFL to sell or deliver a firearm to a person he or she knows does not reside in the State in which the FFL's place of business is located. Additionally, under 18 U.S.C. 923(g)(1)(A), each FFL must maintain records of importation, production, shipment, receipt, sale, or other disposition of firearms at its place of business as prescribed by regulation. Federal regulations at Title 27, Code of Federal Regulations (CFR), sections 478.124 and 478.125, require licensed importers, manufacturers, and dealers to record each firearm sale or disposition and complete a Firearms Transaction Record, ATF Form 4473.

Revenue Ruling 69-248, 1969-1 C.B. 360, addresses temporary assignments of firearms to FFL employees, agents, and representatives. That ruling states that a licensee may lawfully ship or transport a firearm to its unlicensed employees, agents, or representatives without violating section 922(a)(2), provided (1) the employee, agent, or representative is not prohibited under section 922(g) from possessing a firearm; and (2) the shipment is for a bona fide business purpose, except to sell or dispose of the firearm. Upon completion of the business purpose, the employee, agent, or representative must return the firearm to the licensee. Rev. Rul. 69-248 is limited to situations in which **"the actual custody of the firearms . . . is transferred for a limited period of time and where title and ultimate control of the firearms . . . remain in such licensee."**

ATF issued Industry Circular 72-23 (dated August 23, 1972) to clarify Rev. Rul. 69-248. Industry Circular 72-23 explained that professional writers, consultants and evaluators were included within the category of agents and representatives discussed in Rev. Rul. 69-248. The industry circular noted that Rev. Rul. 69-248 is explicitly limited to **"firearms . . . acquired from a licensee for limited lengths of time and where the title to and ultimate control of the firearm remains in the licensee."**

In 1993, over 20 years after issuance of the ruling and circular, Congress enacted the Brady Handgun Violence Prevention Act (the Brady Law), Title 18, U.S.C., section 922(t). The Brady Law amended the GCA, in part, by providing that an FFL may not transfer a firearm to any unlicensed person unless the FFL contacts and successfully completes a background check through NICS, and the FFL has verified the identity of the transferee by examining a valid identification document containing a photograph of the transferee. In enacting the Brady Law, Congress intended to create a system to help ensure that FFLs did not allow prohibited persons to come into possession of firearms.

Neither the GCA nor its implementing regulations define the term **"transfer."** The common legal definition of **"transfer"** broadly encompasses any method of disposing of an asset. A **"transfer"** includes any change in dominion or control of a firearm, whether temporary or permanent, commercial or noncommercial. A change in dominion or control may occur even when such change does not convey title to the firearm.

Businesses carry out operations through their employees. When an FFL temporarily assigns a firearm to an employee for bona fide business purposes, title and control of the firearm remain with the licensee. For this reason, no transfer or disposition occurs for purposes of the GCA. Accordingly, no NICS check, disposition record entry, or ATF Form 4473 is required for the temporary assignment of a firearm by an FFL to its employee. Bona fide business purposes, in this context, are purposes integral to the FFL's business operations, and do not include permanently assigning a firearm to a specific employee, or loaning or renting a firearm to an employee for personal use. Those are considered transfers or other dispositions that would trigger recordkeeping and NICS requirements. Because FFLs are accountable for their firearms inventories, ATF strongly recommends that FFLs ensure accountability for firearms temporarily assigned to employees. In addition, ATF reminds FFLs that they may not knowingly make available or assign a firearm to any person whose receipt or possession of firearms is prohibited under the GCA.

Businesses may also support their operations through contractors or volunteers. When an FFL temporarily assigns a firearm to a non-employee, even for bona fide business purposes, a transfer occurs for purposes of the GCA. Temporary firearms assignments to employees are different from temporary firearms assignments to non-employee contractors, agents, and representatives because the FFL exerts a higher level of control over its employees than its contractors or agents. Unlike contractors and agents, employees work for wages or salaries under direct supervision. In an employer-employee relationship, the employer controls not only the result of the employee's work, but also the manner, training, and hours in which the work will be carried out. In independent contractor or non-employee agency relationships, the independent contractor or non-employee agent has control of the manner, training, and hours of performing the work, and is only responsible for the result. Because an FFL relinquishes control over a firearm by temporarily assigning it to a non-employee, the GCA requirements apply.

If the FFL and the unlicensed person do not wish to enter into an employer-employee relationship, certain unlicensed persons may obtain a Federal firearms license from ATF. As discussed above, the Brady Law does not apply to transfers between FFLs; therefore, no

NICS check would be required to transfer a firearm from one FFL to another. In ATF Ruling 73-19, for example, ATF determined that firearms consultants and experts may be licensed as dealers to receive firearms from unlicensed persons for testing and evaluation. ATF Rul. 73-19, 1973-ATF C.B. 93. Licensed dealers who engage solely in firearms consulting or expert services for firearms testing or evaluation need not maintain regular business hours or open their business premises to the general public, although they must comply with all other applicable GCA requirements.

Held, the temporary assignment of a firearm by an FFL to its unlicensed employees

for bona fide business purposes, where the actual custody of the firearm is transferred for a limited period of time, and where title and control of the firearm remain with the FFL, is not a transfer or disposition for purposes of the Gun Control Act, as amended, and, accordingly, the FFL need not contact NICS for a background check, record a disposition entry, nor complete an ATF Form 4473.

Held further, the temporary assignment of a firearm by an FFL to its unlicensed agents, contractors, volunteers, or any other person who is not an employee of the FFL, even for bona fide business purposes, is a transfer or disposition for purposes of the Gun Control Act, as amended, and, accordingly, the FFL must contact NICS for a background check, record a disposition entry, and complete an ATF Form 4473. As a reminder, an FFL may not transfer a firearm other than a rifle or shotgun to an unlicensed person who does not reside in the FFL's State. An FFL may transfer a firearm to another FFL in the unlicensed person's State of residence, and that FFL may subsequently transfer the firearm to the unlicensed person, after contacting NICS for a background check, recording a disposition entry, and completing an ATF Form 4473.

Rev. Rul. 69-248, 1969-1 C.B. 360, is hereby superseded and ATF Rul. 73-19, 1973-ATF C.B. 93, is modified. In addition, Industry Circular 72-23, dated August 23, 1972, is no longer in effect.

Date approved: May 20, 2010

18 U.S.C. 921(a)(23): DEFINITIONS (MACHINEGUN)

18 U.S.C. 923(i): IDENTIFICATION OF

FIREARMS

26 U.S.C. 5842(a): IDENTIFICATION OF FIREARMS

26 U.S.C. 5845(a): DEFINITIONS (FIREARM)

26 U.S.C. 5845(b): DEFINITIONS (MACHINEGUN)

27 CFR 478.11: DEFINITIONS (FIREARM FRAME OR RECEIVER, MACHINEGUN)

27 CFR 478.92: IDENTIFICATION OF FIREARMS

27 CFR 479.11: DEFINITIONS (FIREARM, FRAME OR RECEIVER, MACHINEGUN)

27 CFR 479.102: IDENTIFICATION OF FIREARMS

The right side-plate of a Vickers/Maxim-type firearm, manufactured with its camming lobe affixed in the proper location, and without an ATF approved block that prevents installation of machinegun fire control components, is a machinegun receiver, and therefore, a "machinegun" as defined by the Gun Control Act of 1968, 18 U.S.C. 921(a)(23), the National Firearms Act, 26 U.S.C. 5845(b), and their implementing regulations, 27 CFR 478.11 and 479.11. Provided it has not been disassembled into its component parts, a complete Vickers/Maxim-type machinegun that is currently registered in the National Firearms Registration and Transfer Record with its serial number located on a component part of the receiver box other than the right side- plate has been lawfully registered under the National Firearms Act and may be lawfully possessed by the registrant under 18 U.S.C. 922(o).

ATF Rul. 2010-3

The Bureau of Alcohol, Tobacco, Firearms and Explosives (ATF) has received requests for clarification from persons asking what part of Vickers/ Maxim-type machineguns constitutes the **"frame or receiver"** that must be marked with a serial number and required identifying information and registered in the National Firearms Registration and Transfer Record (NFRTR).

ATF examined various types of Vickers/Maxim-type firearms in both semi-automatic and machinegun configurations. The design of these firearms, in either configuration, consists of forward and rear sections. The forward section is made of a reciprocating barrel assembly. The majority of the barrel assembly is contained inside a jacket. The jacket has a rear ring (often integrally formed with the trunnion), an outer body, and a forward ring or cap. The forward ring incorporates a mounting point for an accelerator and guide for the forward part of the barrel. The rear section is made of the receiver- box, which contains the internal fire control components of the firearm and provides housing for the rear portion of the barrel assembly and the ammunition feed tray. The receiver-box consists of a left and a right side-plate, bottom channel, and trunnion that are held together by a number of rivets. The receiver-box also provides mounting for the handle block, top cover, and recoil mechanism. Camming lobes, which are permanently affixed to the inside portion of each side-plate, hold the lock piece in place and guide the movement of the extractor, which is attached to the lock piece. Vickers/Maxim-type side-plates are recognizable and properly classified as side-plates for this weapon when the camming lobe has been affixed. The right side-plate of the Vickers/Maxim-type machinegun provides housing for the hammer, breechblock, and firing mechanism, and attaches to the trunnion, which holds the rear portion of the barrel. The right side-plate typically contains a mounting point for the cocking-lever assembly.

ATF evaluated Vickers/Maxim-type firearms and parts in the following configurations:

1. A completely assembled Vickers/Maxim-type firearm;

2. A complete receiver-box;

3. A left side-plate;

4. A right side-plate (both automatic and semi-automatic configurations);

5. A trunnion;

6. A parts kit containing a right side-plate; and

7. A parts kit containing a left side-plate.

During the evaluations, complete, original models of Vickers and German

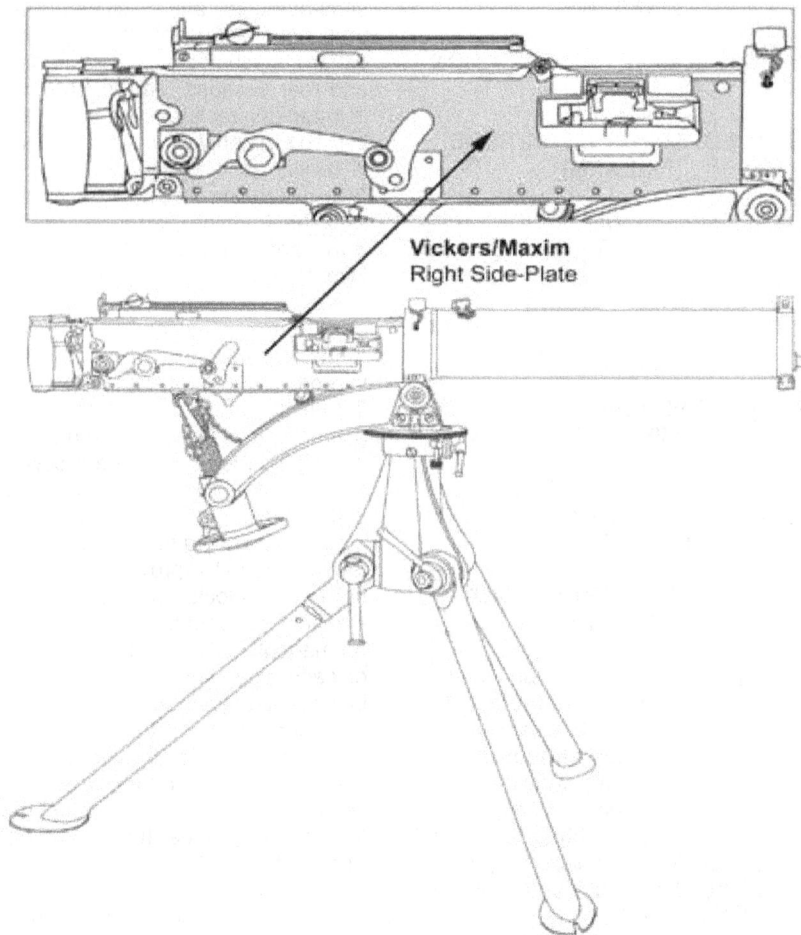

**Vickers/Maxim
Right Side-Plate**

Maxim-type machinegun is a **"machinegun"** as defined by the GCA, 18 U.S.C. 921(a)(23), the NFA, 26 U.S.C. 5845(b), and their implementing regulations, 27 CFR 478.11 and 479.11.

Moreover, though not necessary for classification, only with a complete right side-plate could the Vickers/Maxim-type machinegun shoot automatically more than one shot, without manual reloading, by a single function of the trigger. Therefore, the right side-plate of a Vickers/Maxim-type machinegun, either stand-alone or assembled as part of the complete receiver-box, is a machinegun receiver, and therefore, a **"machinegun,"** as defined by the GCA, NFA, and their implementing regulations. The right side-plate of the Vickers/Maxim-type machinegun is illustrated below.

Because the right side-plate of a Vickers/Maxim-type machinegun is the machinegun receiver, any kit or collection of parts of a Vickers/Maxim-type firearm containing the right side-plate is also a machinegun. In contrast, any individual part of a Vickers/Maxim-type firearm other than the machinegun's right side-plate, or any kit or collection of parts of a Vickers/Maxim-type firearm that does not contain the machinegun right side-plate is not a machinegun receiver, and therefore, not a **"machinegun,"** as defined by the GCA, NFA, and their implementing regulations.

Machineguns are required to be identified by a serial number, pursuant to 18 U.S.C. 923(i), 26 U.S.C. 5842, 27 CFR 478.92, and 479.102. Under 27 CFR 478.92(a)(1) and (2) and 479.102(a)(1) and (e), the serial number must be placed on the frame or receiver of a machinegun regardless of whether it is part of a complete firearm. Thus, the right side-plate, either unattached or assembled as part of a complete Vickers/Maxim-type machinegun, must be identified with a serial number and registered in the NFRTR.

ATF is aware that, without the benefit of this ruling, some complete Vickers/Maxim-type machineguns registered in the NFRTR were identified with serial numbers located on parts other than the right side-plate. Federal regulations at 27 CFR 478.92(a)(4) and 479.102(c) provide that the Director of ATF may authorize other means of identification of firearms if the other method is reasonable and will not hinder the effective administration of the regulations. Provided that it has not been disassembled into its component parts, ATF finds that a complete Vickers/Maxim-type machinegun identified with a serial

MG-08 Maxim-type machineguns were tested and found to fire automatically more than one round of ammunition by a single function of the trigger. Additionally, a reference model of a Vickers-type machinegun was tested for function; first, by partially removing the right side-plate, and then after completely removing the right side-plate. Although the sample firearm was capable of firing one shot with a partial right side-plate, once the entire right side-plate was removed, testing demonstrated the sample was not capable of functioning without the right side-plate. Accordingly, it was determined that neither a Vickers nor any Maxim-type machinegun would be capable of firing with the entire right side-plate removed. Nonetheless, ATF has previously examined semi-automatic versions of the Vickers/Maxim firearm where the right side-plate has an approved block that prevents installation of machinegun fire control components. These semi-automatic firearms have been classified by ATF as **"firearms"** under the Gun Control Act of 1968 (GCA), but not as **"machineguns"** under the GCA or the National Firearms Act (NFA).

The NFA, 26 U.S.C. 5845(a)(6), and its implementing regulation, 27 CFR 479.11, define the term **"firearm,"** in part, as **"a machinegun."** The term **"machinegun"** is defined by the GCA, 18 U.S.C. 921(a)(23), the NFA, 26 U.S.C. 5845(b), and their implementing regulations, 27 CFR 478.11 and 479.11, as **"any weapon which shoots, is designed to shoot, or can be readily restored to shoot, automatically more than one shot, without manual reloading, by a single function of the trigger. The term shall also include the frame or receiver of any such weapon ..."** The term **"frame or receiver"** is defined in 27 CFR 478.11 and 479.11 as **"[t]hat part of a firearm which provides housing for the hammer, bolt or breechblock and firing mechanism, and which is usually threaded at its forward portion to receive the barrel."**

A completely assembled Vickers/Maxim-type machinegun is a weapon that shoots automatically more than one shot, without manual reloading, by a single function of the trigger. Therefore, a completely assembled Vickers/

number on a component part of the receiver box other than the right side-plate has been registered in the NFRTR in a manner that will not hinder the effective administration of the NFA or its regulations.

However, if the complete machinegun is disassembled, such as for resale of the right side-plate, the right side-plate must be marked with its registered serial number and name of the manufacturer, as required by 26 U.S.C. 5842(a) and 27 CFR 479.102. If disassembled without its required markings, the unmarked right side-plate of a Vickers/Maxim-type firearm would be a machinegun receiver not identified or registered in the NFRTR, and therefore, unlawful to transfer and possess, pursuant to 18 U.S.C. 922(o) and 26 U.S.C. 5861.

Held, the right side-plate of a Vickers/Maxim-type firearm, manufactured with its camming lobe affixed in the proper location, and without an ATF approved block that prevents installation of machinegun fire control components, is a machinegun receiver, and therefore, a "**machinegun**" as that term is defined by the Gun Control Act of 1968, 18 U.S.C. 921(a)(23), the National Firearms Act, 26 U.S.C. 5845(b), and their implementing regulations, 27 CFR 478.11 and 479.11.

Held further, provided it has not been disassembled into its component parts, a complete Vickers/Maxim-type machinegun that is currently registered in the National Firearms Registration and Transfer Record with its serial number located on a component part of the receiver box other than the right side-plate, has been lawfully registered under the National Firearms Act and may be lawfully possessed by the registrant under 18 U.S.C. 922(o).

To the extent this ruling is inconsistent with any prior classifications, they are hereby superseded.

Date approved: September 28, 2010

18 U.S.C. 922(a)(3): PROHIBITED TRANSPORTATION OR RECEIPT

18 U.S.C. 922(a)(5): PROHIBITED TRANSFER BY NON-LICENSEE

18 U.S.C. 922(a)(9): PROHIBITED RECEIPT BY NON-RESIDENT

18 U.S.C. 922(b)(3): PROHIBITED TRANSFER BY LICENSEE

18 U.S.C. 922(t): REQUIREMENTS

TO TRANSFER FIREARMS

27 CFR 478.11: DEFINITIONS (STATE OF RESIDENCE)

27 CFR 478.11: DEFINITIONS (IDENTIFICATION DOCUMENT)

27 CFR 478.124: FIREARMS TRANSACTION RECORD

For the purpose of acquiring firearms under the Gun Control Act of 1968, a United States citizen who temporarily resides in a foreign country, but who also demonstrates the intention of making a home in a particular State, is a resident of the State during the time period he or she actually resides in that State. The intention of making a home in a State must be demonstrated to a Federal firearms licensee by presenting valid identification documents. Such documents include, but are not limited to, driver's licenses, voter registration, tax records, or vehicle registration.

ATF Rul. 2010-6

The Bureau of Alcohol, Tobacco, Firearms and Explosives (ATF) has received inquiries seeking clarification as to whether, under Federal law, United States citizens who maintain residences in both a foreign country and a particular State may purchase firearms while in the State.

The Gun Control Act of 1968 (GCA), as amended, Title 18, United States Code (U.S.C.), section 922(a)(3) prohibits any person not licensed under the GCA (non-licensees) from receiving any firearm purchased or otherwise obtained by such person outside the State in which the person resides. Furthermore, section 922(a)(9) provides, in part, that it is unlawful for any unlicensed person who does not reside in any State to receive any firearm unless such receipt is for lawful sporting purposes. Pursuant to section 922(b)(3), the GCA also prohibits licensees from selling or delivering firearms to any person who the licensee knows or has reasonable cause to believe does not reside in the State in which the licensee's place of business is located. Section 922(a)(5) extends the prohibition, with some limited exceptions, to any unlicensed person transferring a firearm to anyone who the transferor knows or has reasonable cause to believe does not reside in the transferor's State of residence.

A person's **"State of residence"** is defined by regulation in 27 CFR 478.11

as **"the State in which an individual resides. An individual resides in a State if he or she is present in a State with the intention of making a home in that State."** Ownership of a home or land within a given State is not sufficient, by itself, to establish a State of residence. However, ownership of a home or land within a particular State is not required to establish presence and intent to make a home in that State. Furthermore, temporary travel, such as short-term stays, vacations, or other transient acts in a State are not sufficient to establish a State of residence because the individual demonstrates no intention of making a home in that State.

To ensure compliance with this residency requirement, section 922(t) of the GCA requires licensees to examine a valid **"identification document"** (as defined in 18 U.S.C. 1028(d) and 27 CFR 478.11) of a firearm transferee. This document must contain the residence address of the transferee so that the licensee may verify the identity of the transferee and discern whether the transferee has the intention of making a home in a particular State. Licensees transferring a firearm to a person not licensed under the GCA are required, pursuant to 27 CFR 478.124, to record the firearm transaction on an ATF Form 4473, which requires, among other things, the transferee's residence address, including the transferee's State of residence as it appears on the valid identification document.

The term **"identification document"** is defined by 18 U.S.C. 1028(d)(3) as **"a document made or issued by or under the authority of the United States Government, a State, political subdivision of a State . . . which, when completed with information concerning a particular individual, is of a type intended or commonly accepted for the purpose of identification of individuals."** The regulations, 27 CFR 478.11, define the term **"identification document"** as **"[a] document containing the name, residence address, date of birth, and photograph of the holder and which was made or issued by or under the authority of the United States Government, a State, political subdivision of a State . . . which, when completed with information concerning a particular individual, is of a type intended or commonly accepted for the purpose of identification of individuals."** Identification documents include, but are not limited to, a driver's license, voter registration, tax records, or vehicle registration. As explained in ATF Ruling 2001-5

(ATFQB 2001-4, 37), a combination of valid government documents may be used to satisfy the GCA's State residency requirement.

ATF has previously addressed the eligibility of individuals to acquire firearms who maintain residences in more than one State. Federal regulations at 27 CFR 478.11 (definition of State of Residence), Example 2, clarify that a U.S. citizen with homes in two States may, during the period of time the person actually resides in a particular State, purchase a firearm in that State. See also ATF Publication 5300.4 (2005), Question and Answer B12, page 179. Similarly, in ATF Ruling 80-21 (ATFB 1980-4, 25), ATF held that, during the time college students actually reside in a college dormitory or at an off-campus location, they are considered residents of the State where the on-campus or off-campus housing is located.

The same reasoning applies to citizens of the United States who reside temporarily outside of the country for extended periods of time, but who also maintain residency in a particular State. Where a citizen temporarily resides outside of the country, but also has the intention of making a home in a particular State, the citizen is a resident of the State during the time he or she actually resides in that State. In acquiring a firearm, the individual must demonstrate to the transferor-licensee that he or she is a resident of the State by presenting valid identification documents.

Held, for the purpose of acquiring firearms under the Gun Control Act of 1968, a United States citizen who temporarily resides in a foreign country, but who also demonstrates the intention of making a home in a particular State, is a resident of the State during the time period he or she actually resides in that State.

Held further, the intention of making a home in a State must be demonstrated to a Federal firearms licensee by presenting valid identification documents. Such documents include, but are not limited to, driver's licenses, voter registration, tax records, or vehicle registration.

Date approved: November 10, 2010

18 U.S.C. 923(g)(1)(A): RECORDS REQUIRED

27 CFR 478.22: ALTERNATE METHODS OR PROCEDURES

27 CFR 478.121: RECORDS RE-QUIRED

27 CFR 478.123: RECORDS MAINTAINED BY MANUFACTURERS

27 CFR 478.125: RECORDS OF FIREARMS RECEIPT AND DISPOSITION

27 CFR 478.129: RECORD RETENTION

ATF authorizes an alternate method or procedure from the firearms acquisition and disposition recordkeeping requirements of 27 CFR 478.123. Specifically, ATF authorizes licensed manufacturers to consolidate their records of manufacture or other acquisition of firearms and their separate firearms disposition records, provided all of the conditions in this ruling are met.

ATF Rul. 2010-8

The Bureau of Alcohol, Tobacco, Firearms and Explosives (ATF) has received requests from licensed manufacturers for permission to consolidate their records of firearms manufacture or other acquisition and their separate records of firearms disposition.

The Gun Control Act of 1968 (GCA), at Title 18, United States Code, section 923(g)(1)(A), provides, in part, that each licensed manufacturer must maintain records of importation, production, shipment, receipt, sale, or other disposition of firearms at his place of business for such period, and in such form, as the Attorney General may by regulations prescribe. Federal regulations at Title 27, Code of Federal Regulations (CFR), section 478.123(a), require licensed manufacturers to record the type, model, caliber or gauge, and serial number of each complete firearm manufactured or otherwise acquired, and the date such manufacture or other acquisition was made, not later than the seventh day following the date such manufacture or other acquisition was made. The records of manufacture or other acquisition must be retained by the manufacturer on the licensed premises permanently, pursuant to 27 CFR 478.121(a) and 478.129(d).

Federal regulations at 27 CFR 478.123(b) require licensed manufacturers to record the disposition of firearms to other licensees showing the quantity, type, model, manufacturer, caliber, size or gauge, serial number of the firearms transferred, the name and license num-ber of the licensee to whom the firearms were transferred, and the date of the transaction. This information must be entered in the proper record book not later than the seventh day following the date of the transaction, and such information must be recorded under the format prescribed by 27 CFR 478.122, except that the name of the manufacturer need not be recorded if the firearm is of the manufacturer's own manufacture. Under 27 CFR 478.129(d), the manufacturer's records of the sale or other disposition of firearms to licensees must be retained by the manufacturer for 20 years.

In addition, under 27 CFR 478.123(d) licensed manufacturers must maintain separate records of the sales or other dispositions of firearms made to nonlicensees. These records must be maintained in the form and manner prescribed by regulations at 27 CFR 478.124, 478.125(e), and 478.125(i), with regard to firearms transaction records and records of firearms disposition. Under 27 CFR 478.129(d), the manufacturer's records of the sale or other disposition of firearms to nonlicensees must be retained for 20 years.

Licensed manufacturers may seek approval from ATF to use an alternate method or procedure to record the acquisition and disposition of firearms. Federal regulations at 27 CFR 478.123(c) provide that ATF may authorize alternate records of the disposal of firearms when it is shown by the licensed manufacturer that the alternate records will accurately and readily disclose the information required to be maintained. Additionally, under 27 CFR 478.22, the Director may approve an alternate method or procedure in lieu of a method or procedure specifically prescribed in the regulations when he finds that: (1) good cause is shown for the use of the alternate method or procedure; (2) the alternate method or procedure is within the purpose of, and consistent with the effect intended by, the specifically prescribed method or procedure and that the alternate method or procedure is substantially equivalent to that specifically prescribed method or procedure; and (3) the alternate method or procedure will not be contrary to any provision of law and will not result in an increase in cost to the Government or hinder the effective administration of 27 CFR Part 478.

ATF recognizes that, provided certain conditions are met, the consolidation of records of manufacture or other acquisition of firearms by a licensed manufacturer with corresponding firearms disposition records will accurately and readily disclose the information required to be maintained. It will also make it easier for manufacturers and ATF to account for and trace a manufacturer's firearms inventory. ATF therefore finds that there is good cause to authorize a variance to the separate acquisition and disposition records requirements of the Federal firearms regulations. Further, this alternate method is not contrary to any provision of law, will not increase costs to ATF, and will not hinder the effective administration of the Federal regulations.

Licensed firearms manufacturers are authorized to consolidate their records of manufacture or other acquisition of firearms and their separate firearms disposition records provided the following conditions are met:

1. Within seven (7) days of the date of manufacture or other acquisition, the licensed manufacturer records the following information for each firearm:

    a. Date of manufacture or other acquisition;

    b. Name of the person from whom the firearm was acquired;

    c. Address of the person from whom the firearm was acquired if the transferor is a nonlicensee or the complete 15-digit license number of the licensed manufacturer or other licensee from whom the firearm was acquired;

    d. Name of the manufacturer (to include the licensed manufacturer) and licensed importer (if applicable);

    e. Model;

    f. Serial number;

    g. Type; and

    h. Caliber, size or gauge.

2. Within seven (7) days of the date of sale or other disposition, beside the corresponding line item record of manufacture or other acquisition, the licensed manufacturer records the following information for each firearm:

    a. Date of sale or other disposition;

    b. Name of the person to whom the firearm was transferred (to include the licensed manufacturer); and

    c. Address of the person to whom the firearm was transferred if the transferee is a nonlicensee or the ATF Form 4473 serial number if the Forms 4473 are filed numerically, or if transferred to a licensee, the transferee's complete 15-digit license number.

3. A manufacturer intending to make any change to the model, type, caliber, size or gauge of a frame, receiver, or assembled firearm must log the firearm out of the acquisition and disposition record as a disposition to the licensee. Once the change has been made, the firearm must be recorded as a new firearm manufactured on a separate line of the acquisition and disposition record. As the manufacturer, the licensee should record his or her name and license number.

4. For firearms dispositions to a licensee, the commercial record of the transaction shall be retained separately from other commercial documents maintained by the licensed manufacturer until the transaction is recorded, and be readily available for inspection on the licensed premises.

5. For firearms dispositions to a nonlicensee, the Firearms Transaction Record, ATF Form 4473, shall be retained separately from the licensee's Form 4473 file, and be readily available for inspection on the licensed premises until the transaction is recorded. After that time, the Form 4473 shall be retained alphabetically (by name of purchaser), chronologically (by date of sale or other disposition), or numerically (by transaction serial number) as part of the licensed manufacturer's required records.

6. By using this variance, a line item will be recorded for each firearm manufactured or otherwise acquired and sold or otherwise disposed of by a licensed manufacturer. The quantity of firearms manufactured or otherwise acquired of the same type, model, and caliber or gauge must be able to be readily determined by adding all associated line items.

7. All consolidated firearms acquisition and disposition records must be maintained permanently by the licensed manufacturer. Additionally, as provided by 27 CFR 478.127, upon discontinuance of business all required records must be forwarded to the ATF Out-of-Business Records Center.

Licensees are reminded of their responsibility to ensure the accuracy and completeness of all required records, and to maintain such records on their licensed premises available for inspection. Additionally, this approval does not relieve licensees of any requirements of State, local, or other Federal government agencies. If acquisition and disposition records are maintained in electronic form, licensees must comply with ATF Ruling 2008-2 (approved August 25, 2008).

Held, pursuant to 27 CFR 478.22 and 478.123(c), ATF authorizes an alternate method or procedure from the firearms acquisition and disposition recordkeeping requirements of 27 CFR 478.123. Specifically, ATF authorizes licensed manufacturers to consolidate their records of manufacture or other acquisition of firearms and their separate firearms disposition records, provided all of the conditions in this ruling are met.

Held further, if ATF finds that a licensee has failed to abide by the conditions of this ruling, or uses any procedure that hinders the effective administration of the Federal firearms laws or regulations, ATF may notify the licensee that the licensee is no longer authorized to consolidate his acquisition and disposition records under this ruling.

Date approved: December 6, 2010

22 U.S.C. 2778(b): REQUIREMENTS FOR IMPORTATION

27 CFR 447.52: REQUIREMENTS FOR IMPORTATION

Persons holding a valid Federal firearms license and/or who are registered as importers of articles on the U.S. Munitions Import List importing surplus military defense articles importable as curios or relics, may submit photocopies of the original supporting statements and documents with ATF Form 6, if they certify, under penalties provided by law, that the supporting documentation is true, correct, and complete. Such persons who are authorized under ATF Ruling 2003-6 to file eForm 6, may submit digitally scanned copies of the original supporting statements and documents with eForm 6, if they certify, under penalties provided by law, that the supporting documentation is true, correct, and complete.

ATF Rul. 2010—9

This ruling was superseded by ATF Rul. 2013-1.

18 U.S.C. 921(a): DEFINITIONS

18 U.S.C. 922(a)(1)(A): LICENSES REQUIRED

18 U.S.C. 923(a): LICENSES RE-QUIRED

18 U.S.C. 923(i): IDENTIFICATION OF FIREARMS

27 CFR 478.11: DEFINITIONS

27 CFR 478.41(a): LICENSES RE-QUIRED

27 CFR 478.92: IDENTIFICATION OF FIREARMS

Any person licensed as a dealer-gunsmith who repairs, modifies, embellishes, refurbishes, or installs parts in or on firearms (frames, receivers, or otherwise) for, or on behalf of a licensed importer or licensed manufacturer, is not required to be licensed as a manufacturer under the Gun Control Act, provided the firearms for which such services are rendered are: (I) not owned, in whole or in part, by the dealer-gunsmith; (2) returned by the dealer-gunsmith to the importer or manufacturer upon completion of the manufacturing processes, and not sold or distributed to any person outside, the manufacturing process; and (3) already properly identified/marked by the importer or manufacturer in accordance with Federal law and regulations.

ATF Rul. 2010-10

The Bureau of Alcohol, Tobacco, Firearms and Explosives (ATF) has received inquiries from firearms industry members asking whether licensed dealer-gunsmiths who would be engaged in the business of repairing, modifying, embellishing, refurbishing, or installing parts in or on firearms for, or on behalf of a licensed importer or manufacturer are required to be licensed as manufacturers and abide by the requirements imposed on manufacturers.

In recent years, licensed firearms importers and manufacturers have contracted certain firearms manufacturing activities on their behalf to specialized licensed firearms manufacturers. Such activities include applying special coatings and treatments to firearms (e.g., bluing, anodizing, powder-coating, plating, polishing, heat/chemical treating). This has caused confusion over which importers and manufacturers are required to identify/mark firearms and maintain permanent records of importation or manufacture. For this reason, licensed importers and manufacturers have asked whether licensed dealergunsmiths, who are not required to mark firearms and keep production records, may engage in such manufacturing activities on their behalf.

The Gun Control Act of 1968 (GCA), Title 18, United States Code (U.S.C.), section

923(a), provides, in part, that no person shall engage in the business of importing, manufacturing, or dealing in firearms until he has filed an application with and received a license to do so from the Attorney General. A **"firearm"** is defined by 18 U.S.C. 921(a)(3) to include any weapon (including a starter gun) which will or is designed to or may readily be converted to expel a projectile by the action of an explosive, and the frame or receiver of any such weapon. The term **"manufacturer"** is defined by 18 U.S.C. 921(a)(10) as any person engaged in the business of manufacturing firearms or ammunition for purposes of sale or distribution. As applied to a manufacturer of firearms, the term **"engaged in the business"** is defined by 18 U.S.C. 921(a)(2l)(A) and 27 CPR 478.11, as a **"person who devotes time, attention, and labor to manufacturing firearms as a regular course of trade or business with the principal objective of livelihood and profit through the sale or distribution of the firearms manufactured."** The term "dealer" is defined by 18 U.S.C. 921(a)(11)(B) and 27 CPR 478.11 to include **"any person engaged in the business of repairing firearms or of making or fitting special barrels, stocks, or trigger mechanisms to firearms ... "** (i.e., a gunsmith). **As applied to a gunsmith, the term "engaged in the business"** is defined by 18 U.S.C. 92l(a)(21)(D) and 27 CFR478.11 as a **"person who devotes time, attention, and labor to engaging in such activity as a regular course of trade or business with the principal objective of livelihood and profit ..."**

In Revenue Ruling 55-342 (C.B. 1955-1, 562), ATF's predecessor agency interpreted the meaning of the terms **"manufacturer"** and **"dealer"** for the purpose of firearms licensing under the Federal Firearms Act, the precursor statute to the GCA. It was determined that a licensed dealer could assemble firearms from component parts on an individual basis, but could not engage in the business of assembling firearms from component parts in quantity lots for purposes of sale or distribution without a manufacturer's license. Since then, ATF has similarly and consistently interpreted the term **"manufacturer"** under the GCA to mean any person who engages in the business of making firearms, by casting, assembly, alteration, or otherwise, for the purpose of sale or distribution. Such persons must have a manufacturer's license under the GCA, maintain permanent records of manufacture, and submit annual manufacturing reports. The Revenue Ruling did not address whether dealergunsmiths who engage in the business of repairing, modifying, embellishing, refurbishing, or installing parts in or on firearms for, or on behalf of an importer or manufacturer are engaged in the business of manufacturing firearms requiring a manufacturer's license.

Manufacturing

ATF's long-standing position is that any activities that result in the making of firearms for sale or distribution, to include installing parts in or on firearm frames and receivers, and processes that primarily enhance a firearm's durability, constitute firearms manufacturing that may require a manufacturer's license. In contrast, some activities are not firearms manufacturing processes, and do not require a manufacturer's license. For example, ATF Ruling 2009-1 (approved January 12, 2009) explained that performing a cosmetic process or activity, such as camouflaging or engraving, that primarily adds to or changes the appearance or decoration of a firearm is not manufacturing. Likewise, ATF Ruling 2009-2 (approved January 12, 2009) stated that installing **"drop-in"** replacement parts in or on existing, fully assembled firearms does not result in any alteration to the original firearms. Persons engaged in the business of these activities that do not constitute firearms manufacturing need only obtain a dealer's license.

Although installing parts in or on firearms, and applying special coatings and treatments to firearms are manufacturing activities, the definition of **"manufacturer"** in 18 U.S.C. 921(a)(10) and 27 CFR 478.11 also requires that a person be **"engaged in the business"** before the manufacturer's license requirement of section 923(a) applies. Thus, a person who manufactures a firearm will require a manufacturer's license if he/she devotes time, attention, and labor to such manufacture as a regular course of trade or business with the principal objective of livelihood and profit through the sale or distribution of the firearms manufactured. If the person is performing such services only for a customer on firearms provided by that customer, and is not selling or distributing the firearms manufactured, the person would be a **"dealer"** as defined by 18 U.S.C. 921(a)(l1)(B) and 27 CFR 478.11, requiring a dealer's license, assuming the person is **"engaged in the business"** as defined in 18 U.S.C. 921(a)(21)(D) and 27 CFR 478.11 (i.e., **"gunsmithing"**).

Gunsmithing

A dealer is **"engaged in the business"** of gunsmithing, as defined in 18 U.S.C. 92l(a)(21)(D) and 27 CFR 478.11, when he/she receives firearms (frames, receivers, or otherwise) provided by a customer for the purpose of repairing, modifying, embellishing, refurbishing, or installing parts in or on those firearms. Once the work is completed, the gunsmith returns the firearms, and charges the customer for labor and parts. As with an individual customer, a licensed dealer-gunsmith may receive firearms (properly identified with a serial number and other information required by 27 CFR 478.92) and conduct gunsmithing services for a customer who is a licensed importer or manufacturer. A dealergunsmith is not **"engaged in the business"** of manufacturing firearms because the firearms being produced are not owned by the dealer-gunsmith, and he/she does not sell or distribute the firearms manufactured. Once the work is completed, the dealer-gunsmith returns the firearms to the importer or manufacturer upon completion of the manufacturing processes, and does not sell or distribute them to any person outside the manufacturing process. Under these circumstances, the licensed dealer-gunsmith is not **"engaged in the business"** of manufacturing firearms requiring a manufacturer's license.

In contrast, a dealer-gunsmith may make or acquire his/her own firearms, and repair, modify, embellish, refurbish, or install parts in or on those firearms. If the dealer-gunsmith then sells or distributes those firearms for livelihood and profit, the dealer-gunsmith is engaged in his/her own business of manufacturing firearms. A person engaged in the business of manufacturing firearms for sale or distribution is required to be licensed as a manufacturer, identify/mark all firearms manufactured, maintain permanent records of manufacture, submit annual manufacturing reports, and pay any taxes imposed on firearm manufacturers. A licensed dealer-gunsmith who becomes licensed

as a manufacturer must also segregate all firearms manufactured for that business separately from firearms for which gunsmithing services are being performed.

To facilitate inspection and ensure that ATF can determine that a licensed dealer-gunsmith is not engaged in the business of manufacturing firearms for his own sale or distribution without a manufacturer's license, licensees may take the following steps:

   (1) maintain a copy of the current, active license of all contracted licensees;

   (2) maintain a copy of the contract and all instructions for gunsmithing services rendered;

   (3) maintain a copy of the invoices for gunsmithing services;

   (4) timely and accurately reflect all firearms acquisitions and dispositions consistent with the contract for gunsmithing services rendered; and

   (5) in the case of a licensed dealer-gunsmith, maintain required bound acquisition and disposition records for all gunsmithing activities separate from other dealer's records.

Unless licensees take these steps, ATF may presume that a particular dealer-gunsmith is engaged in his own business of manufacturing firearms for sale or distribution without a manufacturer's license, and take corrective administrative or other enforcement action.

Identification of Firearms

The GCA at 18 U.S.C. 923(i) provides, in part, that licensed manufacturers and importers must **"identify"** each firearm manufactured or imported by a serial number in the manner prescribed by regulation. Federal regulations at 27 CFR 478.92(a)(l) further require importers and manufacturers to identify each firearm by engraving, casting, stamping (impressing), or otherwise conspicuously placing the individual serial number and certain additional information - the model (if designated), caliber/gauge, manufacturer's name, and place of origin on the frame, receiver, or barrel - at a minimum depth. Section 478.92(a)(2) specifies that a **"firearm frame or receiver that is not a component part of a complete weapon at the time it is sold, shipped, or otherwise disposed of ... must be identified as required by this section."**

Because dealer-gunsmiths are not required to identify firearms manufactured, it is incumbent upon the importer or manufacturer, prior to shipping firearms to a dealer gunsmith for gunsmithing services, to mark them with a serial number and other required information. With regard to frames and receivers shipped separately, section 478.92(a)(2) provides, in part, that the manufacturer or importer must mark all frames and receivers prior to shipment with all information required by section 478.92 (i.e., serial number, model (if designated), caliber/gauge, manufacturer's name, and place of origin). This will ensure that the frames and receivers can be traced by ATF in the event they are lost or stolen during the manufacturing process.

Held, any person licensed as a dealer-gunsmith who repairs, modifies, embellishes, refurbishes, or installs parts in or on firearms (frames, receivers, or otherwise) for, or on behalf of a licensed importer or licensed manufacturer, is not required to be licensed as a manufacturer under the Gun Control Act, provided the firearms for which such services are rendered are: (1) not owned, in whole or in part, by the dealer-gunsmith; (2) returned by the dealer-gunsmith to the importer or manufacturer upon completion of the manufacturing processes, and not sold or distributed to any person outside the manufacturing process; and (3) already properly identified/marked by the importer or manufacturer in accordance with Federal law and regulations.

This ruling is limited to an interpretation of the requirements imposed upon importers, manufacturers, and dealer-gunsmiths under the Gun Control Act of 1968, and does not apply to persons making or manufacturing firearms subject to the National Firearms Act, 26 U.S.C. 5801 et. seq.

Revenue Ruling 55-342, C.B. 1955-1, 562, is hereby clarified. To the extent this ruling may be inconsistent with any prior letter rulings, they are hereby superseded.

Date approved: December 27, 2010

18 U.S.C. 923(g)(l)(A): RECORDS REQUIRED

27 CFR 478.22; ALTERNATE METHODS OR PROCEDURES

27 CFR 478.U1: RECORDS REQUIRED

27 CFR 478.122: RECORDS MAINTAINED BY IMPORTERS

27 CFR 478.125: RECORDS OF FIREARMS RECEIPT AND DISPOSITION

27 CFR 478.U9: RECORD RETENTION

ATF authorizes an alternate method or procedure from the firearms acquisition and disposition recordkeeping requirements of 27 CFR 478.122. Specifically, ATF authorizes licensed importers to consolidate their records of importation or other acquisition of firearms and their separate firearms disposition records, provided all of the requirements stated in this ruling are met.

ATF Rul. 2011-I

The Bureau of Alcohol, Tobacco, Firearms and Explosives (ATF) has received requests from licensed importers for permission to consolidate their records of firearms importation or other acquisition and their separate records of firearms disposition.

The Gun Control Act of 1968 (GCA), at Title 18.United States Code, section

923(g)(I)(A), provides, in part, that each licensed importer must maintain records of importation, production, shipment, receipt, sale, or other disposition of firearms at his place of business for such period, and in such form, as the Attorney General may by regulations prescribe. Federal regulations at Title 27, Code of Federal Regulations (CFR), section 478.122(a), require licensed importers, within 15 days of importation or other acquisition, to record the type, model, caliber or gauge, manufacturer, country of manufacture, and the serial number of each firearm imported or otherwise acquired, and the date such importation or other acquisition was made. The records of importation or other acquisition must be retained by the importer on the licensed premises permanently, pursuant to 27 CPR 478.121(a) and 478.129(d).

Federal regulations at 27 CFR 478.122(b) require licensed importers to record the disposition of firearms to other licensees showing the quantity, type, manufacturer, country of manufacture, caliber or gauge, model, serial number of the firearms transferred, the name and license number of the licensee to whom the firearms were transferred, and the date of the transaction. This in-

formation must be entered in the proper record book not later than the seventh day following the date of the transaction, and such information must be recorded under the format prescribed by 27 CFR 478.122. Under 27 CFR 478.129(d), the importer's records of the sale or other disposition of firearms to licensees must be retained by the importer for 20 years.

In addition, Wider 27 CFR 478.122(d) licensed importers must maintain separate records of the sales or other dispositions of firearms made to nonlicensees. These records must be maintained in the form and manner prescribed by regulations at 27 CFR 478.124, 478.125(e), and 478.125(i), with regard to firearms transaction records and records of firearms disposition. Under 27 CFR 478.129(d), the importer's records of the sale or other disposition of firearms to nonlicensees must be retained for 20 years.

Licensed importers may seek approval from ATF to use an alternate method or procedure to record the acquisition and disposition of firearms. Federal regulations at 27 CFR

478.122(c) provide that ATF may authorize alternate records of the disposal of firearms when it is shown by the licensed importer that the alternate records will accurately and readily disclose the information required to be maintained. Additionally, under 27 CFR 478.22, the Director may approve an alternate method or procedure in lieu of a method or procedure specifically prescribed in the regulations when he finds that: (1) good cause is shown for the use of the alternate method or procedure; (2) the alternate method or procedure is within the purpose of, and consistent with the effect intended by, the specifically prescribed method or procedure and that the alternate method or procedure is substantially equivalent to that specifically prescribed method or procedure; and (3) the alternate method or procedure will not be contrary to any provision of law and will not result in an increase in cost to the Government or hinder the effective administration of 27 CFR Part 478.

ATF recognizes that, provided certain conditions are met, the consolidation of records of importation or other acquisition of firearms by a licensed importer with corresponding firearms disposition records will accurately and readily disclose the information required to be maintained. It will also make it easier for importers and ATF to account for and trace an importer's .firearms inventory. ATF therefore finds that there is good cause to authorize a

variance to the separate acquisition and disposition records requirements of the Federal firearms regulations. Further, this alternate method is not contrary to any provision of law, will not increase costs to ATF, and will not hinder the effective administration of the Federal regulations.

Licensed firearms importers are not required to maintain separate records of importation or other acquisition and disposition of firearms, as required by the Federal regulations, provided the following conditions are met:

1. Within 15 days of the date of importation or other acquisition, the licensed importer records the following information for each firearm:

a. Name of the manufacturer (foreign or domestic) and importer (if any);

b. Country of manufacture;

c. Model;

d. Type;

e. Serial number;

f. Caliber or gauge;

g. Date of importation or other acquisition;

h. Name of the person (nonlicensee or other Federal firearms licensee (FFL))from whom the firearm was received (to include a person outside of the United States, if applicable); and

i. Address (if a non licensee) or complete 15-digit license number of the person from whom the firearm was received (to include the address of a person outside of the United States, if applicable).

2. Within seven (7) days of the date of sale or other disposition, beside the corresponding line item record of importation or other acquisition, the licensed importer records the following information for each firearm:

a. Date of sale or other disposition;

b. Name of the person to whom the firearm was transferred; and

c. Address or license number of the person to whom the f1rearm was transferred: if the transferee is a nonlicensee, the address of the transferee or the ATF Form 4473 Firearms Transaction Record serial number if the Forms 4473 are filed numerically; if the transferee is a licensee, the complete 15digit license number.

3. For firearms dispositions to a licensee, the commercial record of the transaction shall be retained separately from other commercial documents maintained by the licensed importer until the transaction is recorded, and be readily available for inspection on the licensed premises.

4. For firearms dispositions to a nonlicensee, the ATF Form 4473 shall be retained separately from the licensee's Form 4473 file and be readily available for inspection on the licensed premises until the transaction is recorded. After that time, the Form 4473 shall be retained alphabetically (by name of purchaser), chronologically (by date of sale or other disposition), or numerically (by transaction serial number) as part of the licensed importer's required records.

5. By using this variance, a line item will be recorded for each firearm imported or otherwise acquired and sold or otherwise disposed of by a licensed importer. The quantity of firearms of the same type, model, and caliber or gauge imported or otherwise acquired and disposed of must be able to be readily determined by adding all associated line items.

6. All consolidated firearms acquisition and disposition records must be maintained permanently by the licensed importer. Additionally, as provided by 27 CFR 478.127, upon discontinuance of business, all required records must be forwarded to the ATF Out-of-Business Records Center.

Licensees are reminded of their responsibility to ensure the accuracy and completeness of all required records, and to maintain such records on their licensed premises available for inspection. Additionally, this approval does not relieve licensees of any requirements of State, local, or other Federal government agencies. If acquisition and disposition records are maintained in electronic form, licensees must comply with ATF Ruling 2008-2 (approved August 25, 2008).

Held, pursuant to 27 CFR 478.22 and 478.122(c), ATF authorizes an alternate method or procedure from the firearms acquisition and disposition recordkeeping requirements of 27 CFR 478.122. Specifically, ATF authorizes licensed importers to consolidate their records of importation or other acquisition of firearms and their separate firearms disposition records, provided all of the requirements stated in this ruling are met.

Held further, if ATF finds that a licensee has failed to abide by the conditions of this ruling, or uses any procedure that hinders the effective administration of the Federal firearms laws or regulations, ATF may notify the licensee that the licensee is no longer authorized to consolidate his acquisition and disposition records under this ruling.

Date Approved: January 26, 2011

Editors Note:

Reference to ATF Rul. 2008-2 is replaced by reference to ATF Rul. 2013-5 approved December 17, 2013.

26 U.S.C. 5845(a)(3): DEFINITIONS (FIREARM)

26 U.S.C. 5845(a)(4): DEFINITIONS (FIREARM)

26 U.S.C. 5845(c): DEFINITIONS (RIFLE)

27 CFR 479.11: DEFINITIONS (RIFLE)

27 CFR 479.11: DEFINITIONS (PISTOL)

A firearm, as defined by the National Firearms Act (NFA), 26 U.S.C. 5845(a) (3), is made when unassembled parts are placed in close proximity in such a way that they: (a) serve no useful purpose other than to make a rifle having a barrel or barrels of less than 16 inches in length; or (b) convert a complete weapon into such an NFA firearm. A firearm, as defined by 26 U.S.C. 5845(a)(3) and (a)(4), is not made when parts within a kit that were originally designed to be configured as both a pistol and a rifle are assembled or re-assembled in a configuration not regulated under the NFA (e.g., as a pistol, or a rifle with a barrel or barrels of 16 inches or more in length). A firearm, as defined by 26 U.S.C. 5845(a)(3) and (a)(4), is not made when a pistol is attached to a part or parts designed to convert the pistol into a rifle with a barrel or barrels of 16 inches or more in length, and the parts are later unassembled in a configuration not regu-

lated under the NFA (e.g., as a pistol). A firearm, as defined by 26 U.S.C. 5845(a)(4), is made when a handgun or other weapon with an overall length of less than 26 inches, or a barrel or barrels of less than 16 inches in length, is assembled or produced from a weapon originally assembled or produced only as a rifle.

ATF Rul. 2011-4

The Bureau of Alcohol, Tobacco, Firearms and Explosives (ATF) has received requests from individuals to classify pistols that are reconfigured into rifles, for personal use, through the addition of barrels, stocks, and other parts and then returned to a pistol configuration by removal of those components. Specifically, ATF has been asked to determine whether such a pistol, once returned to a pistol configuration from a rifle, becomes a **"weapon made from a rifle"** as defined under the National Firearms Act (NFA).

Some manufacturers produce firearm receivers and attachable component parts that are designed to be assembled into both rifles and pistols. The same receiver can accept an interchangeable shoulder stock or pistol grip, and a long (16 or more inches in length) or short (less than 16 inches) barrel. These components are sold individually, or as unassembled kits. Generally, the kits include a receiver, a pistol grip, a pistol barrel less than 16 inches in length, a shoulder stock, and a rifle barrel 16 inches or more in length. Certain parts or parts sets are also designed to allow an individual to convert a pistol into a rifle without removing a barrel or attaching a shoulder stock to the pistol. These parts consist of an outer shell with a shoulder stock into which the pistol may be inserted. When inserted, the pistol fires a projectile through a rifled extension barrel that is 16 inches or more in length, and with an overall length of 26 inches or more. Other parts sets require that certain parts of the pistol, such as the pistol barrel and the slide assembly, be removed from the pistol frame prior to attaching the parts sets. Typically, a separate barrel is sold with the parts set, which is 16 inches or greater in length. The barrel is installed along with an accompanying shoulder stock. The resulting firearm has a barrel of 16 inches or more in length, and an overall length of 26 inches or more.

The NFA, Title 26, United States Code (U.S.C.), Chapter 53, requires that persons manufacturing, importing, transferring, or possessing firearms as defined in the NFA comply with the Act's licensing, registration, and taxation requirements. The NFA defines the term **"firearm"** at 26 U.S.C. 5845(a) to include **"(3) a rifle having a barrel or barrels of less than 16 inches in length;"** (**"short-barreled rifle"**) and **"(4) a weapon made from a rifle if such weapon as modified has an overall length of less than 26 inches or a barrel or barrels of less than 16 inches in length"** (**"weapon made from a rifle"**). The term **"rifle"** is defined by 26 U.S.C. 5845(c) and 27 CFR 479.11 as **"a weapon designed or redesigned, made or remade, and intended to be fired from the shoulder and designed or redesigned and made or remade to use the energy of the explosive in a fixed cartridge to fire only a single projectile through a rifled bore for each single pull of the trigger, and shall include any such weapon which may be readily restored to fire a fixed cartridge."** Although not defined in the NFA, the term **"pistol"** is defined by the Act's implementing regulations, 27 CFR 479.11, as **"a weapon originally designed, made, and intended to fire a projectile (bullet) from one or more barrels when held in one hand, and having (a) a chamber(s) as an integral part(s) of, or permanently aligned with, the bore(s); and (b) a short stock designed to be gripped by one hand and at an angle to and extending below the line of the bore(s)"** (emphasis added).

Unassembled Parts Kits

In United States v. Thompson/Center Arms Company, 504 U.S. 505 (1992), the United States Supreme Court examined whether a short-barreled rifle was **"made"** under the NFA when a carbine-conversion kit consisting of a single-shot **"Contender"** pistol was designed so that its handle and barrel could be removed from its receiver, and was packaged with a 21-inch barrel, a rifle stock, and a wooden fore-end. The Court held that, where aggregated parts could convert a pistol into either a regulated short-barreled rifle, or an unregulated rifle with a barrel of 16 inches or more in length, the NFA was ambiguous and applied the **"rule of lenity"** (i.e., ambiguities in criminal statutes should be resolved in favor of the defendant) so that the pistol and carbine kit, when packaged together, were not considered a **"short-barreled rifle"** for purposes of the NFA.

However, the Court also explained that an NFA firearm is made if aggregated parts are in close proximity such that they: (a) serve no useful purpose other

than to make an NFA firearm (e.g., a receiver, an attachable shoulder stock, and a short barrel); or (b) convert a complete weapon into an NFA firearm (e.g., a pistol and attachable shoulder stock, or a long-barreled rifle and attachable short barrel). Id. at 511-13.

Assembly of Weapons from Parts Kits

The Thompson/Center Court viewed the parts within the conversion kit not only as a Contender pistol, but also as an unassembled **"rifle"** as defined by 26 U.S.C. 5845(c). The inclusion of the rifle stock in the package brought the Contender pistol and carbine kit within the **"intended to be fired from the shoulder"** language in the definition of rifle at 26 U.S.C. 5845(c). Id. at 513 n.6. Thompson/Center did not address the subsequent assembly of the parts. United States v. Ardoin, 19 F.3d 177, 181 (5th Cir. 1994). Based on the definition of **"firearm"** in 26 U.S.C. 5845(a)(3), if parts are assembled into a rifle having a barrel or barrels of less than 16 inches in length, a regulated short-barreled rifle has been made. See, e.g., United States v. Owens, 103 F.3d 953 (11th Cir. 1997); United States v. One (1) Colt Ar-15, 394 F. Supp. 2d 1064 (W.D.Tenn. 2004). Conversely, if the parts are assembled into a rifle having a barrel or barrels 16 inches in length or more, a rifle not subject to the NFA has been made.

Therefore, so long as a parts kit or collection of parts is not used to make a firearm regulated under the NFA (e.g., a short-barreled rifle or **"any other weapon"** as defined by 26 U.S.C. 5845(e)), no NFA firearm is made when the same parts are assembled or re-assembled in a configuration not regulated under the NFA (e.g., a pistol, or a rifle with a barrel of 16 inches or more in length). Merely assembling and disassembling such a rifle does not result in the making of a new weapon; rather, it is the same rifle in a knockdown condition (i.e., complete as to all component parts). Likewise, because it is the same weapon when reconfigured as a pistol, no **"weapon made from a rifle"** subject to the NFA has been made.

Nonetheless, if a handgun or other weapon with an overall length of less than 26 inches, or a barrel or barrels of less than 16 inches in length is assembled or otherwise produced from a weapon originally assembled or produced only as a rifle, such a weapon is a **"weapon made from a rifle"** as defined by 26 U.S.C. 5845(a)(4). Such a weapon would not be a **"pistol"** because the weapon was not

originally designed, made, and intended to fire a projectile by one hand.

Held, a firearm, as defined by the National Firearms Act (NFA), 26 U.S.C. 5845(a)(3), is made when unassembled parts are placed in close proximity in such a way that they:

(a)Serve no useful purpose other than to make a rifle having a barrel or barrels of less than 16 inches in length (e.g., a receiver, an attachable shoulder stock, and barrel of less than 16 inches in length); or

(b) Convert a complete weapon into such an NFA firearm, including –

(1) A pistol and attachable shoulder stock; and

(2) A rifle with a barrel of 16 inches or more in length, and an attachable barrel of less than 16 inches in length.

Such weapons must be registered and are subject to all requirements of the NFA.

Held further, a firearm, as defined by 26 U.S.C. 5845(a)(3) and (a)(4), is not made when parts in a kit that were originally designed to be configured as both a pistol and a rifle are assembled or re-assembled in a configuration not regulated under the NFA (e.g., as a pistol, or a rifle with a barrel of 16 inches or more in length).

Held further, a firearm, as defined by 26 U.S.C. 5845(a)(3) and (a)(4), is not made when a pistol is attached to a part or parts designed to convert the pistol into a rifle with a barrel of 16 inches or more in length, and the parts are later unassembled in a configuration not regulated under the NFA (e.g., as a pistol).

Held further, a firearm, as defined by 26 U.S.C. 5845(a)(4), is made when a handgun or other weapon with an overall length of less than 26 inches, or a barrel or barrels of less than 16 inches in length, is assembled or produced from a weapon originally assembled or produced only as a rifle. Such weapons must be registered and are subject to all requirements of the NFA.

To the extent this ruling may be inconsistent with any prior letter rulings, they are hereby superseded.

Date approved: July 25, 2011

18 U.S.C. 921(a)(3): DEFINITIONS (FIREARM)

18 U.S.C. 923(i): IDENTIFICATION OF FIREARMS

27 CFR 478.11: DEFINITIONS (FIREARM)

27 CFR 478.92: IDENTIFICATION OF FIREARMS

27 CFR 478.123(a): RECORDS MAINTAINED BY MANUFACTURERS

All firearms manufactured, to include firearm frames and receivers that are to be sold, shipped, or otherwise disposed of separately, must be identified by a licensed manufacturer in the manner and with the markings required by 18 U.S.C. 923(i) and 27 CFR 478.92 during the manufacturing process. It is reasonable for a licensed manufacturer to have seven (7) days following the date of completion (to include a firearm in knockdown condition, i.e., complete as to all component parts, or a frame or receiver to be sold, shipped, or disposed of separately) in which to mark a firearm manufactured, and record its identifying information in the manufacturer's permanent records. A firearm frame or receiver that is not a component part of a complete weapon at the time it is sold, shipped, or otherwise disposed of must be marked with all of the required markings; provided, that an alternate means of identification may be approved (i.e., a "non-marking variance") under section 478.92(a)(4); provided further, that the model designation and caliber or gauge may be omitted without a variance if that information is unknown at the time the frames or receivers are marked.

ATF Rul. 2012-1

The Bureau of Alcohol, Tobacco, Firearms and Explosives (ATF) has received inquiries from manufacturers asking when they must mark the firearms they manufacture. They ask specifically whether the requirement to record firearms in their records of manufacture within seven days also requires marking within that time period.

Some licensed firearms manufacturers produce and assemble complete firearms within their licensed business premises. Others manufacture only frames or receivers that are not parts of complete weapons when transferred to other licensees for further manufacturing.

In some instances, a manufacturer may possess firearm frames or receivers for an extended period of time without selling, shipping, or otherwise disposing of them. Such manufacturers may intend to sell these components as frames or receivers without further manufacturing, or intend to manufacture complete weapons from them in the future.

The Gun Control Act (GCA) at 18 U.S.C. 921(a)(3) and its implementing regulation, 27

CFR 478.11, define the term **"firearm,"** in part, to mean: **"any weapon (including a starter gun) which will or is designed to or may readily be converted to expel a projectile by the action of an explosive,"** and includes **"the frame or receiver of any such weapon."** The term **"firearm frame or receiver"** is defined by section 478.11 as **"[t]hat part of a firearm which provides housing for the hammer, bolt or breechblock, and firing mechanism, and which is usually threaded at its forward position to receive the barrel."** Under 18 U.S.C. 923(i) and 27 CFR 478.92, licensed manufacturers and licensed importers must **"identify, by means of a serial number engraved or cast on the receiver or frame of the weapon, in such manner as the Attorney General shall by regulations prescribe, each firearm imported or manufactured ...."** Federal regulations at 478.92(a)(1) require manufacturers to identify each firearm by engraving, casting, stamping (impressing), or otherwise placing certain additional information (the model, caliber/gauge, manufacturer's name, and place of origin) on the frame, receiver, or barrel, at a minimum depth. Paragraph 478.92(a)(2) specifies that a **"firearm frame or receiver that is not a component part of a complete weapon at the time it is sold, shipped, or otherwise disposed of ... must be identified as required by this section."** As to recordkeeping, section 478.123(a) provides that **"[e]ach licensed manufacturer shall record the type, model, caliber or gauge, and serial number of each complete firearm manufactured or otherwise acquired ... not later than the seventh day following the date such manufacture or acquisition was made."** (emphasis added)

Congress did not specify in section 923(i) a time period for a manufacturer to mark firearms manufactured. However, it is unlikely Congress intended to allow manufacturers to stockpile completed firearms (including finished frames or receivers of such firearms to be sold or shipped separately) without markings

for an indefinite period of time. The plain language in section 923(i) requiring that the serial number be **"engraved or cast on"** the receiver or frame of the firearm demonstrates Congressional intent to require marking as an integral part of the manufacturing process. Otherwise, fully assembled firearms, or finished frames and receivers to be sold, shipped, or disposed of separately - stockpiled indefinitely without required markings - would be susceptible to theft or loss without any means to trace them.

Under the regulations, section 478.92, firearms required to be marked at the time of manufacture include both **"complete weapons,"** and complete frames or receivers of such weapons that are to be sold, shipped, or otherwise disposed of separately. Because identifying firearms is an integral part of the manufacturing process, and sections 923(i) and 478.92 do not specify a time period in which to identify firearms, licensed manufacturers are required to mark them during the manufacturing process. Reading the marking requirement for complete weapons in section 478.92(a) together with the seven day recordkeeping requirement for complete firearms in section 478.123(a), ATF concludes that it is reasonable for a manufacturer to have seven (7) days following the date of completion (to include a firearm in knockdown condition, i.e., complete as to all component parts, or a frame or receiver to be sold, shipped, or disposed of separately) in which to mark the firearm and record its identifying information in the manufacturer's permanent records. Further, because firearm frames and receivers to be sold, shipped, or disposed of separately do not have a barrel at the time they are marked, pursuant to 478.92(a)(2), all of the information required by 478.92(a)(1) must be placed on the frame or receiver, unless an alternate means of identification is approved (i.e., a **"non-marking variance"**) under section 478.92(a)(4). The model designation and the caliber or gauge may be omitted without a variance if that information is unknown at the time the firearm frames or receivers are marked.

Nonetheless, ATF recognizes that a manufacturer may require more than seven days to finish the manufacturing process from beginning to end with the required markings, depending on the nature of the process involved. Some firearms may take more time due to differences in the type and capability of the firearm, availability of materials and components, and complexity of the assembly and finishing processes. ATF also recognizes that the equipment necessary to identify the firearms must be available and in working order. However, once the entire manufacturing process has ended, manufacturers must ensure that the firearms have been marked in the manner required by section 478.92.

To facilitate inspection and ensure that ATF can determine that a licensed manufacturer has not unreasonably held completed firearms (to include finished frames and receivers to be sold, shipped, or disposed of separately) after seven (7) days from the date of completion without their required markings, licensees may take the following steps:

(1) maintain a copy of the current, active license of all contracted licensees;

(2) maintain records of firearms production;

(3) maintain work orders, contracts, and related instructions for services rendered that describe the various firearm manufacturing processes;

(4) maintain orders for firearm parts that have yet to be received; and

(5) maintain invoices to repair non-functioning machines.

Held, all firearms manufactured, to include firearm frames and receivers that are to be sold, shipped, or otherwise disposed of separately, must be identified by a licensed manufacturer in the manner and with the markings required by 18 U.S.C. 923(i) and 27 CFR 478.92 during the manufacturing process.

Held further, it is reasonable for a licensed manufacturer to have seven (7) days following the date of completion (to include a firearm in knockdown condition, i.e., complete as to all component parts, or a frame or receiver to be sold, shipped, or disposed of separately) in which to mark a firearm manufactured, and record its identifying information in the manufacturer's permanent records.

Held further, a firearm frame or receiver that is not a component part of a complete weapon at the time it is sold, shipped, or otherwise disposed of must be marked with all of the required markings; provided, that an alternate means of identification may be approved (i.e., a "non-marking variance") under section 478.92(a)(4);provided further, that the model designation and caliber or gauge may be omitted without a variance if that

information is unknown at the time the frames or receivers are marked.

This ruling is limited to an interpretation of the requirements imposed upon licensed manufacturers under the Gun Control Act of 1968, 18 U.S.C. 921 et. seq., and does not apply to persons making or manufacturing firearms subject to the National Firearms Act,26 U.S.C. 5801 et. seq. To the extent this ruling may be inconsistent with any prior letter rulings, they are hereby superseded.

Date Approved: January 12, 2012

18 U.S.C. 923(g)(5)(A): RECORDS

The Bureau of Alcohol, Tobacco, Firearms and Explosives (ATF) authorizes an alternate method or procedure to the requirement at Title 18, United States Code (U.S.C.), section 923(g)(5)(A), that licensees submit a form containing certain required record information. Specifically, licensed manufacturers may submit the Annual Firearms Manufacturing and Exportation Report (AFMER), ATF Form 5300.11, electronically using ATF eForms, provided all of the conditions set forth in this ruling are met.

ATF Rul. 2012-3

The Gun Control Act of 1968 (GCA), 18 U.S.C. 923(g)(5)(A), provides, in relevant part, that each licensed importer, manufacturer, and dealer shall, when required by letter issued by the Attorney General, and until notified to the contrary, submit on a form specified by the Attorney General, for periods and at the times specified in such letter, all record information required to be kept by 18 U.S.C. Chapter 44 or such lesser record information as the Attorney General in such letter prescribes. The authority to administer and enforce the provisions of the GCA was delegated to the Attorney General, who further delegated these responsibilities to the Director of ATF. See 28 CFR 0.130. Pursuant to 18 U.S.C. 923(g)(5)(A), ATF requires licensed manufacturers to submit an Annual Firearms Manufacturing and Exportation Report (AFMER), ATF Form 5300.11. Currently, the form may either be mailed or faxed to ATF.

The Government Paperwork Elimination Act (GPEA), 44 U.S.C. 3504, requires executive agencies to provide for the option of the electronic maintenance, submission, or disclosure of information, as a substitute for paper, and for the use and acceptance of electronic signatures,

when practicable, by October 2003. In accordance with the GPEA's mandate, ATF developed the eForms online electronic filing system for Federal firearms licensees (FFLs) and others to submit a variety of ATF forms.

The eForms online electronic filing system now enables licensed manufacturers to submit the AFMER and meet the submission requirements set forth in 18 U.S.C. 923(g)(5)(A). The ATF Form 5300.11 is accepted and processed by ATF provided the manufacturer has furnished all of the required information. The manufacturer may obtain through the internet electronic copies of submitted ATF Forms 5300.11, which can be printed and preserved as documentation of compliance with the filing requirement.

To register to use the eForms system, applicants must access the ATF website at http://www.atfonline.gov/. This site contains the instructions necessary to access eForms and register to use the system. To register, the applicant must provide his or her name, business name (if applicable), address, telephone number, and email address. Upon proper registration, ATF issues each registrant a user identification (ID), and the registrant creates a password allowing access to the eForms system. Each individual registrant is issued a unique user ID that can be applied to multiple Federal firearms licenses. After registering with the unique user ID, persons filing the AFMER form on behalf of an FFL must request to be associated with the applicable Federal firearms license using the **"My Profile"** function of the eForms system.

The eForms system requires applicants to attest that the information submitted is true and correct, subject to penalties provided by law, and confirm their ATF-issued electronic credentials to complete the application process. Specifically, to complete the application process, an applicant is required to declare first that all statements contained in the application are true and correct, and that he or she has read, understood, and complied with the conditions and instructions for the form being submitted. Second, the applicant is required to declare that he or she authorizes the transmittal via the eForms system of data that may constitute tax return information, as defined in section 6103 of the Internal Revenue Code, Title 26, U.S.C. Applicants are not required to use the eForms system, and licensees may continue to file ATF Form 5300.11 on paper in accordance with the instructions on the form.

In addition to the mandate of GPEA, ATF recognizes that, provided certain conditions are met, the use of electronic AFMERs may more accurately and readily disclose the same information required to be submitted. It will also make it easier, less costly, and faster for manufacturers to submit their reports, and for ATF to process those reports. ATF therefore finds that there is good cause to authorize a variance from the ATF Form 5300.11 submission requirement of the Federal firearms laws. ATF also finds that, provided certain conditions are met, the alternate method set forth in this ruling is within the purposes of, and consistent with the requirements at 18 U.S.C. 923(g)(5)(A) because the same required information is captured on the applicable eForm, which is electronically signed under penalties provided by law. Further, this alternate method is not contrary to any provision of law, will not increase costs to ATF, and will not hinder the effective administration of the regulations.

Held, licensed manufacturers may submit the AFMER, ATF Form 5300.11, electronically using ATF eForms, provided the following conditions are met:

1. The applicant has registered with ATF by completing the on-line registration process;

2. The applicant has received a unique user ID and created a password;

3. The applicant has agreed that the electronic signature assigned to them is intended as their original signature for eForms submissions; and

4. The applicant has agreed to be bound by the Notices and Agreements governing the use of the eForms system.

Held further, if ATF finds that an eForms user has failed to abide by the conditions of this ruling, uses any procedure that hinders the effective administration of the Federal firearms laws or regulations, or any legal or administrative difficulties arise due to the use of eForms, ATF may notify the person that he or she is no longer authorized to use eForms.

Date approved: June 14, 2012

18 U.S.C. 925(d): EXCEPTIONS

22 U.S.C. 2778: CONTROL OF ARMS EXPORTS AND IMPORTS

26 U.S.C. 5844: IMPORTATION

27 CFR 447.42: APPLICATION FOR PERMIT

27 CFR 447.45: IMPORTATION

27 CFR 478.22: ALTERNATE METHODS OR PROCEDURES

27 CFR 478.111-478.113, 478.116, and 478.119: IMPORTATION

27 CFR 479.26: ALTERNATE METHODS OR PROCEDURES

27 CFR 479.111-479.113: IMPORTATION

The Bureau of Alcohol, Tobacco, Firearms and Explosives (ATF) authorizes an alternate method or procedure from the regulations that require the filing of applications to import firearms, ammunition, and other defense articles in paper form. Specifically, individuals, Federal firearms licensees (FFLs), and registered importers of articles enumerated on the US. Munitions Import List (USMIL) may file ATF Form 6-Part I and Form 6A electronically using ATF eForms, provided all of the conditions set forth in this ruling are met. Persons authorized to file eForms may submit digitally scanned copies of the original supporting statements and documents with eForms, if they certify, under penalties provided by law, that the supporting documentation is true, correct, and complete. This ruling supersedes ATF Rulings 2003-6 and 2010-9.

ATF Rul. 2013-1

The Gun Control Act of 1968 (GCA), Title 18, United States Code (U.S.C.) Chapter 44, and the National Firearms Act (NFA), 26 U.S.C. Chapter 53, provide that, with certain exceptions, no firearm, firearm barrel, or ammunition shall be imported or brought into the United States unless the Attorney General has authorized its importation. See 18 U.S.C. 925(d); 26 U.S.C. 5844. The Arms Export Control Act of 1976 (AECA), 22 U.S.C. 2778, gives the President the authority to control the export and import of defense articles and defense services in furtherance of world peace and the security and foreign policy of the United States. The authority to administer and enforce the provisions of the GCA and

NFA, and the permanent import provisions of the AECA, were delegated to the Attorney General, who further delegated at 28 CFR 0.130 these responsibilities to the Director, ATF.

The regulations at 27 CFR 447.42, 478.112, 478.113, 478.113a, 478.116,478.119, 479.111, 479.112, and 479.113 require persons requesting to import firearms or ammunition to file with the Director an ATF Form 6- Part I, Application and Permit for Importation of Firearms, Ammunition and Implements of War. Under 27 CFR 478.112, 478.113, 478.113a, and 478.119, if the Director approves the application, the approved application serves as the import permit. Under 27 CFR 447.45,478.112, 478.113, 478.113a, and 478.119, the importer must then furnish a copy of ATF Form 6A, Release and Receipt of Imported Firearms, Ammunition, and Implements of War, to U.S. Customs and Border Protection (CBP), and must forward a copy of the ATF Form 6A to ATF, to complete the release of the imported articles.

In accordance with the mandate of the Government Paperwork Elimination Act (GPEA), 44 U.S.C. 3504, ATF issued ATF Ruling 2003-6 (approved July 11, 2003), which conditionally authorized a variance from the requirements of 27 CFR 447.42, 478.111,

478.112, 478.113, 479.111, 479.112, and 479.113 only for FFLs and registered importers of articles enumerated on the USMIL who file eForms. In addition, ATF issued ATF Ruling 2010-9 (approved November 30, 2010), allowing only those persons holding a valid Federal firearms license and/or who are registered as importers of articles on the USMIL who are importing surplus military defense articles importable as curios or relics, and who are authorized under ATF Ruling 2003-6 to file eForm 6, to submit digitally scanned copies of the original supporting statements and documents with eForm 6, if they certified, under penalties provided by law, that the supporting documentation is true, correct, and complete.

The eForms online electronic filing system now enables individuals, FFLs, and registered importers under the AECA, to file the ATF Form 6- Part I electronically and obtain an approved import permit from ATF electronically via the Internet. ATF processes eForm online applications and approves, partially approves, denies, returns without action (if incomplete), or withdraws the application (at the request of the applicant). While the applicant may also file the Form 6A electronically, it is not required and the Form 6A may be submitted in the paper version even if the Form 6 was submitted electronically. The applicant may obtain through the Internet electronic copies of the submitted applications with ATF's final determination, which can be printed and furnished to CBP to affect release of the imported articles.

To register to use the eForms system, applicants must access the ATF website at http://www.atfonline.gov/. This site contains the instructions necessary to access the eForms applications and to register to use the system. To register, the applicant must provide his or her name, business name (if applicable), address, telephone number, and email address. Upon proper registration, ATF issues each registrant a user identification (ID), and the registrant creates a password allowing access to the eForms system. Each individual registrant is issued a unique user ID that can be applied to multiple Federal firearms licenses or AECA registrations. After registering with the unique user ID, persons filing permit applications on behalf of an FFL or an import registrant under the AECA must request to be associated with the applicable Federal firearms license or AECA registration number using the **"My Profile"** function of the eForms system.

The eForms system requires individuals, FFLs and registered importers to attest that the information submitted is true and correct, subject to penalties provided by law, and confirm their ATF-issued electronic credentials to complete the application process. Specifically, to complete the application process, an applicant is required to declare first that all statements contained in the application are true and correct and that he or she has read, understood, and complied with the conditions and instructions for the application being submitted. Second, the applicant is required to declare that he or she authorizes the transmittal via the eForms system of data that may constitute tax return information, as defined in section 6103 of the Internal Revenue Code, Title 26, U.S.C. Applicants are not required to use the eForms system and, in certain circumstances, may not be able to participate. Applicants may continue to submit ATF forms on paper in accordance with the instructions on the form.

Under 27 CFR 478.22 and 479.26, the Director may approve an alternate method or procedure in lieu of a meth-

od or procedure specifically prescribed in the regulations when he finds that: (1) good cause is shown for the use of the alternate method or procedure; (2) the alternate method or procedure is within the purpose of, and consistent with the effect intended by, the specifically prescribed method or procedure and that the alternate method or procedure is substantially equivalent to that specifically prescribed method or

procedure; and (3) the alternate method or procedure will not be contrary to any provision of law and will not result in an increase in cost to the Government or hinder the effective administration of 27 CFR Parts 478 or 479.

In addition to the mandate of GPEA, ATF recognizes that, provided certain conditions are met, the use of electronic forms to import firearms and other defense articles will more accurately and readily disclose the information required to be submitted to process import applications. It will also make it easier, less costly, and faster for importers to submit their applications, and for ATF to process those applications. ATF therefore finds that there is good cause to authorize a variance from the paper importation application requirements of the Federal firearms regulations. ATF also finds that, provided certain conditions are met, the alternate method set forth in this ruling is within the purposes of, and consistent with the provisions of 27 CFR 447.42, 447.45, 478.112, 478.113, 478.113a, 478.116, 478.119, 479.111, 479.112, and 479.113, because the same required information is captured on the applicable eForm, which is electronically signed under penalties provided by law. Further, this alternate method is not contrary to any provision of law, will not increase costs to ATF, and will not hinder the effective administration of the regulations.

Held, pursuant to 27 CFR 478.22 and 479.26, ATF authorizes an alternate method or procedure from the provisions of 27 CFR 447.42, 447.45, 478.112, 478.113, 478.113a, 478.116, 478.119, 479.111, 479.112, and 479.113 that require the filing of applications to import firearms, ammunition, and other defense articles in paper form. Specifically, individuals, Federal firearms licensees, and registered importers of articles enumerated on the USMIL may file ATF Form 6-Part I and Form 6A electronically using ATF eForms, provided the following conditions are met:

1. The applicant has registered with ATF by completing the on-line

registration process;

2. The applicant has received a unique user ID and created a password;

3. The applicant has agreed that the electronic signature assigned to them is intended as their original signature for eForms submissions; and

4. The applicant has agreed to be bound by the Notices and Agreements governing the use of the eForms system.

Held further, persons authorized to file eForms may submit digitally scanned copies of the original supporting statements and documents with eForms, if they certify, under penalties provided by law, that the supporting documentation is true, correct, and complete; provided, ATF will determine whether the documentation submitted is acceptable, and may require the submission of the original or additional documentation if necessary; provided further, ATF may deny a permit application when an applicant fails to submit legible copies of original statements or documents, or fails to provide original and/or additional supporting statements or documents upon request.

Held further, if ATF finds that an eForms user has failed to abide by the conditions of this ruling, uses any procedure that hinders the effective administration of the Federal firearms laws or regulations, or any legal or administrative difficulties arise due to the use of eForms, ATF may notify the person that he or she is no longer authorized to use eForms.

Date approved: July 10, 2013

26 U.S.C. 5811 - 5812: TRANSFERS

26 U.S.C. 5821 - 5822: MAKING

26 U.S.C. 5841: REGISTRATION OF FIREARMS

26 U.S.C. 5844: IMPORTATION

26 U.S.C. 5851 - 5854: EXEMPTIONS

27 CFR 479.26: ALTERNATE METHODS OR PROCEDURES

27 CFR 479.61 - 479.64: TAX ON MAKING FIREARMS

27 CFR 479.68 - 479.70: EXCEPTIONS TO TAX ON MAKING FIRE-

ARMS

27 CFR 479.71: REGISTRATION

27 CFR 479.81 - 479.86: TRANSFER TAX

27 CFR 479.88 - 479.91: EXEMPTIONS RELATING TO TRANSFERS OF FIREARMS

27 CFR 479.101, 479.103, 479.104, 479.112: REGISTRATION OF FIREARMS

27 CFR 479.114 - 116: EXPORTATION

The Bureau of Alcohol, Tobacco, Firearms and Explosives (ATF) authorizes an alternate method or procedure from the provisions of Title 27, Code of Federal Regulations (CFR), sections 479.61-479.64, 479.68-479.71, 479.81-479.86, 479.88-479.91, 479.101, 479.103, 479.104, and 479.112-479.116 that require the filing of paper forms to make, manufacture, import, transfer, export, or register National Firearms Act (NFA) firearms. Specifically, Federal firearms licensees (FFLs), legal entities, and government agencies may file ATF Forms 1, 2, 3, 4, 5, 9, and 10 using ATF eForms, provided all of the conditions set forth in this ruling are met. Persons authorized to file eForms may submit digitally scanned copies of the original supporting statements and documents with eForms, if they certify, under penalties provided by law, that the supporting documentation is true, correct, and complete.

ATF Rul. 2013-2

The NFA, Title 26, United States Code (U.S.C.), Chapter 53, requires the registration in the National Firearms Registration and Transfer Record (NFRTR) of certain firearms, such as machineguns, short-barreled rifles, short-barreled shotguns, and silencers. The NFA regulates the making, manufacture, importation, transfer, and exportation of these firearms. Any making, transfer, importation, or exportation of an NFA firearm must be approved by ATF prior to the activity taking place.

The regulations at 479.62, 479.70, 479.84, 479.88, 479.90, 479.103, 479.104, 479.112, and 479.114 require persons and governmental entities who request to make, manufacture, import, transfer, export, or register NFA firearms to file certain forms with the information specified in Federal regulations at 27

CFR Part 479. The following forms are used by persons to register NFA firearms in the NFRTR, and pay associated taxes:

1. ATF Form 1 (5320.1), Application to Make and Register a Firearm.

2. ATF Form 2 (5320.2), Notice of Firearms Manufactured or Imported.

3. ATF Form 3 (5320.3), Application for Tax-Exempt Transfer of Firearm and Registration to Special Occupational Taxpayer (National Firearms Act).

4. ATF Form 4 (5320.4), Application for Tax Paid Transfer and Registration of Firearm.

5. ATF Form 5 (5320.5), Application for Tax Exempt Transfer and Registration of Firearm.

6. ATF Form 9 (5320.9), Application and Permit for Permanent Exportation of Firearms.

7. ATF Form 10 (5320.10), Application for Registration of Firearms Acquired by Certain Governmental Entities.

In accordance with the mandate of the Government Paperwork Elimination Act (GPEA), 44 U.S.C. 3504, ATF has expanded the forms that may be electronically filed to include the above stated NFA forms. ATF's eForms program enables applicants, with certain restrictions depending on the form, to electronically file these forms and download the finalized applications or notices from ATF electronically via the Internet. ATF processes eForms online applications, and approves, denies, or withdraws the application (at the request of the applicant). The electronic copy with ATF's final approval must be printed by the applicant as proof of registration as required by 26 U.S.C. 5841(e) and 27 CFR 479.10l(e). Applicants may attach supporting documentation with their submitted eForms applications, notices, or returns. The supporting documentation will be treated as an original document with the electronic filing.

Persons engaged in the business of importing, manufacturing or dealing in firearms, or a government agency or legal entity (e.g., a trust), may use eForms in accordance with the instructions for the applicable eForm. Currently, the eForms system is not able to accept fingerprint cards or photographs or to verify

the completion of the Law Enforcement Certification that is required with the submission of an ATF Form 1, 4, or 5 by an individual. Therefore, individuals may not submit ATF Forms 1, 4, and 5 electronically to ATF.

To register to use the eForms system, applicants must access the ATF website at http://www.atfonline.gov/. This site contains the instructions necessary to access eForms and register to use the system. To register, the applicant must provide his or her name, business name (if applicable), address, telephone number, and email address. Upon proper registration, ATF issues each registrant a user identification (ID), and the registrant creates a password allowing access to the eForms system. Each individual registrant is issued a unique user ID that can be applied to multiple Federal firearms licenses. After registering with the unique user ID, persons filing NFA fo1ms on behalf of an FFL must request to be associated with the applicable Federal firearms license using the **"My Profile"** function of the eForms system.

The eForms system requires applicants to attest that the information submitted is true, correct, and complete, subject to penalties provided by law, and confirm their ATF-issued electronic credentials to complete the application process. Specifically, to complete the application process, an applicant is required to declare first that all statements contained in the application are true and correct, and that he or she has read, understood, and complied with the conditions and instructions for the form being submitted. Second, the applicant is required to declare that he or she authorizes the transmittal via the eForms system of data that may constitute tax return information, as defined in section 6103 of the Internal Revenue Code, Title 26, U.S.C. Applicants are not required to use the eForms system, and in certain circumstances, may not be able to participate. As noted, NFA forms that require photographs, fingerprints, or the Law Enforcement Certification cannot be submitted electronically. Applicants may continue to file ATF fo1ms on paper in accordance with the instructions on the form.

As provided in 27 CFR 479.26, the Director may approve an alternate method or procedure in lieu of a method or procedure specifically prescribed in the regulations when he finds that: (1) good cause is shown for the use of the alternate method or procedure; (2) the alternate method or procedure is within the purpose of, and consistent with the effect intended by, the specifically prescribed method or procedure and that the alternate method or procedure is substantially equivalent to that specifically prescribed method or procedure; and (3) the alternate method or procedure will not be contrary to any provision of law and will not result in an increase in cost to the Government or hinder the effective administration of 27 CFR Part 479.

In addition to the mandate of GPEA, ATF recognizes that, provided certain conditions are met, the use of electronic forms to make, manufacture, import, transfer, export, or register NFA firearms or pay SOT will more accurately and readily disclose the information required to be submitted to process such applications. It will also make it easier, less costly, and faster for persons to submit their NFA applications and for ATF to process those applications. ATF therefore finds that there is good cause to authorize a variance from the paper filing requirements of the Federal firearms regulations. ATF also finds that, provided certain conditions are met, the alternate method set forth in this ruling is within the purposes of, and consistent with the provisions of 27 CFR, 479.61-479.64, 479.68-479.71, 479.81-479.86, 479.88-479.91, 479.101, 479.103, 479.104, 479.112 and 479.114-116 because the same required information is captured on the applicable eForm, which is electronically signed under penalties provided by law. Further, this alternate method is not contrary to any provision of law, will not increase costs to ATF, and will not hinder the effective administration of the regulations.

Held, pursuant to 27 CFR 479.26, ATF authorizes an alternate method or procedure from the provisions of 27 CFR 479.61-479.64, 479.68-479.71, 479.81-479.86, 479.88-479.91, 479.101, 479.103, 479.104, 479.112 and 479.114-116 that require the filing of paper forms to make, manufacture, import, transfer, export, or register NFA firearms. Specifically, FFLs, legal entities, and government agencies may file ATF Forms 1, 2, 3, 4, 5, 9, and 10 using ATF eForms, provided the following conditions are met:

1. The applicant has registered with ATF by completing the on-line registration process;

2. The applicant has received a unique user ID and created a password;

3. The applicant has agreed that the electronic signature assigned to them is intended as their original signature for eForms submissions; and

4. The applicant has agreed to be bound by the Notices and Agreements governing the use of the eForms system.

Held further, persons authorized to file eForms may submit digitally scanned copies of the original supporting statements and documents with eForms, if they certify under penalties provided by law, that the supporting documentation is true, correct, and complete; provided, ATF will determine whether the documentation submitted is acceptable, and may require the submission of the original or additional documentation if necessary; provided further , ATF may deny a registration application when an applicant fails to submit legible copies of original statements or documents, or fails to provide original and/or additional supporting statements or documents upon request.

Held further, if ATF finds that an eForms user has failed to abide by the conditions of this ruling, uses any procedure that hinders the effective administration of the Federal firearms laws or regulations, or any legal or administrative difficulties arise due to the use of eForms, ATF may notify the person that he or she is no longer authorized to use eForms.

Date approved: July 10, 2013

18 U.S.C. 922(k): REMOVED, OBLITERATED, OR ALTERED SERIAL NUMBER

18 U.S.C. 923(i): IDENTIFICATION OF FIREARMS

26 U.S.C. 5842: IDENTIFICATION OF FIREARMS

26 U.S.C. 5861(h): OBLITERATED, REMOVED, CHANGED, OR ALTERED SERIAL NUMBER

27 CFR 478.92(a)(1): IDENTIFICATION OF FIREARMS

27 CFR 478.92(a)(4)(i): ALTERNATE MEANS OF IDENTIFICATION

27 CFR 478.112(d): IMPORTATION BY A LICENSED IMPORTER

27 CFR 479.102(a): IDENTIFICATION OF FIREARMS

27 CFR 479.102(c): ALTERNATE MEANS OF IDENTIFICATION

The Bureau of Alcohol, Tobacco, Firearms and Explosives (ATF) au-thorizes licensed manufacturers and licensed importers of firearms, and makers of National Firearms Act (NFA) firearms, to adopt the serial number, caliber/gauge, and/or model already identified on a firearm without seeking a marking variance, provided all of the conditions in this ruling are met. Licensed manufacturers seeking to adopt all of the required markings, including the original manufacturer's name and place of origin, must receive an approved variance from ATF. ATF Ruling 75-28 is superseded, and ATF Industry Circular 77-20 is clarified.

ATF Rul. 2013-3

ATF has received numerous inquiries from manufacturers and importers of firearms, and makers of NFA firearms (makers), asking if they can adopt the existing markings that were placed on a firearm by the original manufacturer. They ask specifically if they may use the existing serial number, caliber/gauge, and model already placed on a firearm instead of marking this information on the firearm when they further manufacture or import the firearm.

Some firearm manufacturers and importers acquire receivers and assemble them into completed firearms for the purpose of sale or distribution. Others, including makers, acquire complete firearms and further manufacture them (e.g., re-barreling or machining the frame or receiver to accept new parts). In either case, the firearm is already marked with a serial number, original manufacturer's name, model (if designated), caliber or gauge (if known), and place of origin. Many manufacturers, importers, and makers assert that marking firearms with their own serial numbers, calibers/gauges, and models, in addition to the existing markings, is costly and burdensome. They also contend that multiple markings are confusing for recordkeeping and tracing purposes.

The Gun Control Act of 1968 at Title 18 United States Code (U.S.C.) 923(i) provides, in part, that licensed manufacturers and licensed importers must identify each firearm manufactured or imported by a serial number in the manner prescribed by regulation. The NFA at 26 U.S.C. 5842 provides, in relevant part, that manufacturers, importers, and anyone making a firearm shall identify each firearm manufactured, imported, or made by a serial number, the name of the manufacturer, importer, or maker, and such other identification prescribed by regulation. Title 27 Code of Federal Regulations (CFR) 478.92(a)(1) and 479.102(a) further require the licensed manufacturer, licensed importer, or maker of a firearm to legibly identify each firearm manufactured, imported, or made by engraving, casting, stamping (impressing), or otherwise conspicuously placing the individual serial number on the frame or receiver, and certain additional information - the model (if designated), caliber/gauge, manufacturer/importer's name, and place of origin - on the frame, receiver, or barrel. Both regulations require the serial number to be at a minimum depth and print size, and the additional information to be at a minimum depth. Further, the serial number must be placed in a manner not susceptible of being readily obliterated, altered, or removed, and not duplicate any serial number placed by the licensed manufacturer, licensed importer, or maker on any other firearm. Under 18 U.S.C. 922(k) and 26 U.S.C. 5861(g), it is unlawful for any person to possess or receive any firearm which has had the serial number removed, obliterated, or altered.

Under 27 CFR 478.112(d), licensed importers must place all required identification data on each firearm imported, and record that information in the required records within fifteen days of the date of release from Customs custody. As explained in ATF Ruling 2012-1 (approved January 12, 2012), it is reasonable for licensed manufacturers to have seven days following the date of completion (to include a firearm in knockdown condition, or a frame or receiver to be sold, shipped, or disposed of separately) in which to mark a firearm manufactured, and record its identifying information in the manufacturer's permanent records.

Licensed manufacturers, licensed importers, and makers may seek approval from ATF to use an alternate means of identification of firearms (marking variance). The regulations at 27 CFR 478.92(a)(4)(i) and 479.102(c) provide that the Director of ATF may authorize other means of identification upon receipt of a letter application showing that such other identification is reasonable and will not hinder the effective administration of 27 CFR Parts 478 and 479.

ATF finds, under the conditions set forth in this ruling, that allowing licensed manufacturers and licensed importers of firearms, and makers to adopt the serial number, caliber/gauge, and/or model already marked on the firearm is reasonable and will not hinder the effective administration of the regulations. Multiple serial numbers are confusing to licensees and law enforcement, and potentially hinder effective tracing of firearms. Licensees use the markings to effectively maintain their firearms inventories and required records. Law enforcement officers use the markings to trace specific firearms involved in crimes from the manufacturer, importer, or maker to individual purchasers. The markings also identify particular firearms that have been lost or stolen and help prove in certain criminal prosecutions that firearms used in a crime have travelled in interstate or foreign commerce.

Often there is little space available on a pistol frame to mark the additional information, and the pistol barrel is enclosed by a slide. Thus, it is reasonable to allow licensed manufacturers and licensed importers to mark the slide with the additional information required by the regulations. Therefore, ATF also finds that marking the additional information (i.e., make, model, caliber/gauge, manufacturer/importer's name, and place of origin) on the slide of a pistol is reasonable and will not hinder the effective administration of the regulations.

Held, pursuant to 27 CFR 478.92(a)(4)(i) and 479.102(c), ATF authorizes licensed manufacturers and licensed importers of firearms, and makers, to adopt the serial number, caliber/gauge, and/or model already identified on a firearm without seeking a marking variance, provided all of the following conditions are met:

1. The manufacturer, importer, or maker must legibly and conspicuously place on the frame, receiver, barrel, or pistol slide (if applicable) his/her own name (or recognized abbreviation) and location (city and State, or recognized abbreviation of the State) as specified under his/her Federal firearms license (if a licensee);

2. The serial number adopted must have been marked in accordance with 27 CFR 478.92 and 479.102, including that it must not duplicate any serial number adopted or placed by the manufacturer, importer, or maker on any other firearm;

3. The manufacturer, importer, or maker must not remove, obliterate, or alter the importers' or manufacturer's serial number to be adopted, except that, within 15 days of the date of release from Customs custody, a licensed importer must add letters, numbers, or a hyphen (as described in paragraph 4) to a foreign manufac-

turer's serial number if the importer receives two or more firearms with the same serial number;

4. The serial number adopted must be comprised of only a combination of Roman letters and Arabic numerals, or solely Arabic numerals, and can include a hyphen, that were conspicuously placed on the firearm; and

5. If the caliber or gauge was not identified or designated (e.g., marked "multi") on the firearm, the manufacturer, importer, or maker must legibly and conspicuously mark the frame, receiver, barrel, or pistol slide (if applicable) with the actual caliber/gauge once the caliber or gauge is known.

Held further, licensed manufacturers seeking to adopt all of the required markings, including the original manufacturer's name and place of origin, must receive an approved variance from ATF.

All prior rulings regarding the adoption of markings on firearms, including ATF Ruling 75-28 (ATF C.B. 1975, 59), are hereby superseded. ATF Industry Circular 77-20 is hereby clarified.

Date approved: July 10, 2013

18 U.S.C. 921(a)(1): DEFINITIONS

18 U.S.C. 923(g)(1)(A): LICENSING

27 CFR 478.22: ALTERNATE METHODS OR PROCEDURES

27 CFR 478.121: RECORDS REQUIRED - GENERAL

27 CFR 478.122: RECORDS MAINTAINED BY IMPORTERS

27 CFR 478.123: RECORDS MAINTAINED BY MANUFACTURERS

27 CFR 478.124: FIREARMS TRANSACTION RECORD

27 CFR 478.125: RECORD OF RECEIPT AND DISPOSITION

27 CFR 478.127: DISCONTINUANCE OF BUSINESS

27 CFR 479.26: ALTERNATE METHODS OR PROCEDURES

27 CFR 479.131: RECORDS

The Bureau of Alcohol, Tobacco, Firearms and Explosives (ATF) authorizes

an alternate method or procedure to the firearms acquisition and disposition recordkeeping requirements contained in Title 27, Code of Federal Regulations (CFR) 478.121, 478.122(a), 478.123(a), 478.125(e), 478.125(f), and 27 CFR 479.131. Specifically, ATF authorizes licensed importers, licensed manufacturers, licensed dealers, and licensed collectors to maintain their firearms acquisition and disposition records electronically instead of in paper format provided the conditions set forth in this ruling are met. This ruling supersedes ATF Rul. 2008-2, Records Required for Firearms Licensees.

ATF Rul. 2013-5

ATF has received inquiries from members of the firearms industry seeking to maintain their required firearms acquisition and disposition records electronically on a computer, rather than in paper form.

The Gun Control Act of 1968 (GCA), Title 18, United States Code (U.S.C.), section

923(g)(1)(A), and the implementing regulations at 27 CFR 478.121, provide, in part, that each licensed importer, licensed manufacturer, licensed dealer, and licensed collector (licensee) must maintain records of importation, production, shipment, receipt, sale, or other disposition of firearms at the licensed premises for such period, and in such form, as the Attorney General by regulations prescribe.

The regulation at 27 CFR 478.122(a) requires a licensed importer to record, within 15 days of the date of importation or other acquisition, the type, model, caliber or gauge, manufacturer, country of manufacture, and the serial number of each firearm imported or otherwise acquired, and the date of importation or other acquisition. Further, 27 CFR 478.122(b) requires a licensed importer to record the disposition of firearms to another licensee showing the quantity, type, manufacturer, country of manufacture, caliber or gauge, model, serial number of the firearms transferred, the name and license number of the licensee to whom the firearms were transferred, and the date of the transaction. This information must be entered in the proper record book not later than the seventh day following the date of the transaction, and such information must be recorded in the format prescribed by 27 CFR 478.122. Also, 27 CFR 478.122(d) provides that a licensed importer must maintain separate records of dispositions of firearms to non-licensees, and

that such records shall be maintained in the form and manner prescribed by 27 CFR 478.124 and 27 CFR 478.125.

The regulation at 27 CFR 478.123(a) requires a licensed manufacturer to record the type, model, caliber or gauge, and serial number of each complete firearm manufactured or otherwise acquired and the date of such manufacture or other acquisition. This information must be recorded not later than the seventh day following the date of such manufacture or other acquisition was made. Further, 27 CFR 478.123(b) requires a manufacturer to record the disposition of firearms to another licensee showing the quantity, type, model, manufacturer, caliber, size or gauge, serial number of the firearms transferred, the name and license number of the licensee to whom the firearms were transferred, and the date of the transaction. This information must be entered in the proper record book not later than the seventh day following the transaction date, and must use the record format prescribed by 27 CFR 478.122, except that the name of the manufacturer need not be recorded if the firearm is of the licensee's own manufacture. Also, 27 CFR 478.123(d) requires that each licensed manufacturer maintain separate records of dispositions of firearms to non-licensees in the format specified by 27 CFR

478.124 and 27 CFR 478.125.

The regulation at 27 CFR 478.125(e) requires a licensed dealer to record, no later than the close of the next business day following the date a firearm was received, the date of receipt, name and address or the name and license number of the person from whom the firearm was received, name of the manufacturer and importer (if any), model, serial number, type, and caliber or gauge. This section further requires a dealer to record the disposition of firearms showing the date of sale or other disposition, and the name and address or the name and license number (if a licensee) of the person to whom the firearm was transferred. This information must be entered not later than seven days following the date of such transaction, and must be maintained in bound form under the format prescribed by 27 CFR 478.125(e).

Additionally, 27 CFR 478.125(f) requires a licensed collector to record, not later than the close of the next business day following the date of purchase or other acquisition of a curio or relic, the date of receipt, the name and address or the name and license number of the person

from whom the curio or relic was received, the name of the manufacturer and importer (if any), the model, serial number, type, and the caliber or gauge of the firearm curio or relic. This section further requires a collector to record the date of the sale or other disposition of each firearm curio or relic, the name and address of the person to whom the firearm curio or relic is transferred, or the name and license number of the person to whom transferred if such person is a licensee, and the date of birth of the transferee if other than a licensee. This information must be entered not later than seven days following the date of such transaction, and must be maintained in bound form under the format prescribed by 27 CFR 478.125(f).

The regulation at 27 CFR 479.131 requires each manufacturer, importer, and dealer in National Firearms Act (NFA), 26 U.S.C., Chapter 53, firearms to keep and maintain records regarding the manufacture, importation, acquisition (whether by making, transfer, or otherwise), receipt, and disposition of NFA firearms as described by 27 CFR Part 478.

Licensees may seek ATF approval to use an alternate method or procedure to record the acquisition and disposition of firearms. Pursuant to 27 CFR 478.122(c), 478.123(c), and 478.125(h), ATF may authorize alternate records of firearms, as well as acquisition, when it is shown by the licensee that the alternate records will accurately and readily disclose the information required to be maintained.

The regulations at 27 CFR 478.22 and 27 CFR 479.26, provide that the Director may approve an alternate method or procedure in lieu of a method or procedure specifically prescribed in the regulations when he or she finds that: (1) good cause is shown for the use of the alternate method or procedure; (2) the alternate method or procedure is within the purpose of, and consistent with the effect intended by, the specifically prescribed method or procedure, and that the alternate method or procedure is substantially equivalent to that specifically prescribed method or procedure; and (3) the alternate method or procedure will not be contrary to any provision of law and will not result in an increase in cost to the Government or hinder the effective administration of 27 CFR Part 478 or 479.

ATF understands that using computers to record and maintain firearms acquisition and disposition records saves time and money in bookkeeping and auditing expenses. Most businesses computerize inventory, sales, customer lists, and other business records. This allows companies to automate inventories, using technology such as bar codes or radio frequency identification (RFID) chips. Furthermore, this technology may facilitate better accountability of inventory, and reduce the potential for accounting errors. Computerized records also facilitate tracing and tracking of firearms through licensee inventories, thus reducing time spent by ATF officials examining records during the inspection process. Additionally, the search capability of electronically stored records makes it easier and faster for licensees to locate specific records and respond to ATF trace requests. Therefore, ATF finds that there is good cause to authorize a variance from the firearms acquisition and disposition recordkeeping requirements of the Federal firearms regulations.

ATF also finds that, provided certain conditions are met, the alternate method set forth in this ruling is within the purpose of and consistent with the provisions of 27 CFR 478.121, 478.122, 478.123, 478.125(e) and 478.125(f), and 27 CFR 479.131, because the same required information will be captured in the electronic acquisition and disposition record. Further, this alternate method is not contrary to any provision of law, will not increase costs to ATF, and will not hinder the effective administration of the regulations.

Held, pursuant to 27 CFR 478.22, 478.122(c), 478.123(c), 478.125(h), and 27 CFR 479.26, ATF authorizes an alternate method or procedure to the paper bound firearms acquisition and disposition recordkeeping requirements of 27 CFR 478.122, 478.123, 478.125(e), and 478.125(f), and 27 CFR 479.131. Specifically, ATF authorizes licensed importers, licensed manufacturers, licensed dealers, and licensed collectors to maintain their firearms acquisition and disposition records electronically, provided all of the following conditions are met.

1. The licensee records in the computer system all of the acquisition and disposition information required by 27 CFR 478.121, 478.122, 478.123, 478.125(e) and 478.125 (f), and 27 CFR 479.131, as applicable. Required information includes a record of both the manufacturer and the importer of foreign-made firearms (if any). Additional columns can be utilized to capture certain additional information (e.g. inventory number, new/used, etc.), so long as the additional information is separate from the required information and the required information is readily apparent. An ATF Form 4473 serial number may be used instead of the address for recording the transfer of a firearm to a non-licensee if such forms are filed numerically.

2. The system must retain any correction of errors as an entirely new entry, without deleting or modifying the original entry (e.g., macro created to track changes). Alternatively, the system may allow for entries in a **"notes"** column to explain any correction and/or track changes (i.e., what was changed, who made the change, why the change was needed). ATF suggests that the recordkeeping system be capable of blocking fields from correction (e.g., protect workbook function).

3. The system cannot rely upon invoices or other paper/manual systems to provide any of the required information.

4. The system must allow queries by serial number, acquisition date, name of the manufacturer or importer, name of the purchaser, and address of purchaser or other transferee.

5. The licensee must print or download all records from the system:

    a. at least semiannually;

    b. upon request of an ATF officer (must be provided within 24 hours);

    c. prior to discontinuance of the database; and

    d. prior to discontinuance of the licensee's firearms business.

The printouts/downloads must include all firearms in inventory, as well as all firearms transferred during the period covered, sequentially by date of acquisition, and must be limited to display only the information required by the applicable regulations. The printouts/ downloads may contain additional columns capturing certain additional information, provided that the required information is readily apparent.

6. If the licensee prints out the records, the printout must be retained until the next printout is prepared.

7. Printouts may include antique firearms, but cannot include other merchandise. However, antique firearms must be identified as such in the **"firearm type"** column.

8. If a licensee downloads the records on a portable storage device (e.g., Compact Disc (CD), Digital Versatile Disc (DVD), or Universal Serial Bus (USB) Flash Drive), the download must be retained on the portable storage device until the next download is prepared. Additionally, the licensee must be able to present the most current version of the requested records in a printed format at ATF's request.

9. Electronic firearms acquisition and disposition records may be stored on a computer server owned and operated solely by the person (as defined by 18 U.S.C. section 921(a) (1)) holding the license, provided that the records are readily accessible through a computer device located at the licensed premises during regular business hours. The server must be located within the United States.

10. The system must back-up the firearms acquisition and disposition records on a daily basis to protect the data from accidental deletion or system failure.

11. Upon discontinuance of a license, the licensee must provide an American Standard Code for Information Interchange (ASCII) text file (in conformity with industry standards) containing all acquisition and disposition records, and a file description, to ATF Out-of-Business Records Center, in accordance with 27 CFR 478.127. The complete printout and ASCII text file (and file description) must contain all information prescribed by regulation.

All laws, regulations, policies, and procedures applicable to the paper form of the firearms acquisition and disposition records also apply to electronic versions. Licensees are not required to use an electronic acquisition and disposition record, and may continue to use a paper record in the format prescribed by regulation. Licensees are reminded of their responsibility to ensure accuracy and completeness of all of their required records.

Held further, if the licensee fails to abide by the conditions of this ruling, uses any

procedure that hinders the effective administration of the Federal firearms laws or regulations, or any legal or administrative difficulties arise due to the use of an electronic acquisition and disposition record, the licensee is no longer authorized to maintain acquisition and disposition records electronically under this ruling until all conditions of this ruling are met.

This ruling supersedes all previous rulings regarding alternate methods or procedures for electronically maintaining firearms acquisition and disposition records, including ATF Rul. 2008-2, Records Required for Firearms Licensees, approved August 25, 2008. Further, the references to ATF Rul. 2008-2 in ATF Rul. 2010-8 Consolidation of Required Records for Manufacturers, approved December 6, 2010, and ATF Rul. 2011-1, Consolidation of Required Records for Importers, approved January 26, 2011, are hereby replaced to reference this ruling.

This ruling replaces and rescinds all previously approved variances covering electronically maintaining firearms acquisition and disposition records. Thus, if a licensee holds a previously approved variance that meets the conditions of this ruling, no ATF variance approval is required to maintain such records.

Date approved: December 17, 2013

18 U.S.C. 921(a)(23): DEFINITIONS (MACHINEGUN)

18 U.S.C. 922(o): TRANSFER AND POSSESSION OF MACHINEGUNS

26 U.S.C. 5845(b): DEFINITIONS (MACHINEGUN)

26 U.S.C. 5845(j): DEFINITIONS (TRANSFER)

27 CFR 478.36: TRANSFER OR POSSESSION OF MACHINE GUNS

27 CFR 479.11: DEFINITIONS (MACHINE GUN AND TRANSFER)

27 CFR 479.103: REGISTRATION OF (NFA) FIREARMS MANUFACTURED

27 CFR 479.105: TRANSFER AND POSSESSION OF MACHINE GUNS

A manufacturer licensed and qualified under the Gun Control Act (GCA) and National Firearms Act (NFA) may manufacture and maintain an inventory of machineguns for future sale to Federal, State, or local government agencies without a specific government contract or official request; provided, the machineguns are properly registered, and their subsequent transfer is conditioned upon and restricted to the sale or distribution of such weapons for the official use of Federal, State, or local

government agencies. A manufacturer may deliver machineguns it has manufactured to another qualified manufacturer for any manufacturing process; provided, the first manufacturer maintains continuous dominion or control over the machineguns. A manufacturer may transfer machineguns it has manufactured for present or future sale to a Federal, State, or local government agency to another qualified manufacturer for any manufacturing processes or storage; provided, the manufacturer has a specific government contract or official written request documenting that it is an agent of the government agency requesting and authorizing such transfer. ATF Ruling 2004-2 is clarified.

ATF Rul. 2014-1

The Bureau of Alcohol, Tobacco, Firearms and Explosives (ATF) has received numerous inquiries from licensed firearms manufacturers asking whether machineguns may be transferred to other licensees involved in the manufacturing process and whether manufacturers may maintain an inventory of post-1986 machineguns for future transfer to Federal, State, or local government agencies.

In recent years, licensed firearms manufacturers have contracted certain machinegun manufacturing processes to other qualified manufacturers. Such activities may include assembly, development, testing, repair, or applying special coatings and

treatments to machineguns. Although some manufacturers transport and remain with the machineguns during the manufacturing process, these manufacturers assert that it would be less costly to leave machineguns with the other qualified manufacturer while such processes are being conducted. In some cases, a manufacturer may not yet have a sales contract with the government, but desires to manufacture and maintain an inventory of machineguns in anticipation of future sales to the government (i.e., stockpiling).

The GCA at Title 18, United States Code (U.S.C.), section 922(o), generally makes it unlawful for any person to transfer or possess a machinegun. The term **"machinegun"** is defined under the GCA, 18 U.S.C. 921(a)(23), and the National Firearms Act (NFA), 26 U.S.C. 5845(b), to include the frame or receiver of any such weapon, any part designed and intended solely and exclusively, or combination of parts designed and intended, for use in converting a weapon into a machinegun, and any combination

of parts from which a machinegun can be assembled if such parts are in the possession or under the control of a person. While the GCA does not define **"transfer,"** case law interprets this term to mean a change in dominion or control of a firearm. See ATF Rul. 2010-1 (approved May 20, 2010). Under the NFA, the term **"transfer"** is defined by 26 U.S.C. 5845(j) to include selling, assigning, pledging, leasing, loaning, giving away, or otherwise disposing of an NFA firearm.

Section 922(o)(2)(A) exempts from section 922(o) transfer[s] to or by, or possession by or under the authority of a Federal, State, or local government agency. The regulations implementing section 922(o) further provide that qualified manufacturers may manufacture machineguns for sale or distribution to a Federal, State, or local government agency so long as they are registered in the National Firearms Registration and Transfer Record (NFRTR), and their subsequent transfer is conditioned upon and restricted to the sale or distribution of such weapons for the official use of Federal, State, or local government entities. See 27 CFR 478.36, 27 CFR 479.103 and 479.105(c). Prior to discontinuance of business, qualified manufacturers must transfer machineguns manufactured to a Federal, State or local governmental entity, qualified manufacturer, qualified importer, or a dealer qualified to possess such machineguns. See 27 CFR 479.105(d) and (f).

Manufacture and Possession of Machineguns

Manufacturers necessarily come to **"possess"** machineguns during the manufacturing process (assuming they manufacture them to completion or to a point where the parts constitute a machinegun). While section 922(o) does not expressly prohibit the manufacture of machineguns, it would normally bar possessing and maintaining an inventory of machineguns unless those activities fall within an exception.

The government exception, section 922(o)(2)(A), does not expressly state that the contemplated transfer of machineguns to governmental entities will include machineguns produced by domestic manufacturers. But the fact that section 922(o)(2)(A)'s companion provision, section 922(o)(2)(B), broadly permits **"any lawful transfer or lawful possession of a machinegun that was lawfully possessed before the date this subsection takes effect"** suggests that section 922(o)(2)(A) is intended to address transfers of machineguns manufactured or imported after the statute's effective date. There is also no evidence that Congress meant to require government agencies to import all machineguns.

In light of the breadth of the phrase **"possession . . . under the authority of"** a government agency, and the fact that the statute contemplates that manufacturers will possess machineguns prior to their transfer to the government, it is reasonable to read section 922(o)(2)(A) as permitting manufacturers to possess machineguns produced for future transfer to the government **"under the authority of"** that government agency. In addition, permitting qualified manufacturers to maintain an inventory of weapons in anticipation of future government sales is necessary to ensure that military and law enforcement personnel have enough machineguns available during times of war or national emergency.

Delivery of Machineguns to Contractors

As stated above, the GCA, section 922(o), makes it generally unlawful for any person to transfer or possess a machinegun. Under the GCA, a **"transfer"** occurs when there is a change in dominion or control of a firearm. Thus, once a machinegun has been manufactured for present or future sale, no transfer in violation of the GCA would occur if the registrant—the manufacturer—maintains continuous dominion and control over the machinegun.

To preclude a transfer under the GCA, a manufacturer, including an authorized employee, may deliver the machineguns to the other licensed manufacturer's business premises, remain with the machineguns during the performance of another process, and return with the machineguns to the first manufacturer's premises. If the process takes longer than a day, the machineguns must be stored in a manner so that only the registrant has access to them during the overnight period. No work may be conducted on the machineguns without the registrant's employees being present to maintain dominion and control over them.

Transfers of Machineguns to Contractors

As stated above, under the GCA, a machinegun transfer would occur if there was a change in dominion or control of the machinegun. Thus, a transfer would occur if the manufacturer relinquished control over a machinegun by leaving it at the contractor's premises for further manufacture, or otherwise disposed of the machinegun to another manufacturer.

Again, while section 922(o) would normally prohibit such a transfer, section 922(o)(2)(A) exempts machinegun transfers **"to or by"** or possession **"by or under the authority of"** a Federal, State, or local government agency. Because the phrase, **"under the authority of"** only modifies **"possession,"** a machinegun can only be transferred **"by"** a government agency if the agency itself transfers the machinegun, or another person authorized to act on that agency's behalf, as a lawful agent, transfers the machinegun.

Under agency common law, the government (as principal) may authorize a manufacturer (as agent) to act on its behalf in accordance with the express terms of a contract or agreement between them. A manufacturer (as agent) may then appoint another manufacturer (as subagent) to perform functions that the manufacturer has consented to perform for, and on behalf of, the government (as principal), so long as the government has granted the first manufacturer authority to do so. Therefore, assuming the machinegun is for future sale to the government, a manufacturer may transfer one or more machineguns to another for further manufacture if it is expressly authorized by a government agency through a specific government contract or official written request. Such documentation would be sufficient to authorize a machinegun **"transfer by"** and subsequent **"possession under the authority of"** the interested government department or agency.

Held, a manufacturer licensed and qualified under the Gun Control Act (GCA) and National Firearms Act (NFA) may manufacture and maintain an inventory of machineguns for future sale to Federal, State, or local government agencies without a specific government contract or official request; provided, the machineguns are properly registered, and their subsequent transfer is conditioned upon and restricted to the sale or distribution of such weapons for the official use of Federal, State, or local government agencies.

Held further, a manufacturer may deliver machineguns it has manufactured to another qualified manufacturer for assembly, anodizing, or other manufacturing processes; provided, the first manufacturer maintains continuous dominion or control over the machineguns.

Held further, a manufacturer may transfer machineguns it has manufactured for present or future sale to a Federal, State, or local government agency to another qualified manufacturer for assembly, development, testing, repair, other manufacturing processes, or storage on behalf of that government agency; provided, the first manufacturer has a specific government contract or official written request stating that it is an agent of the government agency requesting and authorizing such transfer and, in the case of a written request, it is on official government letterhead signed and dated by an authorized government official, includes the official's title and position, and includes the following statements to document government approval:

1. A statement that the firearms to be transferred are machineguns (as defined by Federal law i.e., the GCA, 18 U.S.C. 921(a)(23), and the NFA, 26 U.S.C. 5845(b));

2. A statement that the machineguns to be transferred are particularly suitable for official use by the requesting Federal, State, or local government agency; and

3. A statement that the Federal, State, or local government agency requests and authorizes the manufacturer to transfer the machineguns to and/or from other licensed manufacturers for assembly, repair, development, testing, other manufacturing processes, or storage, as the case may be, for that government agency.

A manufacturer wishing to transfer machineguns under government authority must attach a specific government contract or official written request to the transfer application submitted to ATF's National Firearms Act Branch and receive ATF approval before making the transfer.

ATF Ruling 2004-2 (approved April 7, 2004) is hereby clarified with respect to the documentation required under the GCA for a qualified importer to temporarily transfer a lawfully imported machinegun for inspection, testing, calibration, repair, reconditioning, further manufacture, or incorporation into another defense article. To the extent, this ruling may be inconsistent with any prior letter rulings or marking variances, they are hereby modified.

Date approved: September 4, 2014

PROCEDURES

PART 178: COMMERCE IN FIREARMS

(Also 26 CFR 178.94, 178.12)

Recordkeeping procedures for "drop shipments" of firearms and ammunition are prescribed.

ATF Proc. 75-3

Purpose: This ATF procedure sets forth the recordkeeping procedures for **"drop shipments"** of firearms (other than National Firearms Act firearms as defined in section 5845(a) of Chapter 53, Title 26, U.S.C.) and ammunition between federally licensed firearms dealers, importers, and manufacturers.

Background: The Bureau has experienced difficulty in tracing firearms in instances where drop shipments have been made to third parties and where the recordkeeping procedures employed by the three parties do not lend themselves to easy and fast tracing of firearms and ammunition. For this reason, the Bureau has prescribed recordkeeping procedures for **"drop shipments"** as set forth below.

Procedures: (1) Where licensee **"A"** places an order for firearms or ammunition with licensee **"B"** and **"B"** transmits the order to licensee **"C"** for direct shipment (drop shipment) to **"A,"** a certified copy of the license of **"A"** must be forwarded to **"C"** prior to shipment of the order. On shipment of the order to **"A,"** **"C"** shall enter in his bound record the disposition of the firearms or ammunition to **"A."** On receipt of the shipment by **"A,"** he shall enter the acquisition of the firearms or ammunition in his bound record. Both licensees shall make such entries in the manner prescribed by regulations. Since the actual movement of the firearms or ammunition is between **"C"** and **"A"** and since **"B"** does not take physical possession of them, **"B"** will make no entry in his bound record. However, **"B"** should make appropriate entries or notations in his commercial records to reflect the transaction.

(2) For example, where a licensed dealer orders firearms from a wholesaler and the wholesaler requests drop shipment from the manufacturer to the dealer, a certified copy of the dealer's license shall accompany the wholesaler's order to the manufacturer. The manufacturer shall enter in his bound record the disposition of the firearms to the dealer, and the dealer shall enter the acquisition of the firearms in his bound record reflecting receipt from the manufacturer. The wholesaler, although a part of the business transaction, neither acquires nor disposes of the firearms and would, therefore, enter nothing of the transaction in his bound record.

NFA Firearms: Transfer of National Firearms Act firearms may be accomplished only pursuant to the manner outlined in Subpart F, Part 179, Title 27, Code of Federal Regulations.

Inquiries: Inquiries concerning this procedure should refer to its number and be addressed to the Firearms Industry Programs Branch.

Editor's Note:

The above procedure no longer applies to ammunition other than shipments of armor piercing ammunition or ammunition for destructive devices.

[75 ATF C.B. 78]

27 CFR 179.35: EMPLOYER IDENTIFICATION NUMBER

(Also see 179.34, 179.84 179.88, 179.90, 179.103 and 179.112)

Identification Number for Special (Occupational) Taxpayer

ATF Proc. 90-1

Purpose: The purpose of this ATF procedure is to inform Federal Firearms licensees who have paid the special (occupational) tax to import, manufacture, or deal in National Firearms Act (NFA) firearms of the discontinuance of the use of the ATF Identification Number and the replacement with the use of the Employer Identification Number (EIN) on all NFA transaction forms.

Background: Section 5801 of Title 26, U.S.C. provides that on first engaging in business, and thereafter on or before the first day of July of each year, every importer, manufacturer, and dealer in NFA firearms shall pay the appropriate special (occupational) tax. In addition, section 5802 requires each importer, manufacturer, and dealer to register with the Secretary his name and the address of each location where he will conduct business. The filing of ATF Form 5630.5, with payment of the appropriate tax required by section 5801, also accomplishes registration requirements under section 5802.

The regulations at 27 CFR 179.34 require that the special tax be paid by return (ATF Form 5630.5, Special Tax Registration and Return) and require that all information called for on the return be provided, including the Employee Identification Number. 27 CFR 179.35 provides the instruction for applying for an EIN.

The regulations in 27 CFR 179.84, 179.88, and 179.90 require that the application to transfer an NFA firearm identify the special tax stamp, if any, of the transferor and transferee. The regulations in 27 CFR 179.103 and 179.112 require that the notice submitted to register NFA firearms identify the special tax stamp of the manufacturer or importer respectively. Identification of the tax stamp is necessary to ensure the tax liability has been satisfied, that the parties are qualified to import, manufacture, or deal in NFA firearms, and, in certain instances, is necessary to ensure that both parties in a transfer application are entitled to an exemption from the transfer tax.

In 1980, because of delays in the issuance of special tax stamps resulting in the inability of special taxpayers to conduct business operations, ATF Procedure 80-6 was implemented. This procedure notified taxpayers that they could obtain an ATF identification number which should be used in lieu of the IRS special tax stamp number on all NFA transaction forms. This procedure was established to facilitate the processing of NFA forms and to eliminate the delay caused by the time period required for IRS processing of the special tax stamp.

ATF has recently taken over the collection of special tax from the Internal Revenue Service, and is now issuing the special tax stamps. The number used to identify the special tax stamp is the EIN.

Identification: Because the number used to identify the special tax stamp is the EIN, this number must appear on all forms (applications, notices, and returns) involving NFA firearms. The problems that caused the implementation of the procedure in 1980 have been resolved. In fact, the assignment of an ATF identification number is now duplicative and requires more paperwork of the taxpayer. Accordingly, the use of the ATF identification number is no longer necessary and is discontinued.

ATF Procedure 80-6 is cancelled.

Inquiries: Inquiries regarding this ATF procedure should refer to its number and be addressed to the Bureau of Alcohol, Tobacco and Firearms, Chief, National Firearms Act Branch, Martinsburg, West Virginia.

Editor's Note:

ATF Form 5630.5 has been replaced by ATF Form 5630.7, Special Tax Registration and Return National Firearms Act (NFA).

[ATFB 1990-1 55]

PART 478: COMMERCE IN FIRE-ARMS

(18 U.S.C. 922(t), 922(b)(2), 922(z); 27 CFR 478.58, 478.102, 478.122, 478.123, 478.124, 478.125, 478.126a, 478.129)

Recordkeeping and background check procedures for facilitation of private party firearms transfers.

ATF Proc. 2013-1

Purpose: The purpose of this ATF procedure is to set forth the recordkeeping and National Instant Criminal Background Check System (NICS) procedures for

Federal firearms licensees (FFLs) who facilitate the transfer of firearms between private unlicensed individuals. This procedure does not apply to pawn transactions, consignment sales, or repairs.

Background: On January 16, 2013, ATF issued an Open Letter to all FFLs encouraging them to facilitate the transfer of firearms between private individuals to enhance public safety and assist law enforcement. This procedure expands and modifies the guidance provided in that letter with respect to private firearms transfers facilitated by FFLs. As explained in the letter, unlicensed persons do not have the ability to use NICS to conduct background checks on prospective transferees (buyers) and, consequently, have no comprehensive way to confirm whether or not the transferee (buyer) is prohibited from receiving or possessing a firearm. In addition, several states have laws which prohibit the transfer of firearms between private individuals unless a NICS check is conducted on the transferee (buyer).

Title 18, United States Code (U.S.C.). section 922(t)(l)(A) requires FFLs to contact NICS before completing the transfer of a firearm to an unlicensed person. Title 27, Code of Federal Regulations (CFR) sections 478.102, 478.122, 478.123, 478.124, and 478.125 set forth the recordkeeping and background check requirements with which FFLs must comply to conduct firearms transactions. As provided by 18 U.S.C. 922(t)(1), an FFL may only conduct a NICS check in connection with a proposed firearm transfer by the FFL. For an FFL to lawfully complete the transfer of a firearm, the FFL must first take the firearm into inventory, and record it as an acquisition in the acquisition and disposition (A&D) record.

The procedures below must be followed when the private party (seller) takes a firearm to an FFL with the prospective transferee (buyer) to conduct a transaction. In all cases-

(1) the prospective transferee (buyer) must complete Section A of a Firearms Transaction Record , ATF Form 4473;

(2) the FFL must complete Section B of the Form 4473, conduct a NICS check on the prospective transferee (buyer), and record the response;

(3) the FFL must record the words **"Private Party Transfer"** on the ATF Form 4473, Item 30c; and

(4) the FFL must maintain the Form 4473 in accordance with 27 CFR 478.129(b).

The FFL must record **"Private Party Transfer"** on the form to ensure it can be determined which transaction records correspond with private party transfers in the FFL's A&D record. If **"Private Party Transfer"** is not noted on the form, and there is no corresponding entry in the A&D record, ATF may conclude that the FFL has transferred a firearm without making appropriate entries in the A&D record.

Procedure - Immediate "Proceed" Transactions:

• If the FFL receives an immediate **"proceed"** response from NICS, the FFL enters the firearm into the A&D record as an acquisition from the private party (seller) in accordance with 27 CFR 478.1 22, 478.123, and 478.125.

• The FFL completes Section D of the Form 4473 and transfers the firearm to the transferee (buyer).

• The FFL records the disposition of the firearm out of the A&D record to the transferee (buyer) no later than seven days following the transaction.

Procedure - "Denied" or "Cancelled" Transactions:

• If the FFL receives a **"denied"** or "cancelled" response from NICS, the firearm cannot be transferred to the prospective transferee (buyer). If the private party (seller) has not relinquished possession, he or she may leave the business premises with the firearm. The FFL would not enter the firearm as an acquisition into the A&D record.

• If, for whatever reason, the private party (seller) leaves the firearm in the exclusive possession of the FFL at the FFL's business premises, the FFL must: (1) record the firearm as an acquisition in the A&D record from the private party (seller), in accordance with 27 CFR 478.122, 478.123, and 478.125; (2) complete a Form 4473 to return the firearm to the private party (seller);

(3) conduct a NICS background check on the private party (seller), and receive either a **"proceed"** response or no response after three business days (or appropriate State waiting period), prior to returning the firearm; and

(4) record the return as a disposition in the A&D record no later than seven days following the transaction.

Procedure— "Delayed" Transactions Without a Subsequent Denial:

• If the FFL receives a **"delayed"** response from NICS, the private party (seller) has the option to:

(1) leave the FFL·s business premises with the firearm; or

(2) allow the FFL to retain the firearm at the business premises (with the FFL 's consent) pending a response from NICS, or until the passage of three business days or applicable State waiting period.

• If the private party (seller) chooses to leave the FFL's premises with the firearm, the FFL does not need to record the acquisition into the A&D record. However, the private party (seller) must return the firearm to the FFL's business premises prior to transfer of the firearm to the prospective transferee (buyer).

• If the private party (seller) chooses to allow the FFL to retain the firearm at the FFL's business premises, the FFL is required to take the firearm into inventory, and record the firearm as an acquisition in the A&D record in accordance with 27 CFR 478.122, 478.123, and 478.125. If NICS later issues a "proceed,., or no response after three business days (or appropriate State waiting period), and the FFL chooses to proceed with the transfer, the private party (seller) need not return to the FFL's business premises for transfer of the firearm to the transferee (buyer).

• If the FFL does not receive a response after three business days (or appropriate State waiting period) and NICS (or State point-of-contact) has not advised the FFL that the prospective transferee's (buyer's) receipt or possession of the firearm would be in violation of law, the FFL may, but is not required to, transfer the firearm to the prospective transferee (buyer).

• A transfer to a prospective transferee (buyer) must be completed within 30 calendar days from the date NICS was initially contacted. If the prospective transferee (buyer) does not return during this period, the FFL must conduct a new NICS check prior to the transfer.

• If the transfer of the firearm takes place on a different day from the date that the prospective transferee (buyer) signed Section A, the FFL must again check the photo identification of the prospective transferee (buyer) at the FFL's licensed business premises, and the prospective transferee (buyer) must complete the recertification in Section C of the Form 4473 immediately prior to the transfer of the firearm.

• The FFL must complete Section D of the Form 4473 prior to transferring the firearm to the prospective transferee. The FFL must record the words **"Private Party Transfer"** on the Form 4473, Item 30c and file the form as prescribed by 27 CFR 478.124(b). In addition, the FFL must record in the A&D record the disposition of the firearm to the transferee (buyer) no later than seven days following the transaction.

• If the private parties do not return to complete the transfer, the FFL must retain the Form 4473 in accordance with section 478.129(b).

Procedure - Delayed Transactions With a Subsequent Denial: If the FFL receives a **"denied"** response from NICS prior to transfer of the firearm, it cannot be transferred to the prospective transferee (buyer). If the private party (seller) has chosen to allow the FFL to retain the firearm pending NICS approval, the FFL and the private party (seller) must complete a Form 4473 prior to returning the firearm to the private party (seller). The FFL must also conduct a NICS background check on the private party (seller) and receive a **"proceed"** response, or no response after three business days (or appropriate State waiting period), prior to returning the firearm. The FFL must also record the return as a disposition in the A&D record no later than seven days following the transaction.

Procedure -Secure Handgun Storage or Safety Devices: The FFL must provide the transferee (buyer) or private party (seller) with a secure gun storage or safety device for each handgun he or she transfers, pursuant to 18 U.S.C. 922(z). The FFL is not required to provide the private party (seller) with a secure gun storage or safety device if the private party (seller) does not relinquish exclusive dominion or control of the firearm.

Procedure- Reports of Multiple Sale or Other Disposition of Pistols

and Revolvers: As provided by 27 CFR 478.126a, the FFL must complete an ATF Form 3310.4 and report all transactions in which an unlicensed person acquires, at one time or during five consecutive business days, two or more pistols or revolvers. The form is not required when the pistols or revolvers are returned to the same person from whom they were received.

Procedure - Reports of Multiple Sale or Other Disposition of Certain Rifles: All applicable FFLs and pawnbrokers located in Arizona, California, New Mexico, and Texas must complete an ATF Form 3310.12 and repo11all transactions in which an unlicensed person acquires, at one time or during five consecutive business days, two or more semi-automatic rifles larger than .22 caliber (including .223/5.56 caliber) with the ability to accept a detachable magazine.

The form is not required when the rifles are returned to the same person from whom they were received.

NFA Firearms: Transfers of National Firearms Act firearms may be accomplished only pursuant to the manner outlined in Subpart F, Part 479, Title 27, Code of Federal Regulations.

State and Local Law: An FFL facilitating private party firearms transfers must comply with all State laws and local ordinances. As provided by 27 CFR 478.58, a Federal Firearms License confers no right or privilege to conduct business or activity contrary to State or other law. It is unlawful for an FFL to sell or deliver any firearm to any person in any State where the purchase or possession by such person of such firearm would be in violation of any State law or published ordinance applicable at the place of sale, delivery, or other disposition. See 18 U.S.C. 922(b)(2). Compliance with the provisions of any State or other law affords no immunity under Federal law or regulations.

Procedure-Out-of-State Transactions: Private party firearm transfers conducted with out-of-State FFLs must comply with all interstate controls and requirements under the Gun Control Act, 18 U.S.C. 921 et. seq. Except in limited circumstances, FFLs cannot transfer firearms to out-of-State residents. See, e.g.,18 U.S.C. 922(b)(3).

Date approved: March 15, 2013

18 U.S.C. 922(c): NON–OVER–THE–COUNTER SALES OF FIREARMS

18 U.S.C. 922(t): NICS REQUIREMENTS AND EXCEPTIONS

27 CFR 478.96: MAIL ORDER SALES OF FIREARMS (IN–STATE, NICS EXEMPT)

27 CFR 478.102: NICS REQUIREMENTS AND EXCEPTIONS

Recordkeeping procedure for non–over–the–counter firearm sales by licensees to unlicensed in-state residents that are NICS exempt.

ATF Proc. 2013-2

Purpose: This ATF procedure gives guidance to licensed importers, manufacturers, and dealers (licensees) on how to record the sale of a firearm to an unlicensed person who:

a) Has a valid alternate permit or otherwise is exempt from National Instant Criminal Background Check System (NICS) requirements;

b) Resides in the same State as the licensee; and

c) Does not appear in person at the licensee's business premises.

Background: Title 18, United States Code (U.S.C.), section 922(c)(1), states that, in any case not otherwise prohibited by 18 U.S.C. chapter 44, a licensed importer, licensed manufacturer, or licensed dealer may sell a firearm to a person (other than another licensed importer, manufacturer, or dealer) who does not appear in person at the licensee's business premises only if the transferee submits to the transferor a sworn statement in the following form:

Subject to penalties provided by law, I swear that, in the case of any firearm other than a shotgun or a rifle, I am twenty–one years or more of age, or that, in the case of a shotgun or a rifle, I am eighteen years or more of age; that I am not prohibited by the provisions of chapter 44 of title 18, United States Code, from receiving a firearm in interstate or foreign commerce; and that my receipt of this firearm will not be in violation of any statute of the State and published ordinance applicable to the locality in which I reside. Further, the true title, name, and address of the principal law enforcement officer of the locality to which the firearm will be delivered are.

Signature _____

Date _____

The statement must contain blank spaces to attach a true copy of any permit or other information required pursuant to such State statute or published ordinance.

The corresponding regulation at 27 CFR 478.96(b) states, in relevant part, that a licensee may sell a firearm that is not subject to the provisions of 478.102(a) to a nonlicensee who does not appear in person at the licensee's business premises if the nonlicensee is a resident of the same State in which the licensee's business premises are located, and the nonlicensee furnishes to the licensee the firearms transaction record, Form 4473 as required by 478.124. The regulation further requires the nonlicensee to attach to the Form 4473 a true copy of any permit or other information required pursuant to any statute of the State and published ordinance applicable to the locality in which he/she resides.

Furthermore, 18 U.S.C. 922(t)(1) and its implementing regulations at 27 CFR 478.102(a) require a licensee to contact NICS for a background check prior to completion of a firearm transfer to an unlicensed person, and verify the identity of that person by examining a valid identification document. The statute at subsection 922(t)(3) and implementing regulation at subsection

478.102(d) provide exceptions to these requirements: (1) if the transferee has presented a permit or license (**"alternate permit"**) that: (i) allows the transferee to possess, acquire, or carry a firearm; (ii) was issued not more than 5 years earlier by the State in which the transfer is to take place; and (iii) the law of the State provides that such a permit or license is to be issued only after an authorized government official has verified that the information available to such official does not indicate that possession of a firearm by the transferee would be in violation of Federal, State, or local law, and includes completion of a NICS background check; (2) the firearm is subject to the provisions of the National Firearms Act and has been approved for transfer under 27 CFR

Part 479; or (3) on application of the licensee, in accordance with 27 CFR 478.150, the ATF Director has certified that running a NICS background check is impracticable due to the geographical location of the licensee.

ATF recognizes that Form 4473, Part I–Over–the–Counter, does not include the sworn statement required by 18 U.S.C. 922(c)(1), and that licensees who sell firearms subject to section 478.96(b) requirements without that statement would not be in compliance with Federal law. For this reason, ATF authorizes this procedure to use Form 4473, Part I, for non–over–the–counter firearm sales by licensees to unlicensed in-state residents that are NICS exempt.

Procedure: A licensed importer, manufacturer, or dealer may record and conduct the sale of a firearm to a resident of the same state who does not appear in person at the licensed business premises using ATF Form 4473, Part I, provided the transfer is exempt from the NICS requirements, pursuant to 18 U.S.C. 922(t)(3) and 27 CFR 478.102(d), and the procedures below are followed:

1. The buyer must properly complete and execute Section A of ATF Form 4473, Part I, as if the firearm was being transferred over-the-counter;

2. Pursuant to 18 U.S.C. 922(c)(1) and 27 CFR 478.96(b), the buyer must execute and attach to the Form 4473 a sworn statement in the format prescribed by 18 U.S.C. 922(c)(1) (set forth above), and, as applicable, a true copy signed and dated by the buyer of the valid alternate permit that qualifies as a NICS exception, and any other documentation required pursuant to State statute or published ordinance;

3. The buyer must send the original ATF Form 4473 and attachment(s) to the licensee. The licensee must then complete the remaining portions of the form as if the firearm was being transferred over–the–counter;

4. The licensee must document the applicable NICS exception in Section B of the ATF Form 4473;

5. The licensee is permitted to omit recording identifying information (type of identification, number on identification, expiration date of identification) in Section B. However, if the buyer is a nonimmigrant alien, the licensee must record the type of documentation showing an exception to the nonimmigrant alien prohibition. This supporting documentation must be attached to the ATF Form 4473;

6. Neither the licensee nor the purchaser should complete Section C of the ATF Form 4473;

7. The licensee must complete Section D. In addition, the licensee must record the words "Non over the counter transaction" at the top of the first page of the ATF Form 4473;

8. Pursuant to 18 U.S.C. 922(c)(2) and 27 CFR 478.96(b), the licensee must, prior to shipment or delivery of the firearm, forward by registered or certified mail (return receipt requested) a copy of the Form 4473, sworn statement, and valid alternate permit or other required information to the chief law enforcement officer (CLEO) named on such statement;

9. Pursuant to 18 U.S.C. 922(c)(3) and 27 CFR 478.96(b), the licensee must delay shipment or delivery of the firearm for at least seven days following receipt by the licensee of either the return receipt evidencing delivery of the copy of the Form 4473,

to the CLEO, or the return of the copy of the Form 4473 to the licensee due to refusal of the CLEO to accept the same in accordance with U.S. Postal Service regulations;

10. The licensee must retain the original Form 4473, Part I, sworn statement, NICS exemption documentation, and evidence of receipt or rejection of delivery of the information sent to the CLEO as a part of the records required to be kept by the licensee under the provisions of 27 CFR Subpart H, including 478.129 and 478.131;

11. The licensee must record the firearms disposition in the licensee's acquisition and disposition record in accordance with 27 CFR 478.122, 478.123, or 478.125 (as applicable); and

12. The licensee must report any multiple sales or other disposition of pistols or revolvers on ATF Form 3310.4 in accordance with 27 CFR

478.126a. In addition, the licensee must report any multiple sales or other disposition of certain rifles on ATF Form 3310.12 in accordance with 18 U.S.C. 923(g)(5)(A), as applicable.

Licensees are reminded of their responsibility to ensure the accuracy and completeness of all required records, and to maintain such records on their licensed premises available for inspection. Failure to abide by any of these procedures may result in a violation of 18 U.S.C. 922(b)(5), 922(c), 922(m), 923(g), and/or 924(a)(3)(B), and corresponding regulations.

Inquiries: Inquiries concerning this procedure should refer to its number and be addressed to the Firearms Industry Programs Branch at (202) 648-7190 or fipb@atf.gov.

Date approved: August 15, 2013

B. Todd Jones Director

INDUSTRY CIRCULARS

Industry Circular 72-23

SHIPMENT OR DELIVERY OF FIRE-ARMS BY LICENSEES TO EMPLOY-EES, AGENTS, REPRESENTATIVES, WRITERS, AND EVALUATORS

This circular is no longer in effect. See ATF Rul. 2010-1.

Industry Circular 72-30

IDENTIFICATION OF PERSONAL FIREARMS ON LICENSED PREMIS-ES NOT OFFERED FOR SALE

Purpose. The purpose of this circular is to urge licensed firearms dealers to identify their personal collection of fire-arms kept at the business premises.

Scope. The provisions of Section 923(g), 18 U.S.C. Chapter 44, and Sub-part H of the regulations (27 CFR 178) require all licensed firearms dealers to maintain records of their receipt and disposition of all firearms at the licensed premises. Section

178.121(b) of the regulations and the law further provide for the examination and inspection during regular business hours or other reasonable times of fire-arms kept or stored on business premis-es by licensees and any firearms record or document required to be maintained.

Guidelines for Identifying Personal Firearms on the Business Premises of Li-censed Dealers. A presumption exists that all firearms on a business premises are for sale and accordingly must be entered in the records required to be maintained under the law and regulations. However, it is recognized that some dealers may have personal firearms on their business premises for purposes of display or dec-oration and not for sale. Firearms deal-ers who have such personal firearms on licensed premises should not intermingle such firearms with firearms held for sale. Such firearms should be segregated from firearms held for sale and appropriately identified (for example, by attaching a tag) as being **"not for sale"**. Personal

firearms on licensed premises which are segregated from firearms held for sale and which are appropriately identified as not being for sale need not be entered in the dealer's records.

There may be occasions where a firearms dealer utilizes his license to acquire firearms for his personal collec-tion. Such firearms must be entered in his permanent acquisition records and subsequently be recorded as a dispo-sition to himself in his private capacity. If such personal firearms remain on the licensed premises, the procedures de-scribed above with respect to segrega-tion and identification must be followed.

The above procedures will facilitate the examination and inspection of the records of firearms dealers and result in less inconvenience to licensees.

Industry Circular 74-13

VERIFICATION OF IDENTITY AND LI-CENSED STATUS OF TRANSFEREE

Purpose. The purpose of this circular is to remind firearms dealers, manufac-turers, and importers of the provisions of 18 U.S.C. Chapter 44, and Subpart F of the Regulations thereunder (26 CFR 178) pertaining to sales or deliveries of firearms or ammunition between licensees.

Background. Regulations, 26 CFR 178.94, provide as follows: "A licensed importer licensed manufacturer, or li-censed dealer selling or otherwise dis-posing of firearms or ammunition and a licensed collector selling or otherwise disposing of curios or relics, to another licensee shall verify the identify and li-censed status of the transferee prior to making the transaction…Such verifica-tion should be established by the trans-feree (buyer) furnishing to the transferor (seller) a certified copy of the transfer-ee's license and by such means as the transferor deems necessary…

Recent reports involving thefts of firearms licenses and of the purchas-ing copies thereof have brought to our

attention the need to emphasize the im-portance of verifying the identity of trans-ferees involved in such transactions.

Guidelines for Verifying Identity and License Status of Transferee. A licens-ee who appears in person at another licensee's business premises for the purpose of acquiring firearms or ammu-nition should be required to furnish, to the transferor, positive identification in addition to a certified copy of his license. Such identification should prove to the satisfaction of the transferor that the per-son receiving the firearm or ammunition is, in fact, the same person to whom the license furnished has been issued.

With respect to mail order sales be-tween firearms licensees, where the shipment is to be made to an address other than the transferee's premises as listed on his license, it is suggested that the transferor verify the address as be-ing that of the transferee.

The Bureau urges all firearms licens-ees to require whatever information they deem necessary and within reason in order to verify the identity and licensed status of the transferee licensees with whom they do business.

Inquiries. Inquiries regarding this cir-cular should refer to its number and be addressed to the Chief, Firearms Indus-try Programs Branch, Washington D.C. 20226.

Editor's Note:

This procedure no longer applies to distributions of ammunition, other than armor piercing ammunition or ammuni-tion for destructive devices.

Industry Circular 77-20

DUPLICATION OF SERIAL NUM-BERS BY LICENSED IMPORTERS

ATF has noted cases where some licensed importers have adopted the same serial number for more than one firearm. These instances of duplication have generally occurred when firearms are received from more than one source.

Title 27, CFR section 178.92 requires that the serial number affixed to a firearm must not duplicate the number affixed to any other firearm that you import into the United States. Those of you who import destructive devices are under the same requirement due to the inclusion of destructive devices in the definition of firearm as used in 27 CFR 178.11. ATF Ruling 75-28 stated that a serial number affixed by the foreign manufacturer may be adopted to fulfill this unique serial number requirement. However, the manufacturer's serial number must be affixed in the manner set forth in 27 CFR 178.92 and must not duplicate a number previously adopted by you for another firearm.

If you receive two or more firearms with the same serial number, it is your responsibility to affix additional markings to make each serial number unique.

ATF Ruling 75-28 also reminds you of the other identifying marks required by 27 CFR 178.92. In addition to a unique serial number, each firearm must be marked to show the model (if any); the caliber or gauge; the name of the manufacturer and importer, or recognizable abbreviations; the country of manufacture; and the city and State (or recognized abbreviations) in which your licensed premises are located.

Editors Note:

This Industry Circular is further clarified with ATF Rul. 2013-3.

GENERAL INFORMATION

TABLE OF CONTENTS

GENERAL INFORMATION

1. INFORMATION CONCERNING M16 PARTS AND AR–15 TYPE RIFLES

ATF has encountered various AR–15 type rifles such as those manufactured by Colt, E.A. Company, SGW, Sendra and others, which have been assembled with fire control components designed for use in M16 machineguns. The vast majority of these rifles which have been assembled with an M16 bolt carrier, hammer, trigger, disconnector and selector will fire automatically merely by manipulation of the selector or removal of the disconnector. Many of these rifles using less than the 5 M16 parts listed above also will shoot automatically by manipulation of the selector or removal of the disconnector.

Any weapon which shoots automatically more than 1 shot without manual reloading, by a single function of the trigger, is a machinegun as defined in 26 U.S.C. 5845(b), the National Firearms Act (NFA). The definition of a machinegun also includes any combination of parts from which a machinegun may be assembled, if such parts are in possession or under the control of a person. An AR–15 type rifle which fires more than 1 shot by a single function of the trigger is a machinegun under the NFA. Any machinegun is subject to the NFA and the possession of an unregistered machinegun could subject the possessor to criminal prosecution.

Additionally, these rifles could pose a safety hazard in that they may fire automatically without the user being aware that the weapon will fire more than 1 shot with a single pull of the trigger.

In order to avoid violations of the NFA, M16 hammers, triggers, disconnectors, selectors and bolt carriers should not be used in assembly of AR–15 type semiautomatic rifles, unless the M16 parts have been modified to AR–15 Model SP1 configuration. Any AR–15 type rifles which have been assembled with M16 internal components should have those parts removed and replaced with AR–15 Model SP1 type parts which are available commercially. The M16 components also may be modified to AR–15 Model SP1 configuration.

It is important to note that any modification of the M16 parts should be attempted by fully qualified personnel only.

Should you have any questions concerning AR–15 type rifles with M16 parts, please contact your nearest ATF office. A list of ATF field offices may be found at http://www.atf.gov.

2. FEDERAL AGE RESTRICTIONS

Federal law prohibits Federal firearms licensees from selling or delivering any firearm or ammunition to any individual who the licensee knows or has reasonable cause to believe is less than 18 years of age, and, if the firearm is other than a shotgun or rifle, or ammunition for a shotgun or rifle, to any individual who the licensee knows or has reasonable cause to believe is less than 21 years of age. (18 U.S.C. 922(b)(1), 27 CFR 478.99(b)(1).) Ammunition interchangeable between rifles and handguns (such as .22 caliber rimfire) may be sold to an individual 18 years of age, but less than 21, if the licensee is satisfied that the ammunition is being acquired for use in a rifle.

It is generally unlawful for a juvenile (a person less than 18 years of age) to possess a handgun, or for any person to transfer a handgun to a juvenile. However exceptions are provided for the transfer and possession of a handgun for the purposes of employment, ranching, farming, target practice or hunting. (18 U.S.C. 922(x))

3. SALES TO LAW ENFORCEMENT OFFICERS

Most provisions of the GCA do not apply to the sale of firearms to law enforcement agencies or law enforcement officers for official duty. However, the GCA does prohibit the receipt or possession of firearms by an officer who has been convicted by a misdemeanor crime of domestic violence.

To qualify for this exemption, the law enforcement officer must present to the licensee a certification letter on the agency's letterhead, signed by a person in authority within the agency stating: the officer will use the firearm in performance of official duties and that a records check reveals that the purchasing officer has not been convicted of a misdemeanor crime of domestic violence. ATF considers the following as persons having authority to certify that law enforcement officers purchasing firearms will use the firearms in performance of official duties:

a. In a city or county police department, the director of public safety or the chief or commissioner of police.

b. In a sheriff's office, the sheriff.

c. In a State police or highway patrol department, the superintendent or the supervisor in charge of the office to which the State officer or employee is assigned.

d. In Federal law enforcement offices, the supervisor in charge of the office to which the Federal officer or employee is assigned.

The Bureau would also recognize someone signing on behalf of a person in authority, provided there is a proper delegation of authority and overall responsibility has not changed in any way. If the purchasing officer is a supervisory officer, the certification must be made by that officer's supervisor. In other words, the purchasing officer and the certifying officer may not be the same person.

Licensees who receive a qualifying letter are not required to prepare an ATF Form 4473 or conduct a NICS background check. However, disposition to the officer is to be entered in the licensee's permanent records and the certification letter kept on file. The permanent records should show the residence address of the purchasing officer, not the address of the officer's employing agency. The officer specified in the certification may purchase a firearm from a licensee regardless of the State in which the officer resides, or in which the agency is located.

Law enforcement officers who do not have a certification letter may purchase firearms in the same manner as any other unlicensed person, including completing an ATF Form 4473 and undergoing a NICS background check.

4. SALES TO ALIENS IN THE UNITED STATES FOR EXPORT

Removal of a firearm or ammunition from the U.S. by anyone is an exportation. With few exceptions, the firearms licensee must obtain an export license (Form DSP–5) from the State Department's Directorate of Defense Trade Controls (DDTC) or the Commerce Department's Bureau of Industry and Security (BIS) prior to exportation. When a licensee exports firearms directly to an alien's residence outside the United States, the licensee need only record the name and address of the foreign customer in his or her bound book. ATF Form 4473 need not be completed.

Exportation of firearms other than sporting shotguns is regulated by the Department of State, DDTC. Any person who intends to export or temporarily export firearms, ammunition or components as defined under the United States Munitions List (22 CFR 121) must obtain the approval of the DDTC prior to export unless the export qualifies for an exemption under the provisions of the International Traffic in Arm Regulations (22 CFR 120–130).

For further information about the exportation of firearms, ammunition or components, contact DDTC:

U.S. DEPARTMENT OF STATE
DIRECTORATE OF DEFENSE
TRADE CONTROLS
COMPLIANCE &
REGISTRATION DIVISION
2401 E STREET, NW, SA–1,
ROOM H1200
WASHINGTON, DC 20522–0112

TELEPHONE: 202–663–1282
WEBSITE: HTTP://WWW.PMDDTC.STATE.GOV

The Department of Commerce regulates the exportation of shotguns with a barrel length of 18 inches and over, as well as related parts, components, shotgun shells, and certain muzzle loading (black powder) firearms. The Department of Commerce requires a specific license to export or re–export these items to most destinations. For further information, contact an Export Counselor at:

202–482–4811 – Outreach and Educational Services Division (located in Washington, DC)

949–660–0144 – Western Regional Office (located in Newport Beach, CA)
408–998–8806 – Northern California branch (located in San Jose, CA)

WEBSITE: HTTP://WWW.BIS.DOC.GOV/

5. SALES TO DIPLOMATS, EMBASSIES OR CONSULATES

Diplomats, as individuals, are not exempt from Federal, State or local firearms laws. Sales to individuals, including diplomats and embassy personnel, must comply with all requirements of the GCA and the firearms regulations (27 CFR Part 478), including the general prohibition on nonimmigrant aliens unless the diplomat is purchasing the firearm/ammunition for official purposes.

Foreign embassies or consulates may purchase firearms for the purpose of the physical security of embassy or consulate grounds. The firearms become the property of the government whose embassy or consulate made the purchase, not the private property of an individual. ATF views the transaction as an exportation because embassy or consulate grounds are regarded as foreign territory.

To document that the sale is to the foreign mission and not to an individual diplomat, the licensee should obtain the following:

(1) A purchase order or invoice from the foreign mission;

(2) Payment out of government funds rather than from private funds; or

(3) A written statement by the principal officer of the embassy or consulate that the weapons are being purchased by, and will be the property of, the mission.

Once the licensee has documented that a sale is to a foreign mission, he or she may complete the transaction by shipping or delivering the firearms directly to the embassy or consulate. An ATF Form 4473 need not be completed because ATF considers the sale to be exportation, however the licensee must reflect the disposition to the foreign mission in the acquisition and disposition record.

6. CANADIAN FIREARMS INFORMATION

a. General

Implementation of the Firearms Act on December 1, 1998, brought about extensive changes to Canadian firearm regulations. Most changes affecting visitors bringing firearms into Canada came into effect on January 1, 2001.

b. Firearms Users Visiting Canada

An individual must be at least 18 years of age to bring a firearm into Canada. Individuals that are younger than 18 may use a firearm in certain circumstances, but an adult must remain present and responsible for the firearm.

Prohibited firearms (see discussion below on prohibited firearms) may not be brought into Canada or transported through Canada.

Restricted firearms (see discussion below on restricted firearms) may only be imported with prior authorization from the Chief Firearms Officer of the province or territory to which you are traveling.

Anyone entering Canada must declare all firearms to Canadian Customs.

For more information on bringing firearms into Canada, contact the Canadian Firearms Program at 1–800–731–4000.

1. Prohibited Firearms

The following firearms are classified as prohibited firearms and cannot be brought into Canada:

• Short–barreled handguns (handguns with a barrel length equal to or less than 105 mm)

• .25 caliber handguns

• .32 caliber handguns

No handgun listed above is prohibited if it is prescribed by regulation for use in competitions governed by the rules of the International Shooting Union.

- sawed–off rifles or sawed–off shotguns less than 660 mm in overall length

- sawed–off rifles or sawed–off shotguns which have a barrel length of less than 457 mm and are equal to or more than 660 mm in overall length

- all automatic firearms

- automatic firearms that have been converted to semiautomatic or single shot

- firearms prohibited by Criminal Code Regulations.

Some large capacity magazines are prohibited even if the firearms for which the magazines are designed are allowed. As a general rule, the maximum capacity is:

Five cartridges for most magazines designed for a centre–fire semi–automatic long gun; and

Ten cartridges for most handgun magazines.

There is no maximum magazine capacity for other types of long guns, including semi–automatics that discharge only rim–fire ammunition.

Replica firearms, except for replicas of antique firearms, are prohibited and cannot be brought into Canada.

Replica firearms are devices that look exactly or almost exactly like real firearms. As a rule to be prohibited, a device must closely resemble an existing make and model of firearm, not just a generic firearm. Many of these devices have to be assessed on a case–by–case basis.

When a prohibited firearm is declared at Canadian Customs, a customs officer may allow the firearm to be exported back to its country of origin. Firearms that are not immediately exported are forfeited.

For information on firearms prohibited by regulations or on firearms prescribed as International Shooting Union handguns, contact the Canadian Firearms Program at 1–800–731–4000.

2. Restricted Firearms

The following firearms are classified as restricted firearms requiring an Authorization to Transport from a Chief Firearms Officer to bring into Canada:

- all handguns which are not prohibited firearms

- semiautomatic centerfire rifles and shotguns that have a barrel length less than 470 mm and are not prohibited

- rifles and shotguns that can fire after being reduced to an overall length of less than 660 mm, by any temporary means such as folding or telescoping

- firearms restricted by Criminal Code Regulations.

Anyone bringing a restricted firearm into Canada must have an Authorization to Transport for the restricted firearm. This authorization will permit transport of the restricted firearm between specified places within Canada. This authorization must be obtained in advance from the Chief Firearms Officer of the Canadian province or territory to be visited.

An applicant for an Authorization to Transport must have a valid purpose for bringing restricted firearms to Canada, such as for use in target practice, or a target shooting competition at an approved shooting club or range. Restricted firearms cannot be used for hunting.

For more information on Authorizations to Transport, call the Chief Firearms Officer of the Canadian province or territory that you will be visiting. You can obtain the address, telephone and fax numbers from the Canada Firearms Centre.

c. Licensing Requirements

Firearm owners and users in Canada must have firearms licenses for the class of firearms in their possession. A license issued under Canada's Firearms Act is different from a provincial hunting license.

Non–residents have two options for meeting the Canadian licensing requirements:

Option 1

Declare firearms in writing to a customs officer at the point of entry to Canada, using the Non–Resident Firearm Declaration (form RCMP 5589).

If there are more than three firearms, a Non–Resident Firearm Declaration Continuation Sheet (form RCMP 5590) should be added.

The declaration form should be filled out prior to arrival at the point of entry, in order to save time. However, it should not be signed before arriving at the entry point, as a Canada Border Services Agency (CBSA) customs officer must witness the signature.

A confirmed declaration costs a flat fee of $25, regardless of the number of firearms listed on it. It is valid only for the person who signs it and only for those firearms listed on the declaration.

Once the declaration has been confirmed by the CBSA customs officer, it acts as a license for the owner and it is valid for 60 days. The declaration can be renewed for free, providing it is renewed before it expires, by contacting the Chief Firearms Officer (call 1–800–731–4000) of the relevant province or territory.

Option 2

Apply for a five–year Possession and Acquisition License (PAL).

To apply for a PAL, applicants must provide evidence that they have passed the written and practical tests for the Canadian Firearms Safety Course. A course from another country does not meet Canadian legal requirements. However, it may be possible to take the tests without taking the course.

The CFO of the province or territory to be visited can provide information on any other documents that will be required to complete the background security check.

With a Canadian firearms license, there is no need to complete the Non–Resident Firearms Declaration. However, an oral declaration must still be made to the customs officer.

For information on the declaration process, please call the CBSA:

Within Canada: 1–800–461–9999

Outside Canada: 204–983–3500 or 506–636–5064

d. Borrowing Firearms while in Canada

No license is required if the firearms user remains under the direct and immediate supervision of a licensed adult.

Otherwise, one of the following is necessary:

- A PAL (see above), or

- A confirmed Temporary Firearms Borrowing License (for Non–residents) (Form RCMP 5513).

A confirmed Non–Resident Firearms Declaration does NOT currently permit the borrowing of firearms in Canada. A temporary borrowing license permits the following uses:

- hunting under the supervision of an outfitter or other person authorized to organize hunting services in Canada;

- hunting with a Canadian resident who has the proper firearms license and hunting license;

- competing in a shooting competition;

- target shooting at an approved shooting club or range;

- taking part in a historical reenactment or display;

- engaging in a business or scientific activity being carried on in a remote area where firearms are needed to control animal predators;

- taking part in a parade, pageant or other similar event; or

- using firearms for movie, television, video or theatrical productions or publishing activities.

Visitors are advised to apply for this temporary license well in advance of arrival at the Canadian border. License application forms are available from the Canada Firearms Centre.

e. Fees (in Canadian funds)

A confirmed Non–Resident Firearm Declaration costs $25. This fee covers all the firearms listed on the declaration.

An initial PAL costs $60. It is valid for five years. For more information on the current license fee structure, please contact CFP by one of the methods listed at the end of this article.

A Temporary Firearms Borrowing License (for Non–Residents) costs $30.

f. Storage, Display and Transportation

In order to bring a firearm to Cana-da, the Storage, Display, Transportation and Handling of Firearms by Individuals Regulations, must be complied with. For non–restricted firearms:

- A secure locking device, such as a trigger lock or cable lock, should be attached, so firearms cannot be fired; or

- The firearms should be locked in a cabinet, container or room that is difficult to break into.

The ammunition should be stored separately or locked up. It can be stored in the same locked container as the firearms.

If left in an unattended vehicle, firearms should be kept in the trunk, or out of sight. The vehicle should be locked.

More information on requirements for the safe storage, display, transportation and handling of firearms in Canada is available from the Canadian Firearms Program.

g. Canadian Firearms Program

The Canadian Firearms Program (CFP)is an agency of the Government of Canada responsible for the implementation of the Firearms Act. For more information contact the CFP at:

ROYAL CANADIAN MOUNTED POLICE
CANADIAN FIREARMS PROGRAM
OTTAWA, ON K1A OR2

TELEPHONE: 1–800–731–4000
WEBSITE: HTTP://WWW.RCMP.
GC.CA
E–MAIL: CFP–PCAF@RCMP–GRC.
GC.CA

Information in this section was obtained from http://www.rcmp.gc.ca/cfp-pcaf/fs-fd/visit–visite–eng.htm.

7. ANTIQUE FIREARMS

Under section 921(a)(16)of the GCA, the term antique firearm means:

(A) any firearm (including any firearm with a matchlock, flintlock, percussion cap, or similar type of ignition system) manufactured in or before 1898; or

(B) any replica of any firearm described in subparagraph (A) if such replica—

(i) is not designed or redesigned for using rimfire or conventional centerfire fixed ammunition, or

(ii) uses rimfire or conventional centerfire fixed ammunition which is no longer manufactured in the United States and which is not readily available in the ordinary channels of commercial trade; or

(C) any muzzle loading rifle, muzzle loading shotgun, or muzzle loading pistol, which is designed to use black powder, or a black powder substitute, and which cannot use fixed ammunition. For purposes of this subparagraph, the term "antique firearm" shall not include any weapon which incorporates a firearm frame or receiver, any firearm which is converted into a muzzle loading weapon, or any muzzle loading weapon which can be readily converted to fire fixed ammunition by replacing the barrel, bolt, breechblock, or any combination thereof.

Under section 5845(g) of the NFA, antique firearm means:

"...Any firearm not designed or redesigned for using rim fire or conventional center fire ignition with fixed ammunition and manufactured in or before 1898 (including any matchlock, flintlock, percussion cap, or similar type of ignition system or replica thereof, whether actually manufactured before or after 1898) and also any firearm using fixed ammunition manufactured in or before 1898, for which ammunition is no longer manufactured in the United States and is not readily available in the ordinary channels of commercial trade."

To illustrate the distinction between the two definitions of antique firearm under the GCA and NFA, a rifle manufactured in or before 1898 would be an antique firearm under the provisions of the GCA, even though it uses conventional ammunition. However, if such rifle has a barrel of less than 16 inches in length AND uses conventional fixed ammunition which is available in the ordinary channels of commercial trade, it would not be an antique firearm under the NFA.

An antique firearm as defined in both the GCA and NFA is exempt from all of the provisions and restrictions contained in both laws. Consequently, such an antique firearm may be bought, sold, transported, shipped, etc., without regard to the requirements of these laws.

Under the Arms Export Control Act certain **"antique firearms"** are not subject to the import controls under that Act. These **"antique firearms"** are defined as **"muzzle loading (black powder) firearms (including any firearm with a matchlock, flintlock, percussion cap, or similar type of ignition system) or firearms covered by Category I(a) established to have been manufactured in or before 1898."** No all–inclusive list of antique firearms is published by ATF.

The Bureau of Industry and Security (BIS) requires an export license for muzzle loading (black powder) firearms with a caliber less than 20 mm that were manufactured later than 1937 and that are not reproductions of firearms manufactured earlier than 1890. This BIS control does not apply to weapons used for hunting or sporting purposes that were not specially designed for military use and are not of the fully automatic type. This exemption does not apply if the weapon meets certain control requirements. Contact BIS at 202–482–4811 for additional information.

8. IMPORTATION

a. Importation of Rifles

18 U.S.C. § 922(l) prohibits the importation or bringing into the United States or any possession any firearm. Section 925(d) provides limited exceptions to the general prohibition, and subsection (d)(3) requires, among other things, that firearms be **"...particularly suitable for or readily adaptable to sporting purposes...."**

ATF, in 1989, published the Report and Recommendation of the ATF Working Group on the Importability of Certain Semiautomatic Rifles (the **"1989 Study"**). The Study identified eight physical features on rifles that created a military configuration. These features were: 1) ability to accept a detachable magazine, 2) folding/telescoping stocks, 3) separate pistol grips, 4) ability to accept a bayonet, 5) flash suppressors, 6) bipods, 7) grenade launchers, and 8) night sights. The Bureau took the position that when considered with other characteristics, any of these military

configuration features, other than the ability to accept a detachable magazine, would negatively impact the importability of semiautomatic rifles.

In 1998, ATF published an updated report, the Study on the Sporting Suitability of Modified Semiautomatic Assault Rifles (the **"1998 Study"**). The 1998 Study validated the findings of the 1989 Study and also found that **"the ability to accept a detachable large capacity military magazine (LCMM) – a magazine with a capacity of more than ten rounds – originally designed and produced for a military assault weapon should be added to the list of disqualifying military configuration features identified in 1989."** The findings in this regard were based on the observation that LCMMs primarily serve **"combat–functional ends"** and that firearms equipped with LCMMs do not serve a traditional sporting purpose.

Both the 1989 and 1998 Studies were conducted to address a particular type of rifle–the **"semiautomatic assault rifle"**–and therefore the respective Studies did not address rifles outside of this limited rifle **"type."** However, rifles not included in the Studies–those not meeting the definition of semiautomatic assault rifles–are still subject to the sporting purposes test under Section 925(d)(3) before they may legally be imported. Such rifles are not to be considered sporting merely on the basis of their exclusion from the respective Studies. Therefore, ATF must apply the sporting purposes test to these rifles as well.

b. Importation of Shotguns

The purpose of section 925(d)(3) is to provide a limited exception to the general prohibition on the importation of firearms without placing "any undue or unnecessary Federal restrictions or burdens on law–abiding citizens with respect to the acquisition, possession, or use of firearms.

In January 2011, ATF published the **"Study on the Importability of Certain Shotguns"**. This study was held to establish the **"sportability"** criteria for shotguns as the '89 and '98 studies had done for rifles and the '68 study did for handguns.

The working group identified certain features and characteristics that if included on a shotgun, could not be characterized as sporting. Thus, if a shotgun has any of the following enumerated

features, it may not be imported into the United States:

(1) Folding, telescoping or collapsible stock.

(2) Bayonet Lug.

(3) Flash Suppressor.

(4) Magazine over 5 rounds, or a Drum Magazine.

(5) Grenade Launcher Mount.

(6) Integrated Rail Systems.

(7) Light Enhancing Devices.

(8) Excessive Weight. (generally less than 10 pounds fully assembled)

(9) Excessive Bulk.

Sporting shotguns are generally no more than 3 inches in width or more than 4 inches in depth.

(10) Forward Pistol Grip or Other Protruding Part Designed or Used for Gripping the Shotgun with the Shooter's Extended Hand.

c. Assembly of Nonsporting Semi-Automatic Rifles and Shotguns from Imported Parts

Section 922(r), Title 18, U.S.C., makes it unlawful for any person to assemble from imported parts any semiautomatic rifle or any shotgun which is identical to any rifle or shotgun prohibited from importation under section 925(d)(3) of the GCA. Regulations implementing the law in 27 C.F.R. 478.39 provide that a violation of section 922(r) will result if a semiautomatic rifle or shotgun is assembled with more than 10 of the following imported parts:

(1) Frames, receivers, receiver castings, forgings, or stampings

(2) Barrels

(3) Barrel extensions

(4) Mounting blocks (trunnions)

(5) Muzzle attachments

(6) Bolts

(7) Bolt carriers

(8) Operating rods

(9) Gas pistons

(10) Trigger housings

(11) Triggers

(12) Hammers

(13) Sears

(14) Disconnectors

(15) Buttstocks

(16) Pistol grips

(17) Forearms, handguards

(18) Magazine bodies

(19) Followers

(20) Floorplates

Section 922(r) does not prohibit the importation, sale, or possession of parts which may be used to assemble a semiautomatic rifle or shotgun in violation of the statute. However, 18 U.S.C. § 2 provides that a person who aids or abets another person in the commission of an offense is also responsible for the offense. Therefore, a person who sells parts knowing that the purchaser intends to use the parts in assembling a firearm in violation of section 922(r) would also be responsible for the offense.

d. Importation of Handguns

Section 925(d)(3), Title 18, U.S.C., states that the Attorney General shall authorize a firearm or ammunition to be imported or brought into the United States if it is of a type that does not fall within the definition of a firearm in section 5845(a) of the Internal Revenue Code of 1986 and is generally recognized as particularly suitable for or readily adaptable to sporting purposes, excluding surplus military firearms. The statute also prohibits the importation of any frame, receiver, or barrel of a firearm that would be prohibited if assembled.

In 1968, the Secretary of the Treasury established the Treasury Department Firearms Evaluation Panel. The purpose of this panel was to assist in the establishment of guidelines for use in determining if a particular firearm would be importable under the sporting purposes test prescribed by section 925(d)(3). This panel was composed of representatives from the firearms industry, law enforcement, and the military.

The panel developed objective numerical criteria with minimum qualifying scores to determine if a handgun is particularly suitable for or readily adaptable to sporting purposes. The criteria, ATF Form 4590, Factoring Criteria for Weapons, assigns point values to handguns based on dimensions, material used in construction, weight, caliber, safety features and miscellaneous equipment. The criteria also have prerequisite requirements concerning safeties and minimum dimensions.

Form 4590 is divided into two parts. The right side of the form is used to evaluate revolvers and the left side is used to evaluate pistols. The minimum qualifying score for a revolver is 45 points and the minimum qualifying score for a pistol is 75 points. The form also provides that revolvers must pass a safety test.

Any handgun being imported into the United States must pass these criteria. The fact that a particular weapon may be of domestic manufacture or classified as a curio or relic does not exempt it from the factoring criteria.

If you have any questions concerning Form 4590, or the score a particular handgun achieves, please contact the ATF Firearms Technology Branch.

e. Importation of Firearms by Nonlicensed U.S. Residents

A permit from ATF must be obtained to permanently import or bring into the United States any firearm, firearm part, or ammunition. The firearm or ammunition must be generally recognized as particularly suitable for, or readily adaptable to, sporting purposes. Firearm frames or receivers or firearm barrels for non-sporting firearms are generally not importable. Surplus military firearms are generally excluded from importation into the United States.

Unlicensed persons may not permanently import firearms into the United States (18 U.S.C. 922(a)(3); 27 CFR 478.111, 113). A Federal firearms licensee located in the nonlicensee's State of residence may act as an agent to import the nonlicensee's personal firearm. The form to be used by the licensee is ATF Form 6, Part I, Application and Permit for Importation of Firearms, Ammunition and Implements of War, and may be obtained from the ATF website at http://www.atf.gov.

A nonlicensee may obtain a permit to import sporting ammunition for personal use (excluding armor piercing ammunition, or tracer or incendiary ammunition), firearm parts, other than frames or receivers, or barrels for sporting firearms without engaging the services of a Federal firearms licensee. Silencer parts and certain machinegun parts are subject to the NFA and may not be imported. If the nonlicensee chooses to have a licensee handle the importation, the licensee should complete and send to ATF an ATF Form 6, Part I, in accordance with the instructions on the form. The nonlicensee's name, address, and telephone number should appear in Item 9, **"Specific purpose of importation."**

No permit or authorization from ATF is required to bring into the United States a firearm or ammunition that was previously taken out of the U.S. by the person bringing it in. U.S. Customs and Border Protection (CBP) is authorized to release a firearm or ammunition without a permit from ATF upon a proper showing of proof that the firearm or ammunition was taken out of the country by the person bringing it in. This proof is best established by having registered the item or items on CBP Form 4457, Certificate of Registration, at the point and time of departure.

For further information, see ATF Rul. 81–3, ATF Rul. 85–10, and ATF Ruling 2013–1, set out within the Rulings, Procedures, and Industry Circulars portion of this publication.

f. Importation by Non–resident U.S. Citizens Returning to the United States and Non–resident Aliens Immigrating to the United States

A non–resident U.S. citizen returning to the United States, or a non–resident alien lawfully immigrating to the United States, may apply for a permit from ATF to import for personal use, and not for resale, firearms and ammunition without having to utilize the services of a federally licensed firearms dealer. ATF Form 6, Part I application should include a statement, on the application form or on an attached sheet, that:

(1) the applicant is a non–resident U.S. citizen who is returning to the United States from a residence outside of the United States or, in the case of an alien, is lawfully immigrating to the United States from a residence outside of the United States, and

(2) the firearms and ammunition are being imported for personal use and not for resale.

No permit will be issued to import firearms or ammunition which are not generally recognized as particularly suitable for, or readily adaptable to, sporting purposes, surplus military firearms, or National Firearms Act (NFA) firearms (e.g., machineguns, silencers, destructive devices, short–barreled rifles, short–barreled shotguns, etc.).

The firearms must accompany non–resident U.S. citizens, since once a person is in the United States and has acquired residence in a State, he or she may import a firearm only by arranging for the importation through a Federal firearms licensee.

g. Importation by Members of the Armed Forces

(1) Import Permit Requirements

Section 925(a)(4) of the GCA provides that:

When established to the satisfaction of the Attorney General to be consistent with the provisions of this chapter [the GCA] and other applicable Federal and State laws and published ordinances, the Attorney General may authorize the transportation, shipment, receipt or importation into the United States to the place of residence of any member of the United States Armed Forces who is on active duty outside the United States (or who has been on active duty outside the United States within the 60 day period immediately preceding the transportation, shipment, receipt, or importation), of any firearm or ammunition which is:

(A) determined by the Attorney General to be generally recognized as particularly suitable for sporting purposes, or determined by the Department of Defense to be a type of firearm normally classified as a war souvenir, and

(B) intended for the personal use of such member.

Applications to import such firearms are filed on ATF Form 6, Part II and should include a detailed description of each firearm to be imported. Incomplete information will cause return of your application. Applications should be completed in triplicate and mailed to the Bureau of ATF, Firearms and Explosives Imports Branch.

A member of the Armed Forces who does not meet the above criteria must obtain the services of a Federal firearms licensee located in his or her State of residence to import a firearm on behalf of the member. The licensee would submit an application on ATF Form 6, Part I.

If your application is approved, the original will be returned to you. This will be your authorization to import the firearm(s) described on the form. The permit is valid for 1 year from the date of approval. If disapproved, your application will be stamped disapproved and returned to you with the reason for disapproval stated.

A permit must be obtained for all firearms to be imported, regardless of the date purchased. However, this does not apply to a firearm previously taken out of the United States by the person bringing it in (if they can prove they previously took the firearm out of the United States), nor to a firearm shipped by a licensee in the United States to a serviceperson on active duty outside the United States or to an authorized rod and gun club abroad specifically for the serviceperson importing the firearm.

Authorization will not be given to import a machinegun, or any other firearm as defined in the NFA, regardless of the degree of serviceability. Additionally, authorization will not be given to import any surplus military firearm.

(2) Importation of War Souvenirs or War Trophy Firearms

Generally, war souvenirs or war trophies are not importable without specific authorization by the Department of Defense. Active duty service members should contact their respective branch for additional information.

(3) ATF Ruling 74–13 — Importation of Handguns

ATF was informed by State and local authorities that handguns were being transported, shipped, received, or imported into the United States by members of the U.S. Armed Forces to their State of residence without such members having obtained the required permits or other authorizations required by the State for lawful possession or ownership of handguns in that State.

Ruling 74–13 holds that a member of the U.S. Armed Forces who is a resident of any State or territory which requires that a permit or other authorization be issued prior to possessing or owning a handgun shall submit evidence of compliance with State law before an application to import a handgun may be approved.

ATF Rul. 74–13 is set out within the Rulings, Procedures, and Industry Circulars portion of this publication.

9. SPECIAL (OCCUPATIONAL) TAXPAYERS AND NFA FIREARMS

a. General

Anyone wishing to manufacture, import, or deal in firearms as defined in the NFA must:

1. Be properly licensed as a Federal firearms licensee;

2. Have an employer identification number (even if you have no employees); and

3. Pay the Special (Occupational) Tax required of those manufacturing, importing, or dealing in NFA firearms.

Those weapons defined as NFA firearms can be found in sections 5845(a)–(f) of the NFA.

b. What You Need to Proceed

If you do not already have an employer identification number (EIN), you must obtain and complete a Form SS–4 application to obtain such a number. This number must appear on all registration documents when you apply to receive or transfer any NFA firearm.

Federal firearms licensees who wish to engage in the business of importing, manufacturing, or dealing in NFA firearms are required to pay special occupational tax before beginning business. You may file one return (ATF Form 5630.7) to cover several locations or classes of taxable activity. However, you must submit a separate return for each tax period. The special occupational tax period runs from July 1 through June 30 and payment is due annually by July 1. If you begin business any time during the tax year, you are responsible for the full amount of tax for the entire year, i.e., the taxes are not prorated.

Upon receipt of your properly completed ATF Form 5630.7, together with your remittance, a Special Tax Stamp will be mailed directly to you.

c. Changes in Operations

If a licensee intends to change location/address, the licensee must file an Application for an Amended Federal Firearms License, ATF Form 5300.38, with the Chief, Federal Firearms Licensing Center, not less than 30 days prior to the move. The licensee must obtain the amended license before commencing business at the new location.

In addition, for a change of address, location, or trade name, an amended ATF Form 5630.7 must be filed according to the instructions on the form and approved prior to the change.

d. Machineguns

Machineguns produced, imported, or registered after May 19, 1986, the effective date of 18 U.S.C. 922(o), generally are unlawful to possess or transfer except for use by a government agency or for exportation. Licensed dealers may, however, acquire **"sales samples"** of post-1986 machineguns if they obtain proper official documentation as specified in 27 CFR 479.105(d).

e. Going Out Of Business

Machineguns. When a Special (Occupational) Taxpayer does not renew payment of the special tax, all machineguns possessed by that taxpayer that are restricted under 18 U.S.C. 922(o) must be transferred to a Special (Occupational) Taxpayer having a legitimate need for the weapon(s) or be exported. Such transfer must occur before the Federal firearms license and special tax status expires. Otherwise, these firearms may not lawfully be possessed, and must be disposed of in accordance with 27 CFR 479.105(f).

(1) Pre–86 Machineguns. Unless otherwise prohibited by State or local law, when an individual Special (Occupational) Taxpayer goes out of business, the individual may continue to possess machineguns lawfully imported or manufactured prior to May 19, 1986 which are lawfully registered to the individual. These firearms may also be transferred after ATF has approved a proper application for transfer under the NFA.

When a corporation, partnership, or other type of business entity Special (Occupational) Taxpayer goes out of business, the business may continue to possess pre–86 machineguns registered to the business only if the corporation, partnership, or other type of business entity continues to exist under State law and only if the title to the machineguns remains in the business after the Special (Occupational) tax stamp expires. Prior to a corporation, partnership or other type of business entity ceasing to exist under State law, all NFA firearms registered to the entity must have been properly transferred to another person. Transfer applications must be submitted and approved before dissolution occurs to avoid placing the possessors in violation of the NFA. If the registered machineguns are transferred to officers or directors of a corporate registrant or individual partners of a partnership, the transaction is a transfer subject to all applicable provisions of the NFA and GCA, including payment of tax.

(2) Post–86 Machineguns. When a Special (Occupational) Taxpayer goes out of business, machineguns lawfully imported or manufactured on or after May 19, 1986, may not be retained by the business. Such machineguns are subject to the restrictions of 18 U.S.C. 922(o) and must be transferred to a law enforcement agency for official use or to a qualified NFA dealer as dealer sales samples in accordance with 27 C.F.R. 479.105(d).

NFA Firearms Other Than Machineguns. When a individual Special (Occupational) Taxpayer goes out of business, the individual may continue to possess firearms registered to the business. Subsequent transfers of the registered weapons are subject to all the provisions of the NFA and GCA.

When a corporation, partnership, or other type of business entity Special (Occupational) Taxpayer goes out of business, the business may continue to possess firearms other than machineguns registered to the business only if the corporation, partnership, or other type of business entity continues to exist under State law and only if title to the firearms remains in the business after the Special (Occupational) tax stamp expires. Prior to a corporation, partnership or other type of business entity ceasing to exist under State law, all NFA firearms registered to the entity must have been properly transferred to another person. Transfer applications must be submitted and approved before dissolution occurs to avoid placing the possessors in violation of the NFA. If the registered machineguns are transferred to officers or directors of a corporate registrant or individual partners of a partnership, the transaction is a transfer subject to all applicable provisions of the NFA and GCA, including payment of tax.

Any NFA firearms retained by the business that were imported under 26 U.S.C. 5844 for use as samples or for scientific or research purposes may only be transferred to government agencies for official use or to a Federal firearms licensee who has paid the special (occupational) tax to manufacture, import, or deal in NFA firearms.

f. NFA Firearms in Decedents' Estates

Possession of an NFA firearm not registered to the possessor is a violation of Federal law. However, a reasonable time is allowed for transfer of lawfully registered NFA firearms in a decedent's estate.

It is the responsibility of the executor or the administrator of an estate to transfer firearms registered to a decedent. ATF Form 5, Application for Tax Exempt Transfer and Registration of a Firearm, is used to apply for a tax–exempt transfer to a lawful heir. A lawful heir is anyone named in the decedent's will or, in the absence of a will, anyone entitled to inherit under the laws of the State in which the decedent last resided. NFA firearms may be transferred directly interstate to a beneficiary of the estate. However, if any Federal, State or local law prohibits the heir from receiving or possessing the firearm, ATF will not approve the application. When a firearm is being transferred to an individual heir, his or her fingerprints on FBI Forms FD–258 must accompany the transfer application.

ATF Form 4 is used to apply for the tax paid transfer of a serviceable NFA firearm to a person outside the estate (not a beneficiary). ATF Form 5 is also used to apply for the tax–exempt transfer of an unserviceable NFA firearm to a person outside the estate. As noted above, all requirements, such as fingerprint cards for transfers to individuals and compliance with State or local law, must be met before an application could be approved.

If the NFA firearm in the estate was imported for use as a **"sales sample,"** this restriction on the firearm's possession remains. The NFA firearm may only be transferred to a Federal firearms licensee who has paid the special (occupational) tax to deal in NFA firearms or to a government agency.

For further information, contact: Bureau of Alcohol, Tobacco, Firearms and Explosives, National Firearms Act Branch.

10. ARMOR PIERCING AMMUNITION

Under Federal law, it is unlawful for any person (including licensed importers and manufacturers) to import or manufacture armor piercing ammunition unless such manufacture is for a governmental agency, exportation, or testing/experimentation. See 18 U.S.C. 922(a)(7),(8); 27 CFR 478.37, 478.148. Licensed dealers may only transfer armor piercing ammunition to a governmental agency if the ammunition was received and maintained by the dealer as part of its business inventory prior to August 28, 1986. See 18 U.S.C. 923(e); 27 CFR 478.99(e).

The GCA defines the term **"armor piercing ammunition"** as:

(i) a projectile or projectile core which may be used in a handgun and which is constructed entirely (excluding the presence of traces of other substances) from one or a combination of tungsten alloys, steel, iron, brass, bronze, beryllium copper, or depleted uranium; or

(ii) a full jacketed projectile larger than .22 caliber designed and intended for use in a handgun and whose jacket has a weight of more than 25 percent of the total weight of the projectile."

Armor piercing ammunition includes the following:

KTW AMMUNITION, all calibers. Identified by a green coating on the projectile.

ARCANE AMMUNITION, all calibers. Identified by a pointed bronze or brass projectile.

THV AMMUNITION, all calibers. Identified by a brass or bronze projectile and a head stamp containing the letters SFM and THV.

CZECHOSLOVAKIAN manufactured 9mm Parabellum (Luger) ammunition having an iron or steel bullet core. Identified by a cupronickel jacket and a head stamp containing a triangle, star, and dates of 49, 50, 51, or 52. This bullet is attracted to a magnet.

GERMAN manufactured 9mm Parabellum (Luger) having an iron or steel bullet core. Original packaging is marked Pistolenpatronen 08 m.E. May have black colored bullet. This bullet is attracted to a magnet.

MSC AMMUNITION, caliber .25. Identified by a hollowpoint brass bullet. **NOTE:** MSC ammunition, caliber .25 identified by a hollowpoint copper bullet is **not** armor piercing.

BLACK STEEL ARMOR PIERCING AMMUNITION, all calibers, as produced by National Cartridge, Atlanta, Georgia.

BLACK STEEL METAL PIERCING AMMUNITION, all calibers, as produced by National Cartridge, Atlanta, Georgia.

7.62mm NATO AP, identified by black coloring in the bullet tip. This ammunition is used by various NATO countries. The U.S. military designation is M61 AP.

7.62mm NATO SLAP. Identified by projectile having a plastic sabot around a hard penetrator. The penetrator protrudes above the sabot and is similar in appearance to a Remington accelerator cartridge.

PMC ULTRAMAG, .38 Special caliber, constructed entirely of a brass type material, and a plastic pusher disc located at the base of the projectile. **NOTE:** PMC ULTRAMAG 38J late production made of copper with lead alloy projectile is **not** armor piercing.

OMNISHOCK. A .38 Special cartridge with a lead bullet containing a mild steel core with a flattened head resembling a wad cutter. **NOTE:** OMNISHOCK cartridges having a bullet with an aluminum core are **not** armor piercing.

7.62x39mm with steel core. These projectiles have a steel core. **NOTE:** Projectiles having a lead core with steel jacket or steel case are **not** armor piercing.

In addition, the Violent Crime Control and Law Enforcement Act of 1994 added to the definition of armor piercing ammunition the following:

"... a full jacketed projectile larger than .22 caliber designed and intended for use in a handgun and whose jacket has a weight of more than 25 percent of the total weight of the projectile."

Exemptions: The following articles are exempted from the definition of armor piercing ammunition.

5.56 mm (.233) SS 109 and M855 Ammunition, identified by a green coating on the projectile tip.

U.S. .30-06 M2AP, identified by a black coating on the projectile tip.

QUESTIONS & ANSWERS

TABLE OF CONTENTS

A. GENERAL QUESTIONS

(A1) Does Federal law regulate firearms businesses?

Yes. The Gun Control Act (GCA), administered by the Bureau of Alcohol, Tobacco, Firearms and Explosives (ATF) of the Department of Justice, requires a person obtain a Federal license to engage in various firearms businesses (manufacturers, importers, and dealers).

[18 U.S.C. 922(a)(1)(A) and 923(a); 27 CFR 478.41]

(A2) Does the Federal Government issue a license or permit to carry a concealed weapon?

No. Neither ATF nor any other Federal agency issues such a permit or license. Carrying permits may be issued by a State or local government. Please contact your State's Attorney General's Office for information regarding permits to carry firearms.

(A3) Do antique firearms come within the purview of the GCA?

No, assuming the antique firearm is not a replica designed or redesigned for using rimfire or conventional centerfire fixed ammunition. The antique firearm also cannot be a black powder muzzle loading weapon that incorporates a firearm frame or receiver, have been converted into a muzzle loading weapon, or uses fixed ammunition (or readily converted to do so).

[18 U.S.C. 921(a)(3) and (16); 27 CFR 478.11 and 478.141(d)]

(A4) Does the GCA regulate ammunition businesses?

Yes. A license is required to import or manufacture ammunition. However, a license is not required to deal only in ammunition. Ammunition includes cartridge cases, primers, bullets or propellant powder designed for use in any firearm other than an antique firearm.

[18 U.S.C. 921(a)(17), 922(a)(1)(B) and 923(a); 27 CFR 478.11]

(A5) Does the GCA control the sale of firearms parts?

No, except that frames or receivers of firearms are "firearms" as defined in the law and are subject to the same controls as complete firearms. Certain parts of si-

lencers and machineguns are also regulated as firearms under the GCA, as well as under the National Firearms Act (NFA).

[18 U.S.C. 921(a)(3), (23) and (24); 26 U.S.C. 5845; 27 CFR 478.11 and 479.11]

(A6) Does an individual need a license to make a firearm for personal use?

No, a license is not required to make a firearm solely for personal use. However, a license is required to manufacture firearms for sale or distribution. The law prohibits a person from assembling a non–sporting semiautomatic rifle or shotgun from 10 or more imported parts, as well as firearms that cannot be detected by metal detectors or x–ray machines. In addition, the making of an NFA firearm requires a tax payment and advance approval by ATF.

[18 U.S.C. 922(o), (p) and (r); 26 U.S.C. 5822; 27 CFR 478.39, 479.62 and 479.105]

(A7) Are there persons who cannot legally receive or possess firearms and/or ammunition?

Yes, a person who —

(1) Has been convicted in any court of a crime punishable by imprisonment for a term exceeding 1 year;

(2) Is a fugitive from justice;

(3) Is an unlawful user of or addicted to any controlled substance;

(4) Has been adjudicated as a mental defective or has been committed to a mental institution;

(5) Is an alien illegally or unlawfully in the United States or an alien admitted to the United States under a nonimmigrant visa;

(6) Has been discharged from the Armed Forces under dishonorable conditions;

(7) Having been a citizen of the United States, has renounced his or her citizenship;

(8) Is subject to a court order that restrains the person from harassing, stalking, or threatening an intimate partner or child of such intimate partner issued after a hear-

ing at which notice was given to the person and at which the person had an opportunity to participate, and includes a finding that the person subject to the order represents a credible threat to the intimate partner or child or the intimate partner OR explicitly prohibits the use, attempted use, or threatened use of force against the partner; or

(9) Has been convicted of a misdemeanor crime of domestic violence cannot lawfully receive, possess, ship, or transport a firearm or ammunition,is prohibited from shipping, transporting, possessing, or receiving firearms and ammunition.

A person who is under indictment or information for a crime punishable by imprisonment for a term exceeding 1 year cannot lawfully ship, transport, or receive a firearm or ammunition. Such persons may continue to lawfully possess firearms and ammunition obtained prior to the indictment or information, but cannot do so once the conviction becomes final.

[18 U.S.C. 922(g) and (n); 27 CFR 478.32]

(A8) Is there a way for a prohibited person to restore his or her right to receive or possess firearms and ammunition?

Although Federal law provides a means for the relief of firearms disabilities, since October 1992, ATF's annual appropriation has prohibited the expending of any funds to investigate or act upon applications for relief from Federal firearms disabilities submitted by individuals. As long as this provision is included in current ATF appropriations, the Bureau cannot act upon applications for relief from Federal firearms disabilities submitted by individuals.

[18 U.S.C. 925(c); 27 CFR 478.144]

(A9) If a person has been convicted of a disabling offense, is there a means other then relief to have firearms rights restored?

Persons convicted of a Federal offense may apply for a Presidential pardon. 28 CFR Part I specifies the rules governing petitions for obtaining Presidential pardons. You may contact the Pardon Attorney's Office at the U.S. Department of Justice, Washington, DC, to inquire about the procedures for obtaining a Presidential pardon.

Persons convicted of a State offense may contact the State Attorney General's Office in the State of their conviction for information concerning the availability of expungements, set asides, pardons and civil rights restoration.

[18 U.S.C. 921(a)(20) and (a)(33); 27 CFR 478.11 and 478.142]

(A10) Can a person prohibited by law from possessing a firearm own a black powder firearm?

Because black powder firearms are considered antique firearms, the possession of a black powder firearm by a person subject to Federal firearms disabilities is not prohibited by the GCA. However, a person subject to Federal firearms disabilities may not receive and/or possess black powder firearms that can be readily converted to fire fixed ammunition by replacing the barrel, bolt, breechblock, or any combination thereof which are classified as "firearms." Additionally, State law may prohibit the possession of a black powder firearm by persons who are not Federally prohibited from possessing them. Please contact your State Attorney General's Office for information regarding black powder firearms.

[18 U.S.C. 921(a)(3) and (16); 27 CFR 478.11 and 478.141(d)]

B. UNLICENSED PERSONS

(B1) To whom may an unlicensed person transfer firearms under the GCA?

A person may transfer a firearm to an unlicensed resident of his or her State, provided the transferor does not know or have reasonable cause to believe the transferee is prohibited from receiving or possessing firearms under Federal law. There may be State laws that regulate intrastate firearm transactions. A person considering transferring a firearm should contact his or her State Attorney General's Office to inquire about the laws and possible State or local restrictions.

Generally, for a person to lawfully transfer a firearm to an unlicensed person who resides out of State, the firearm must be shipped to a Federal firearms licensee (FFL) within the transferee's State of residence. The transferee may then receive the firearm from the FFL upon completion of an ATF Form 4473 and a NICS background check.

A person may loan or rent a firearm to a resident of any State for temporary use for lawful sporting purposes, if he or she or she does not know or have reasonable cause to believe the person is prohibited from receiving or possessing firearms under Federal law. Another exception is provided for transfers of firearms to non-residents to carry out a lawful bequest or acquisition by intestate succession. This exception would authorize the transfer of a firearm to a nonresident who inherits a firearm under the will of a decedent.

A person may transfer a firearm to a licensee in any State. However, a firearm other than a curio or relic may not be transferred interstate to a licensed collector.

[18 U.S.C 922(a)(5) and 922(d); 27 CFR 478.30, 478.32]

(B2) May an unlicensed person acquire a firearm under the GCA in any State?

Generally, a person may only acquire a firearm within the person's own State. Exceptions include the acquisition pursuant to a lawful bequest, or an over–the–counter acquisition of a rifle or shotgun from a licensee where the transaction is allowed by the purchaser's State of residence and the licensee's State of business. A person may borrow or rent a firearm in any State for temporary use for lawful sporting purposes.

[18 U.S.C 922(a)(3); 27 CFR 478.29]

(B3) How may an unlicensed person receive a firearm in his or her State that he or she or she purchased from an out–of–State source?

An unlicensed person who is not prohibited from receiving or possessing firearms may purchase a firearm from an out–of–State source, provided the transfer takes place through a Federal firearms licensee in his or her State of residence.

[18 U.S.C 922(a)(3) and 922(b)(3); 27 CFR 478.29]

(B4) May an unlicensed person obtain ammunition from an out–of–State source?

Yes, provided he or she or she is not prohibited from possessing or receiving ammunition.

[18 U.S.C. 922(g) and (n)]

(B5) Do law enforcement officers who are subject to restraining orders and who receive and possess firearms for purposes of carrying out their official duties violate the law?

No. Although generally the GCA prohibits the receipt and possession of firearms and ammunition by persons subject to disqualifying restraining orders, the GCA does not prohibit a law enforcement officer subject to a restraining order from receiving or possessing firearms or ammunition for use in performing official duties. Possession of the firearm for official purposes while off duty would be lawful if such possession is required or authorized by law or by official departmental policy. An officer subject to a disabling restraining order would violate the law if the officer received or possessed a firearm or ammunition for other than official use.

[18 U.S.C. 921(a)(32), 922(g)(8) and 925(a)(1); 27 CFR 478.11, 478.32 and 478.141]

(B6) May a nonlicensee mail a firearm through the U.S. Postal Service?

A nonlicensee may mail a shotgun or rifle to a resident of his or her own State or to a licensee in any State. The Postal Service recommends that long guns be sent by registered mail and that no marking of any kind which would indicate the nature of the contents be placed on the outside of any parcel containing firearms. Handguns may not be mailed. A common or contract carrier must be used to ship a handgun.

[18 U.S.C. 1715, 922(a)(5) and 922 (a)(2)(A); 27 CFR 478.31]

(B7) May a nonlicensee ship a firearm by common or contract carrier?

A nonlicensee may ship a firearm by a common or contract carrier to a resident of his or her or her own State or to a licensee in any State. A common or contract carrier must be used to ship a handgun. In addition, Federal law requires that the carrier be notified that the shipment contains a firearm or ammunition, prohibits common or contract carriers from requiring or causing any label to be placed on any package indicating that it contains a firearm and requires obtaining written acknowledgement of receipt.

[18 U.S.C. 922(a)(2)(A), 922(a)(5), 922(e) and (f); 27 CFR 478.30 and 478.31]

(B8) May a nonlicensee ship firearms interstate for his or her use in hunting or other lawful activity?

Yes. A person may ship a firearm to himself or herself in care of another person in the State where he or she or she intends to hunt or engage in any other lawful ac-tivity. The package should be addressed to the owner "in the care of" the out–of–State resident. Upon reaching its destination, persons other than the owner may not open the package or take pos-session of the firearm.

(B9) May a person who is relocating out–of–State move firearms with other household goods?

Yes. A person who lawfully possess-es a firearm may transport or ship the firearm interstate when changing his or her State of residence. If using a moving company, the person must no-tify the mover that firearms are being transported.

Certain NFA firearms must have prior approval from ATF before such firearms may be moved interstate. He or she or she should also check State and local laws where relocating to ensure that movement of firearms into the new State does not violate any State law or local ordinance.

[18 U.S.C. 922(a)(4) and 922(e); 27 CFR 478.28 and 478.31]

(B10) What constitutes residency in a State?

For GCA purposes, a person is a resident of a State in which he or she is present with the intention of making a home in that State. The State of residence for a corporation or other business entity is the State where it maintains a place of business. A member of the Armed Forces on active duty is a resident of the State in which his or her or her permanent duty station is located. If a member of the Armed Forces maintains a home in one State and the member's permanent duty station is in a nearby State to which he or she or she commutes each day, then the member has two States of residence and may purchase a firearm in either the State where the duty station is located or the State where the home is maintained.

[18 U.S.C. 921(b), 922(a)(3), and 922(b)(3); 27 CFR 478.11]

(B11) May a person who resides in one State and owns property in another State purchase a firearm in either State?

If a person maintains a home in 2 States and resides in both States for certain periods of the year, he or she may, during the period of time the person actually resides in a particular State, purchase a firearm in that State. However, simply owning property in another State does not alone qualify the person to purchase a firearm in that State.

[27 CFR 478.11]

(B12) May aliens legally in the United States purchase firearms?

An alien legally in the U.S. is not prohibited from purchasing firearms unless the alien is admitted into the U.S. under nonimmigrant visa and does not meet one of the exceptions as provided in 18 U.S.C. 922(y)(2), such as possession of a valid hunting license or permit.

[18 U.S.C. 922 (d)(5), (g)(5) and (y)(2); 27 CFR 478.11 and 478.32(a)(5)]

(B13) May a parent or guardian purchase firearms or ammunition as a gift for a juvenile (less than 18 years of age)?

Yes. However, persons less than 18 years of age may only receive and possess handguns with the written permission of a parent or guardian for limited purposes, e.g., employment, ranching, farming, target practice or hunting.

[18 U.S.C. 922(x)]

(B14) May an individual between the ages of 18 and 21 years of age acquire a handgun from an unlicensed individual who is also a resident of that same State?

An individual between 18 and 21 years of age may acquire a handgun from an unlicensed individual who resides in the same State, provided the person acquiring the handgun is not otherwise prohibited from receiving or possessing firearms under Federal law. A Federal firearms licensee may not, however, sell or deliver a firearm other than a shotgun or rifle to a person the licensee knows or has reasonable cause to believe is under 21 years of age.

There may be State or local laws or regulations that govern this type of transaction. Contact the office of your State Attorney General for information on any such requirements.

[18 U.S.C. 922(b)(1)]

(B15) Are curio or relic firearms exempt from the provisions of the GCA?

No. Curios or relics are still firearms subject to the provisions of the GCA; however, curio or relic firearms may be transferred in interstate commerce to licensed collectors or other licensees.

[18. U.S.C. 921(a)(3) and (13) and 923(b); 27 CFR 478.11]

(B16) What recordkeeping procedures should be followed when two unlicensed individuals want to engage in a firearms transaction?

When a transaction takes place between unlicensed persons who reside in the same State, the GCA does not require any record keeping. An unlicensed person may sell a firearm to another unlicensed person in his or her State of residence and, similarly, an unlicensed person may buy a firearm from another unlicensed person who resides in the same State. It is not necessary under Federal law for a Federal firearms licensee (FFL) to assist in the sale or transfer when the buyer and seller are "same–State" residents.

There may be State or local laws or regulations that govern this type of transaction. Contact the office of your State Attorney General for information regarding any such requirements.

(B17) How does a person register a firearm or remove a name from a firearms registration?

Only those firearms subject to the National Firearms Act (NFA) (e.g., machineguns, short–barreled rifles and shotguns, silencers, destructive devices, and firearms designated as "any other weapons") must be registered with ATF.

Firearms registration may be required by State or local law. Any person considering acquiring a firearm should contact his or her State Attorney General's Office to inquire about the laws and possible State or local restrictions.

[26 U.S.C. 5841; 27 CFR 479.101]]

C. LICENSING

(C1) Who is eligible for a firearms license?

An application for a Federal firearms license will be approved if the applicant:

- Is 21 years of age or over;

- Is not prohibited from shipping, transporting, receiving or possessing firearms or ammunition, nor in the case of a corporation, partnership, or association, is any individual possessing, directly or indirectly, the power to direct or cause the direction of the management and policies of the corporation, partnership, or association prohibited from shipping, transporting, receiving or possessing firearms or ammunition;

- Has not willfully violated the GCA or its regulations;

- Has not willfully failed to disclose material information or has not made false statements concerning material facts in connection with his or her application;

- Has premises for conducting business or collecting; and

- The applicant certifies that:

    (1) the business to be conducted under the license is not prohibited by State or local law in the place where the licensed premises is located;

    (2) within 30 days after the application is approved the business will comply with the requirements of State and local law applicable to the conduct of the business;

    (3) the business will not be conducted under the license until the requirements of State and local law applicable to the business have been met;

    (4) the applicant has sent or delivered a form to the chief law enforcement officer where the premises is located notifying the officer that the applicant intends to apply for a license; and

* if the applicant is to be a licensed dealer, the applicant certifies that secure gun storage or safety devices will be available at any place in which firearms are sold under the license to persons who are not licensees ("secure gun storage or safety device" is defined in 18 U.S.C. 921(a)(34)).

[18 U.S.C. 923(d)(1); 27 CFR 478.47(b)]

(C2) How does one get a license?

Submit ATF Form 7, Application for License, or ATF Form 7CR, Application for License (Collector of Curios or Relics), with the appropriate fee in accordance with the instructions on the form to ATF. An application packet may be obtained by contacting the ATF Distribution Center.

[18 U.S.C. 923; 27 CFR 478.44]

(C3) May one license cover several locations?

No. A separate license must be obtained for each location. However, separate facilities solely to store firearms are not required to be covered by a separate license, although the records maintained at the licensed premises must reflect all firearms held in the separate storage facility.

[18 U.S.C. 923(a); 27 CFR 478.42 and 478.50]

(C4) Does an importer or manufacturer of firearms also need a dealer's license?

No, as long as the importer or manufacturer is engaged in the business of dealing in firearms at the licensed premises in the same type of firearms authorized by the importer's or manufacturer's license.

[27 CFR 478.41(b)]

(C5) If a person timely files an application for renewal of a license and the present license expires prior to receipt of the new license, may the person continue to conduct the business covered by the expired license?

Yes. A person who timely files an application for renewal of a license may continue operations authorized by the expired license until final action is taken on the application.

If a person does not timely file a license renewal application and the license expires, the person must file a new ATF Form 7, Application for License, or an ATF Form 7CR, Application for License (Collector of Curios or Relics), as required by 27 CFR 478.44.

[27 CFR 478.45]

(C6) Must a licensed importer's, manufacturer's, or dealer's records be surrendered to ATF if the licensee discontinues business?

If the business is being discontinued completely, the licensed dealer, manufacturer or importer is required to submit all records to ATF within 30 days . Records may be delivered to an ATF office or shipped to the following address:

BUREAU OF ATF
ATF OUT–OF–BUSINESS RECORDS
CENTER
244 NEEDY ROAD
MARTINSBURG, WV 25405

A licensee discontinuing business must also notify the Federal Firearms Licensing Center within 30 days.

If someone is taking over the business, the original licensee should underline the final entry in each acquisition and disposition (A&D) record, note the date of transfer, and forward all records and forms to the successor (who must apply for and receive his or her own license before lawfully engaging in business) or forward the records and forms to the ATF Out–of–Business Records Center. If the successor licensee receives records and forms from the original licensee, the successor licensee may choose to forward those records and forms to the ATF Out–of– Business Record Center.

[18 U.S.C. 923(g)(4); 27 CFR 478.57(a) and, 478.127]

(C7) What records are licensees required to forward to ATF upon discontinuance of business?

The records consist of the licensee's acquisition and disposition (A/D) records, ATF Forms 4473, ATF Forms 3310.4 (Report of Multiple Sale or Other Disposition of Pistols and Revolvers), ATF Forms 3310.11 (Federal Firearms Licensee Theft/Loss Report), records of importation (ATF Forms 6 and 6A), and law enforcement certification letters.

[18 U.S.C. 923(g)(4); 27 CFR 478.127]

(C8) Must a person obtain his or her own firearms license if he or she purchases an existing firearms business?

Yes. Each person intending to engage in business as a firearms dealer, importer or manufacturer, or an ammunition importer or manufacturer must obtain the required Federal firearms license prior to commencing business.

[18 U.S.C. 923(a); 27 CFR 478.41]

(C9) May a person obtain a dealer's license to engage in business only at gun shows?

No. A license may only be issued for a permanent premises at which the license applicant intends to do business. A person having such license may conduct business temporarily at gun shows located in the State in which the licensed premises is located, and sell and deliver curio or relic firearms to other licensees at any location.

[18 U.S.C. 923(a) and (j); 27 CFR 478.100]

(C10) Does a person need to obtain a new license if an existing firearms business is moved to a new location?

Yes. To change a business location, a licensee must file an application for an amended license, ATF Form 5300.38, not less than 30 days prior to the move. An amended license must be obtained before commencing business at the new location. Once the new license is obtained, the licensee may no longer conduct business at the former business premises.

[27 CFR 478.52]

D. ATF FORM 4473, FIREARMS TRANSACTION RECORD

(D1) Where can a Federal firearms licensee get ATF Forms 4473?

ATF Forms 4473 are available free of charge from the ATF Distribution Center. Forms may be ordered online at www. atf.gov. Please order a quantity of forms estimated for 6 months use.

(D2) Is an ATF Form 4473 required when an unlicensed person sells or disposes a firearm?

No. The ATF Form 4473 is required only for sales or dispositions by a li-

censed manufacturer, importer, or dealer.

[18 U.S.C. 923(g); 27 CFR 478.124]

(D3) Who signs the ATF Form 4473 for the seller?

The ATF Form 4473 must be signed by the person who verified the identity of the buyer.

(D4) Must a transferee provide his or her social security number on the ATF Form 4473?

No. This information is solicited on an optional basis. However, providing this information will help ensure the lawfulness of the sale and avoid the possibility that the transferee will be incorrectly identified as a felon or other prohibited person.

[27 CFR 478.124]

(D5) Is a Social Security card a proper means of identification for purchasing a firearm from a licensee?

No. An identification document must contain the name, residence address, date of birth and photograph of the holder. A Social Security card does not, by itself, contain sufficient information to identify a firearms purchaser. However, a purchaser may be identified by any combination of government–issued documents which together establish all of the required information.

[27 CFR 478.11 and 478.124(c)]

(D6) Why does ATF Form 4473 ask for race and ethnicity?

The purpose for requiring prospective purchasers of firearms to identify their racial and ethnic background is to aid law enforcement in accurately tracing firearms found in crimes and better enable Federal firearms licensees to identify the purchaser during the background check portion of a firearms transaction. To collect this identifying information, ATF was required to follow the race and ethnicity standards and format for administrative forms and records established by the Office of Management and Budget (OMB). These standards were first published by OMB in the Federal Register on October 30, 1997, and became effective on January 1, 2003.

[62 Fed. Reg. 58782 (October 30, 1997)]

(D7) Are the instructions attached to the ATF Form 4473 required to

be maintained with the Form 4473 as part of a licensee's permanent firearms records?

Yes.

[27 CFR 478.21]

(D8) If a buyer or transferee is unable to read and or write but wants to purchase a firearm, how may the transfer legally be completed?

If the buyer or transferee is unable to read and/ or write, the answers (other than the signature) may be written on the form by another person, excluding the seller. Two persons not directly involved in the firearms transaction (excluding, for example, the licensee and employees of the licensee) must sign as witnesses to the buyer's answers and signature.

[27 CFR 478.21]

(D9) If a licensee notices an error on the ATF Form 4473 after the transfer, what is the procedure to record the correct information?

If the licensee or the buyer/transferee discovers that an ATF Form 4473 is incomplete or improperly completed after the firearm(s) has been transferred, the licensee should photocopy the inaccurate form and make any necessary additions or revisions to the photocopy. The buyer/transferee should only make changes to Sections A and C. The licensee should only make changes to Sections B and D. Whoever made the changes should initial and date the changes. The corrected photocopy should be attached to the original ATF Form 4473 and retained as part of the licensee's permanent records.

[27 CFR 478.21]

(D10) How long are licensees required to maintain ATF Forms 4473?

Licensees shall retain each ATF Form 4473 for a period of not less than 20 years after the date of sale or disposition. Where a licensee has initiated a NICS check for a proposed firearms transaction, but the sale, delivery, or transfer of the firearm is not made, the licensee shall record any transaction number on the Form 4473, and retain the Form 4473 for a period of not less than 5 years after the date of the NICS inquiry.

[18 U.S.C. 923(g)(1)(A); 27 CFR 478.129(b)]

(D11) May an ATF Form 4473 be completed for the transfer of a firearm by a person holding a Power of Attorney?

No. A licensee is required to obtain an ATF Form 4473 from the transferee who must certify that he or she is not prohibited from receiving or possessing a firearm, and whose identity the licensee must verify prior to transfer. A licensee cannot comply with these provisions where a Form 4473 is completed by a person other than the actual transferee.

[18 U.S.C. 923(g); 27 CFR 478.124]

E. RECORDS REQUIRED – LICENSEES

(E1) What is a "bound book?"

The firearms acquisition and disposition (A&D) record, also known as a "bound book", is a permanently bound book or an orderly arrangement of loose–leaf pages which must be maintained at the business premises. The format must follow that prescribed in the regulations and the pages must be numbered consecutively.

[18 U.S.C. 923(g)(1)(A); 27 CFR 478.121 and 478.125]

(E2) May a licensee keep more than one "bound book" at the same time?

Yes. A licensee is not limited to using only one "bound book." It may be convenient for a licensee to account for different brands or types of firearms, or business activities (e.g., retail, pawn, or gunsmith inventories), in separate "bound books."

(E3) Does the Government sell firearms acquisition and disposition record books for licensees to use in recording their receipts and dispositions of firearms?

No, however, some private vendors may have them available for purchase.

(E4) How much time does a licensee have to record the acquisition and disposition of firearms?

1. Dealer Acquisitions:

Generally, for licensed dealers the purchase or other acquisition of a firearm shall be recorded not later than the close of the next business day following the date of the acquisition or purchase. However, if commercial records containing the required information are available for inspection and are separate from other commercial documents, dealers have 7 days from the time of receipt to record the receipt in the acquisition and disposition (A&D) record.

2. Manufacturer Acquisitions:

Each licensed manufacturer shall record each firearm manufactured or otherwise acquired not later than the 7th day following the date of such manufacture or other acquisition.

3. Importer Acquisitions:

Each licensed importer shall within 15 days of the date of importation or other acquisition record the required information of each firearm imported or otherwise acquired.

4. Dispositions:

All licensees shall record sales or other dispositions not later than 7 days following date of the transaction. If a disposition is made before the acquisition has been entered in the A&D record, the acquisition entry must be made at the same time as the disposition entry.

[18 U.S.C. 923(g)(1)(A); 27 CFR 478.122, 123, and 125]

(E5) May a licensee seek an alternate method of maintaining records?

Yes. The licensee may request an alternate method to maintain records.

[27 CFR 478.22, 478.122(c), 478.123(c), and 478.125(h)]

(E6) What is the licensee's responsibility where an alternate method to maintain records has been authorized?

The ATF letter authorizing the alternate method must be kept at the licensed premises and available for inspection. For businesses with more than a single licensed outlet, each outlet covered by the alternate method must have a copy of the letter authorizing the change.

[27 CFR 478.22, 478.122(c), 478.123(c), and 478.125(h)]

(E7) Do licensees need to keep records of acquisition and disposition for ammunition?

Licensees are required to keep records for armor piercing ammunition.

[18 U.S.C. 922(b)(5); 27 CFR 478.122, 478.123, and 478.125]

(E8) Do licensees need to record the rental of a firearm in the acquisition and disposition (A&D) record?

Yes, if the firearm is taken off the premises of the licensee. However, no disposition entry is required for the loan or rental of a firearm for use only on the premises of the licensee.

[27 CFR 478.97]

(E9) How should a licensee record in his or her records the transfer of a frame or receiver to an unlicensed purchaser?

A licensee must record in the acquisition and disposition record and on the

ATF Form 4473 the type of firearm as a "frame" or "receiver" (as applicable). The licensee must also include in any record the make, model, and serial number of the frame or receiver.

As a frame or receiver is neither a shotgun nor a rifle, a licensee is prohibited from selling or delivering a frame or receiver to any individual the licensee knows or has reasonable cause to believe is less than 21 years of age.

[18 U.S.C. 922(b)(5), 27 CFR 478.11]

F. PERSONAL COLLECTIONS– LICENSEES

(F1) May a licensee maintain a personal collection of firearms?

Yes.

[18 U.S.C. 923(c); 27 CFR 478.125a]

(F2) Does a licensee have to record firearms acquired prior to obtaining the license in his or her acquisition and disposition record?

Firearms acquired by a licensee prior to obtaining a license must be recorded as business inventory in the acquisition and disposition record if the licensee intends to sell the firearm. All firearms acquired after obtaining a firearms license must be recorded as an acquisition in the acquisition and disposition record as business inventory.

[18 U.S.C. 923(c); 27 CFR 478.125a]

(F3) How can a licensee maintain a personal collection of firearms?

A licensee may log business firearms out of the acquisition and disposition record as a personal firearm. If that firearm is sold within a year of having been transferred to the licensee's personal collection, the licensee must re–enter the firearm in the business acquisition and disposition record, and sell or transfer the firearm as business inventory, including completing an ATF Form 4473 and a NICS background check if required. If, however, the firearm is held in the personal collection of the licensee for more than one year, the firearm may be sold as the licensee's personal firearm. Although no ATF Form 4473 or NICS background check is required, the licensee must record the disposition of the personal firearm in a personal disposition record.

[18 U.S.C. 923(c); 27 CFR 478.125a]

(F4) May a licensee create a personal collection to avoid the recordkeeping and NICS background check requirements of the GCA?

No. The law presumes that any firearm transferred to a personal collection for the purpose of willfully evading the restrictions placed upon licensees remains business inventory.

[18 U.S.C. 923(c)]

(F5) May a licensee store personal firearms at the business premises?

Yes, however, such firearms must be kept segregated from the licensee's business inventory. A presumption exists that all firearms on a licensee's business premises are for sale and must be entered in the licensee's bound book. Licensees should appropriately identify or tag personal firearms as "not for sale."

G. CONDUCT OF BUSINESS — LICENSEES

(G1) What steps must licensee take prior to transferring a firearm to an unlicensed person?

The following steps must be followed prior to transferring a firearm:

1. The licensee must have the transferee complete and sign ATF Form 4473, Firearms Transaction Record.

2. The licensee must verify the identity of the transferee through a government–issued photo identification.

3. Unless an exception applies, the licensee must contact NICS through either the FBI or a State point of contact (POC). A licensee may not transfer a firearm unless the licensee receives a "proceed" response, or three business days have elapsed since the licensee contacted NICS, and the system has not notified the licensee that the receipt of a firearm by such other person would violate the law. Licensees contacting the FBI directly and receiving a "delayed" response will receive information from the FBI indicating when the 3 business day time period elapses. A licensee may not transfer a firearm where a "denied" response is issued by NICS.

(G2) What form of identification must a licensee obtain from a transferee of a firearm?

The identification document presented by the transferee must have a photograph of the transferee, as well as the transferee's name, residence address, and date of birth. The identification document must also be valid (e.g., unexpired) and have been issued by a governmental entity for the purpose of identification of individuals. An example of an acceptable identification document is a current driver's license.

A combination of government issued documents may be used to meet the requirements of an identification document. For example, a passport which contains the name, date of birth, and photograph of the holder may be combined with a voter or vehicle registration card containing the residence address of the transferee in order to comply with the identification document requirements. A passport issued by a foreign government is also acceptable so long as it has all of the required information.

Whether a hunting license or permit issued by a retailer meets the definition of an identification document is State law specific. This license or permit meets the definition of an identification document if the State in which the retailer is located has authorized the retailer

to supply State issued documents. If the State recognizes the hunting license or permit as government issued, then this license or permit would qualify as being government issued for the purposes of supplementing another government issued identification document.

A description of the location of the residence on an identification document, such as a rural route, is sufficient to constitute a residence address provided the purchaser resides in a State or locality where it is considered to be a legal residence address.

[18 U.S.C. 922(t); 27 CFR 478.11 and 478.124]

(G3) May a licensee transfer a firearm to a nonlicensed individual who does not appear in person at the licensed premises?

Assuming the transfer otherwise complies with Federal and State law, a licensee may transfer a firearm to a nonlicensed person who does not appear in person at the licensed premises only when a background check is not required to transfer the firearm, and both reside in the same State. For example, a licensee may ship firearms to residents within the same State as the licensee where the transferee has a valid permit that has been recognized as an alternative to NICS where the licensee complies with the procedures set forth in 18 U.S.C. 922(c), 27 CFR 478.96(b) and ATF Procedure 2013–2. In any transaction where a NICS check is required, the firearm must be sold over–the–counter.

[18 U.S.C. 922(c) and 922(t); 27 CFR 478.96 and 478.124; ATF Procedure 2013–2]

(G4) Does Federal law require licensees to comply with State laws and local published ordinances when selling firearms?

Yes. It is unlawful for any licensed importer, licensed manufacturer, licensed dealer, or licensed collector to sell or deliver any firearm to any person if the person's purchase or possession would be in violation of any State law or local published ordinance applicable at the place of sale, delivery or other disposition.

[18 U.S.C. 922(b)(2); 27 CFR 478.99(b)(2)]

(G5) May a licensee sell a firearm to a nonlicensee who is a resident of another State?

Generally, a firearm may not lawfully be sold by a licensee to a nonlicensee who resides in a State other than the State in which the seller's licensed premises is located. However, the sale may be made if the firearm is shipped to a licensee whose business is in the purchaser's State of residence and the purchaser takes delivery of the firearm from the licensee in his or her State of residence. In addition, a licensee may sell a rifle or shotgun to a person who is not a resident of the State where the licensee's business premises is located in an over–the–counter transaction, provided the transaction complies with State law in the State where the licensee is located and in the State where the purchaser resides.

[18 U.S.C. 922(b)(3); 27 CFR 478.99(a)]

(G6) May a licensee sell firearms to law enforcement agencies and individual officers?

Yes. Law enforcement officers purchasing firearms for official use who provide a licensee with a certification on agency letterhead that the officer will use the firearm in official duties and that a records check reveals the purchasing officer has no convictions for misdemeanor crimes of domestic violence are not required to complete a ATF Form 4473 or undergo a background check. An officer purchasing a firearm for official duties may purchase the firearm from a licensee in any State, regardless of where the officer resides or the agency is located. Disposition of a firearm to an officer must be entered into the licensee's acquisition and disposition record, and the certification letter used to purchase the firearm must be retained in the licensee's files. Contact your State's Attorney General's Office to ensure there is no State prohibition on such sales.

[18 U.S.C. 925(a)(1); 27 CFR 478.134 and 478.141]

(G7) May a licensee employ an individual who is less than 21 years of age if the licensee sells handguns and ammunition suitable for use in handguns?

Yes. An individual less than 21 years of age may sell handguns and ammunition suitable for use in handguns. How-ever, a person less than 18 years of age must have the prior written consent of a parent or guardian and the written consent must be in the person's possession at all times. Also, the parent or guardian giving the written consent may not be prohibited by law from possessing a firearm. Moreover, State law must not prohibit a person less than 18 years of age from possessing the handguns or ammunition.

[18 U.S.C. 922(x)]

(G8) May a licensee employ an individual who is prohibited from receiving or possessing firearms and ammunition?

A licensee may not allow an individual who is a prohibited person to receive or possess firearms or ammunition, including persons employed by the licensee.

[18 U.S.C. 922(g) or (n), and 2]

(G9) Must a licensee advise ATF if two or more pistols or revolvers are sold or otherwise disposed of to an unlicensed person?

The disposition of two or more pistols or revolvers to any nonlicensee during a period of 5 consecutive business days must be reported on ATF Form 3310.4, Report of Multiple Sale or Other Disposition of Pistols and Revolvers, not later than the close of the business day on the day of disposition of the second pistol or revolver. The licensee must forward a copy of the Form 3310.4 to the ATF office specified thereon, and another copy must be forwarded to the State police or local law enforcement agency where the sale occurred. A copy of the Form 3310.4 must also be attached to the ATF Form 4473 executed upon delivery of the pistols or revolvers.

A business day for purposes of reporting multiple sales of pistols or revolvers is a day that a licensee conducts business pursuant to the license, regardless of whether State offices are open. The application of the term "business day" is, therefore, distinguishable from the term "business day" as used in the NICS context.

Example: A licensee conducts business only on Saturdays and Sundays, days on which State offices are not open. The licensee sells a pistol to an unlicensed person on a Saturday. If that same unlicensed person acquires another handgun the next day (Sunday), the following Saturday or Sunday, or the Saturday after that, the reporting requirement would be triggered, the subsequent acquisition of a pistol or revolver would have to be reported on a Form 3310.4 by the close of the day upon which the second or subsequent pistol or revolver was sold.

[18 U.S.C. 923(g)(3); 27 CFR 478.126a]

(G10) Does a licensee have to prepare a multiple sale report for the return of two or more pistols or revolvers to the same person from whom they were received?

No. A multiple sale report is not required for the return of firearms to the same individual such as a pawn redemption, consignment or repair.

[27 CFR 478.126a]

(G11) Where does a licensee submit the ATF Form 3310.4, Report of Multiple Sale or Other Disposition of Pistols and Revolvers?

ATF Form 3310.4 must be completed in triplicate (3 copies). The original is sent to ATF's National Tracing Center by FAX at 1–877–283–0288, by email at MultipleHandgunSalesForms@atf.gov, or by mail to U.S. Department of Justice, Bureau of Alcohol, Tobacco, Firearms and Explosives, National Tracing Center, P.O. Box 0279, Kearneysville, WV 25430–0279. A copy is to be sent to the designated State police or the local law enforcement agency in the jurisdiction where the sale took place. The remaining copy is to be attached to the corresponding ATF Form 4473 and retained in the licensee's records for a period of not less than 5 years.

[27 CFR 478.126a and 478.129]

(G12) How long may Chief Law Enforcement Officers (CLEOs) retains multiple sale reports?

Federal law, 18 U.S.C. 923(g)(3)(B), provides that Chief Law Enforcement Officers (CLEOs) receiving copies of ATF Form 3310.4 from FFLs for sales to nonprohibited persons may not disclose the contents of the forms and must destroy the forms and any record of the content of the forms within 20 days of receipt.

(G13) Does a customer have to be a certain age to buy firearms or ammunition from a licensee?

Yes. Under the GCA, shotguns and rifles, and ammunition for shotguns or rifles may be sold only to individuals 18 years of age or older. All firearms other than shotguns and rifles, and all ammunition other than ammunition for shotguns or rifles may be sold only to individuals 21 years of age or older. Licensees are bound by the minimum age requirements established by the GCA regardless of State or local law. However, if State law or local ordinances establish a higher minimum age for the purchase or disposition of firearms, the licensee must observe the higher age requirement.

[18 U.S.C. 922(b)(1) and (b)(2); 27 CFR 478.99(b)]

(G14) May a licensee sell interchangeable ammunition such as .22 caliber rimfire ammunition to a person less than 21 years of age?

Yes, provided the buyer is 18 years of age or older, and the licensee is satisfied that the ammunition is for use in a rifle. If the ammunition is intended for use in a handgun, the buyer must be at least 21 years of age.

[18 U.S.C. 922(b)(1); 27 CFR 478.99(b)]

(G15) If the purchaser of a firearm is a licensee, how does the seller verify that the licensed status of the purchaser?

A licensee selling or disposing of firearms to another licensee must verify the licensed status of the transferee by having the transferee furnish to the transferor a certified copy of the transferee's license and by any other means the transferor deems necessary (such as the FFL eZcheck).

[27 CFR 478.94]

(G16) Must a multi–licensed business submit a certified copy of each license when acquiring firearms from another licensee?

No. The multi–licensed business may provide the seller a list, certified to be true, correct and complete, containing the name, address, and license number and expiration date for each location in lieu of a copy of each license.

[27 CFR 478.94]

(G17) May a licensee continue to sell or dispose of firearms to a licensee whose license has expired?

Yes, for a period of 45 days following the date of expiration of the license. After the 45–day period, the transferor is required to verify the licensed status of the transferee with the Chief, Firearms Licensing Center.

[27 CFR 478.94]

(G18) Is a license required to engage in the business of selling small arms ammunition?

No. A license is not required for a dealer in ammunition only. However a license is required to manufacture or import ammunition.

[18 U.S.C. 922 (a)(1)(B); 27 CFR 478.41]]

(G19) May licensees sell firearms at gun shows?

Generally, a licensed manufacturer, importer or dealer may sell firearms temporarily at a qualifying gun show or event located within his or her State of licensure. Licensees may, however, sell curio or relic firearms to another licensee at any location.

[18 U.S.C. 923(j), 27 CFR 478.100]

(G20) What may a licensee do at an out–of–State gun show?

A licensee may only display and take orders for firearms at an out–of–State gun show. In filling any orders for firearms, the licensee must return the firearms to his or her licensed premises and deliver them from that location. Any firearm ordered by a nonlicensee must be delivered or shipped from the licensee's premises to a licensee in the purchaser's State of residence, and the purchaser must obtain the firearm from the licensee located in the purchaser's State. A licensee is prohibited from transferring firearms to another licensee at an out–of–State gun show, except where the firearm being transferred is a curio or relic.

[18 U.S.C. 922(a)(1), (b)(3), 923(a) and (j); 27 CFR 478.100]

(G21) May a licensee mail handguns through the U.S. Postal Service?

Yes. Licensees may mail an unloaded handgun to another licensee in customary trade shipments. Handguns may also be mailed to any officer, employee, agent, or watchman who is eligible under 18 U.S.C. 1715 to receive pistols, revolvers, and other firearms capable of being concealed on the person for use in connection with his or her official duties.

However, postal service regulations must be followed. Any person proposing to mail a handgun must file with the postmaster, at the time of mailing, an affidavit signed by the addressee stating that the addressee is qualified to receive the firearm, and the affidavit must bear a certificate stating that the firearm is for the official use of the addressee. See the current Postal Manual for details.

[18 U.S.C. 1715]

(G22) How does a licensee handle the sale of a consignment firearm?

Firearms received for sale on consignment must be entered in the licensee's acquisition and disposition record. The sale of a consignment firearm is handled in the same manner as other firearm sale.

Return of any consigned firearms by the licensee to the consignor must be entered in the licensee's disposition record. An ATF Form 4473 and a NICS check must be completed prior to the return of such firearms.

[18 U.S.C. 923(g); 27 CFR 478.122, 478.123, 478.124 and 478.125]

(G23) How does a licensee report the theft or loss of firearms?

A theft or loss of firearms must be reported to your local police as well as to ATF within 48 hours after the discovery. Licensees should notify ATF on the 24–hour, 7 days a week toll free line at 1–888–930–9275 and by preparing and submitting ATF Form 3310.11, Federal Firearms Licensee Firearms Inventory-Theft/Loss Report.

The theft or loss of NFA firearms should also be reported to the NFA Branch immediately upon discovery. The NFA Branch can be contacted at (304) 616–4500.

[18 U.S.C. 923(g)(6); 27 CFR 478.39a and 479.141]

(G24) How should a licensee record the theft or loss of a firearm in the acquisition and disposition record?

The licensee should note "stolen" or "lost" and the date in the disposition section of the acquisition and disposition record for each firearm stolen or lost. In addition, the licensee should also record in the bound book the control number provided by ATF upon notification of the theft or loss.

(G25) What is a licensee's responsibility to respond to a request to trace a firearm?

A licensee must provide the requested information immediately and in no event later than 24 hours after receipt of a request by ATF.

[18 U.S.C. 923(g)(7); 27 CFR 478.25a]

(G26) Is a licensee required to provide written notification regarding handguns and juveniles?

Yes. The requirement to give written notification applies when the licensee delivers a handgun to a nonlicensee.

[27 CFR 478.103]

(G27) May firearms be shipped to a licensee at an address different from the business premises address identified on the license?

Yes. Neither the GCA nor its implementing regulations require firearms be shipped only to the licensed business premises Therefore, a licensee may lawfully receive firearms at the licensee's mailing address, storage location, or other address where the licensee intends to ensure safe and secure receipt of the firearms.

(G28) How would a licensee record the disposition of a replacement firearm in the licensee's records?

A licensee who receives a firearm for repair or customizing, and who returns a replacement firearm, must record the disposition in the licensee's acquisition and disposition record. However, no ATF Form 4473 is required if the replacement firearm is returned to the same person from whom the licensee received the firearm being replaced. The replacement firearm must be of the same kind and type.

[18 U.S.C. 922(a)(2)(A); 27 CFR 478.124(a) and 478.147]

(G29) May a licensee lawfully transfer a frame or receiver to an unlicensed individual who is less than 21 years of age?

No. A frame or receiver is a type of firearm "other than a shotgun or rifle" and the transfer by a licensee to an individual less than 21 years of age would be prohibited.

[18 U.S.C. 921(a)(5) and (7) and 922(b)(1); 27 CFR 478.11 and 478.99(b)]

(G30) May a licensee lawfully transfer a frame or receiver to an unlicensed person who resides outside his or her state?

No.

[18 U.S.C. 922(b)(3); 27 CFR 478.99(a)]

(G31) Does a licensee need to complete and submit a multiple sales form after selling, within five consecutive business days, two or more frames or receivers?

No. Multiple sales forms (ATF Forms 3310.4) are not required for sales of frames or receivers as they are not pistols or revolvers.

[18 U.SC 923(g)(3); 27 CFR 478.126a]

(G32) What does the Child Safety Lock Act of 2005 (CSLA) require of a licensee?

When selling, delivering, or transferring a handgun to any person other than another licensee, any licensed importer, licensed manufacturer, or licensed dealer must provide a secure gun storage or safety device to that person for the handgun.

[18 U.S.C. 922(z)]

(G33) What qualifies as a secure gun storage or safety device?

1. A device that, when installed on a firearm, is designed to prevent the firearm from being operated without first deactivating the device;

2. A device incorporated into the design of the firearm that is designed to prevent the operation of the firearm by anyone not having access to the device; or

3. A safe, gun safe, gun case, lock box, or other device that is de-

signed to be or can be used to store a firearm and that is designed to be unlocked only by means of a key, a combination, or other similar means.

Zip ties, rope, and string do not meet this definition.

[18 U.S.C. 921(a) (34)]

H. COLLECTORS

(H1) Is there a specific license which permits a collector to acquire firearms in interstate commerce?

Yes. A person may obtain a collector's license. However, this license applies only to transactions in curio or relic firearms.

[18 U.S.C. 921(a)(13), 923(b); 27 CFR 478.41(c), (d), and 478.93]

(H2) What firearms are considered to be curio and relic firearms?

Curio and relic firearms are defined as firearms which are of special interest to collectors because they possess some qualities not ordinarily associated with firearms intended for sporting use or as offensive or defensive weapons. To be recognized as a curio or relic, firearms must fall within one of the following categories:

(1) Firearms manufactured at least 50 years prior the current date, but not including replicas thereof;

(2) Firearms certified by the curator of a municipal, State, or Federal museum which exhibits firearms to be curios or relics of museum interest; and

(3) Firearms which derive a substantial part of their monetary value from the fact that they are novel, rare, or bizarre or from the fact of their association with some historical figure, period, or event.

ATF has recognized only complete, assembled firearms as curios or relics. ATF's classification of surplus military firearms as curios or relics has extended only to those firearms in their original military configuration. Frames or receivers of curios or relics are not generally recognized as curios or relics.

Collectors wishing to obtain a determination whether a particular firearm qualifies for classification as a curio or relic may submit a written request for a determination to ATF's Firearms Technology Branch. ATF's classifications of curios and relics firearms are published in ATF Publication 5300.11.

(H3) Does a collector's license afford any privileges to the licensee with respect to acquiring or disposing of firearms other than curios or relics in interstate or foreign commerce?

No. A licensed collector has the same status under the GCA as a nonlicensee except for transactions in curio or relic firearms.

[27 CFR 478.93]

(H4) Does a license as a collector of curio or relic firearms authorize the collector to engage in the business of dealing in curios or relics?

No. A dealer's license must be obtained to engage in the business of dealing in any firearms, including curios or relics.

[18 U.S.C. 922(a) and 923(a); 27 CFR 478.41(d)]

(H5) Are licensed collectors required to execute an ATF Form 4473 for transactions in curio or relic firearms?

No. However, licensed collectors are required to keep an acquisition and disposition (A&D) record.

[18 U.S.C. 923(g)(2); 27 CFR 478.125(f)]

(H6) Are transfers of curio or relic firearms by licensed collectors subject to the NICS background check requirements?

No. However, it is unlawful for any person to transfer a firearm to any person knowing or having reasonable cause to believe that such person is a felon or is within any other category of person prohibited from receiving or possessing firearms.

[18 U.S.C. 922(d) and 922(t)(1); 27 CFR 478.32(d) and 478.102]

(H7) Are licensed collectors required to comply with the requirements that written notification be given to handgun transferees and signs be posted on juveniles and handguns?

Licensed collectors are required to comply with written notification to unlicensed individuals upon delivery of a handgun. However, the sign posting requirement does not apply to licensed collectors.

[27 CFR 478.103]

(H8) Are licensed collectors required to turn in their acquisition and disposition records to ATF if they discontinue their collecting activity?

No. Licensed collectors are not required to submit their records to ATF upon discontinuance of their collecting activity.

[18 U.S.C. 923(g)(4); 27 CFR 478.127]

I. MANUFACTURERS

(I1) Must a person who engages in business in both manufacturing firearms and importing firearms have a separate license to cover each type of business?

Yes. A separate license is required to cover each of these types of businesses.

[18 U.S.C. 923; 27 CFR 478.41]

(I2) May a person licensed as a manufacturer of ammunition also manufacture firearms?

No. A person licensed as a manufacturer of ammunition may not manufacture firearms unless he or she obtains a license as a firearms manufacturer.

[18 U.S.C. 923; 27 CFR 478.41]

(I3) May a person licensed as a manufacturer of firearms also manufacture ammunition?

Yes. A manufacturer of firearms may also manufacture ammunition (not including destructive device ammunition or armor piercing ammunition) without obtaining a separate license as a manufacturer of ammunition.

(I4) Is a person who reloads ammunition required to be licensed as a manufacturer?

Yes, if the person engages in the business of selling or distributing reloads for the purpose of livelihood and profit. No, if the person reloads only for personal use.

[18 U.S.C. 922(a) and 923(a); 27 CFR 478.41]

(I5) Must a licensed manufacturer pay excise taxes?

Yes. Licensed manufacturers incur excise tax on the sale of firearms and ammunition manufactured. For additional information on excise taxes contact the Department of Treasury, Alcohol and Tobacco Tax and Trade Bureau.

(I6) Are licensed manufacturers required to report the production of firearms?

Yes. Manufacturers are required to complete the Annual Firearms Manufacturing and Exportation Report (ATF F 5300.11).

[18 U.S.C. 932(g)(5)]

(I7) Are licensed manufacturers required to file the Annual Firearms Manufacturing and Exportation Report even if the licensee did not manufacture any firearms during the reporting period?

Yes, licensees must file a report even if the licensee had no manufacturing activity.

[18 U.S.C. 932(g)(5)]

J. GUNSMITHS

(J1) Is a license needed to engage in the business of engraving, customizing, refinishing or repairing firearms?

Yes. A person conducting such activities as a business is considered to be a gunsmith within the definition of a dealer.

[18 U.S.C. 921(a)(11) and (21); 27 CFR 478.11]

(J2) Does a gunsmith need to enter every firearm received for adjustment or repair into an acquisition and disposition (A&D) record?

If a firearm is brought in for repairs and the owner waits while it is being repaired, or if the gunsmith is able to return the firearm to the owner during the same business day, it is not necessary to list the firearm in the A&D record as an "acquisition." If the gunsmith has possession of the firearm from one business day to another or longer, the firearm must be recorded as an "acquisition" and a "disposition" in the A&D record.

(J3) Is an ATF Form 4473 required when a gunsmith returns a repaired firearm?

No, provided the firearm is returned to the person from whom it was received.

[27 CFR 478.124(a) and 478.147]

(J4) May a gunsmith make immediate repairs at locations other than his or her place of business?

Yes. Licensed gunsmiths may make on–the–spot repairs at skeet, trap, target, and similar organized shooting events.

(J5) May a licensed gunsmith receive an NFA firearm for purposes of repair?

Yes, a licensed gunsmith may receive a National Firearms Act firearm for the sole purpose of repair and subsequent return to its owner. It is suggested that the owner obtain permission from ATF for the transfer by submitting an ATF Form 5, Application for Tax Exempt Transfer and Registration of Firearm, to the NFA Branch and receive approval prior to the delivery. The gunsmith should do the same prior to returning the firearm.

(J6) Is a licensed gunsmith required to comply with the requirements to give written notification to handgun transferees and post signs regarding juveniles and handguns?

Yes. Licensed gunsmiths are required to comply with written notification to unlicensed individuals upon delivery of a handgun. A gunsmith who transfers handguns to nonlicensees must also comply with the sign posting requirement.

[27 CFR 478.103]

(J7) Must a licensed gunsmith conduct a NICS background check on the return of repaired or customized firearms?

No, if the firearm is being returned to the person from whom it was received. If however the firearm is delivered to a person other than the person from whom received, a NICS background check is required.

[18 U.S.C. 922(t); 478.124(a)]

K. PAWNBROKERS

(K1) What is the procedure for a pawnbroker to return a pawned firearm?

The redemption of a pawned firearm is a "disposition" subject to all the record-keeping requirements under the GCA. Dispositions must be properly entered in the pawnbroker's acquisition and disposition record, an ATF Form 4473 must be executed and a NICS background check completed in connection with the redemption.

[18 U.S.C. 922(t), 923(g); 27 CFR 478.102, 478.124 and 478.125]

(K2) To whom may a pawnbroker return a pawned firearm?

(1) If the pawnbroker and non-licensee are residents of the same State, the pawnbroker may return a handgun, long gun, or other firearm to either the person who pawned it or, where State and/or local law allows, a holder of the pawn ticket who resides in the pawnbroker's State.

(2) If the pawnbroker and non-licensee are not residents of the same State:

a. The pawnbroker may return a firearm to the person who pawned it.

b. The pawnbroker may transfer a rifle or shotgun to the holder of a pawn ticket who did not pawn it, provided the transaction complies with the conditions of sale in both the State where the pawnbroker is licensed and the State where the pawn ticket holder resides.

[18 U.S.C. 922(a)(2), 922(a)(3) and 922(b)(3)]

(K3) Are there categories of persons to whom a pawnbroker cannot return firearms?

Yes. Like other licensees, a pawnbroker cannot lawfully return a firearm to a person who is, due to age or other disability, ineligible to receive or possess firearms.

[18 U.S.C. 922(d) and 922(b)(1); 27 CFR 478.99]

(K4) Is a pawnbroker required to comply with the requirements to give written notification to handgun transferees and post signs regarding juveniles and handguns?

Yes. Licensed pawnbrokers are required to comply with written notification to unlicensed individuals upon delivery of a handgun. A pawnbroker who transfers handguns to nonlicensees must also comply with the sign posting requirement.

[27 CFR 478.103]

(K5) May a pawnbroker conduct a NICS background check prior to accepting a firearm for pawn?

Yes. A licensed pawnbroker may conduct a NICS background check on a person at the time the person offers to pawn a firearm. If NICS advises the pawnbroker that receipt or possession of the firearm by the person attempting to pawn the firearm would violate the law, the pawnbroker must advise local law enforcement within 48 hours after receipt of the information.

A pawnbroker who contacts NICS about a person prior to accepting the person's firearm in pawn must still conduct a NICS background check at the time of redemption.

[Public Law 105–277, enacted on October 21, 1998]

L. AUCTIONEERS

(L1) Does an auctioneer who is involved in firearms sales need a dealer's license?

Generally speaking, there are two types of auctions: estate–type auctions and consignment auctions.

In estate–type auctions, the articles to be auctioned (including firearms) are being sold by the executor of the estate of an individual. The firearms belong to and are possessed by the executor. The firearms are controlled by the estate, and the sales of firearms are being made by the estate. The auctioneer is acting as an agent of the executor and assisting the executor in finding buyers for the firearms. In these cases, the auctioneer does not meet the definition of engaging in business as a dealer in firearms and would not need a license. An auctioneer who does have a license may perform this function away from his or her licensed premises.

In consignment–type auctions, an auctioneer often takes possession of firearms in advance of the auction. These firearms are generally inventoried, evaluated, and tagged for identification. The firearms belong to individuals who have entered into a consignment agreement with the auctioneer giving that auctioneer authority to sell the firearms. The auctioneer therefore has possession and control of the firearms. Under these circumstances, an auctioneer would generally need a license. If you are not sure if a license is needed in a particular consignment auction situation, contact your local ATF office.

[ATF Ruling 96–2]

(L2) If a licensed auctioneer is making sales of firearms, where may those sales be made?

In a consignment auction firearms may be displayed at an auction site away from the auctioneer's licensed premises and sales of the firearms can be agreed upon at that location, but the firearms must be returned to the auctioneer's licensed premises prior to transfer. The simultaneous sale and delivery of the auctioned firearms away from the licensed premises would violate the law, i.e., engaging in business at an unlicensed location.

However, if the auctioneer is assisting an estate in disposing of firearms, the estate is the seller of the firearms and the estate is in control and possession of the firearms. In this situation, the firearms may be sold by the estate at the auction site.

[18 U.S.C. 923(a); 27 CFR 478.50]

(L3) Can a licensee conduct background checks and transfer firearms on behalf of an unlicensed auctioneer?

Generally no, as most auctions do not qualify as a gun show or qualifying event and therefore a licensee would not be permitted to conduct business away from the licensed premises.

[18 U.S.C. 923(j); 27 CFR 478.100]

M. IMPORTING AND EXPORTING

(M1) May a licensee who does not have an importer's license make an occasional importation?

Yes. A licensee may make an occasional importation of a firearm for a non-licensee or for the licensee's personal use (not for resale). The licensee must first submit an ATF Form 6, Part I to ATF for approval. The licensee may then present the approved ATF Form 6 and completed ATF Form 6A to U.S. Customs and Border Protection.

[27 CFR 478.113]

(M2) Does a licensee need an export license to export a firearm?

Most firearms and ammunition must be exported in accordance with the provisions of the Arms Export Control Act of 1976. Regulations implementing this Act generally require a license to be obtained from the U.S. Department of State, Directorate of Defense Trade Controls. Additional information may be accessed online at: http://www.pmddtc.state.gov/index.html.

The export of sporting shotguns and ammunition for sporting shotguns is regulated by the U.S. Department of Commerce. An export license is generally needed to export these shotguns and ammunition. For further information, contact the U.S. Department of Commerce, Bureau of Industry and Security. Contact information may be found online at: http://www.bis.doc.gov/

When exporting NFA firearms, an ATF Form 9, Application and Permit for Permanent Exportation of Firearms, must be approved by ATF prior to export.

[22 U.S.C. 2778; 27 CFR 479.114 and 479.116]

N. NATIONAL FIREARMS ACT (NFA)

(N1) What firearms are regulated under the NFA?

(1) a shotgun having a barrel or barrels of less than 18 inches in length;

(2) a weapon made from a shotgun if such weapon as modified has an overall length of less than 26 inches or a barrel or barrels of less than 18 inches in length;

(3) a rifle having a barrel or barrels of less than 16 inches in length;

(4) a weapon made from a rifle if such weapon as modified has an overall length of less than 26 inches or a barrel or barrels of less than 16 inches in length;

(5) any other weapon, as defined in subsection (e);

(6) a machinegun;

(7) any silencer (as defined in section 921 of title 18, United States Code); and

(8) a destructive device.

[26 U.S.C. 5845; 27 CFR 479.11]

(N2) How can a person legally obtain NFA firearms?

A person may make an NFA firearm by filing and receiving an approved ATF Form 1 Application to Make and Register a Firearm.

A person may transfer an NFA firearm to another person by filing and receiving an approved ATF Form 4, Application for Tax Paid Transfer and Registration of Firearm.

Applications to make or transfer a firearm will not be approved if Federal, State, or local law prohibits the making or possession of the firearm.

[26 U.S.C. 5812 and 5822; 27 CFR 479.62 and 479.84]

(N3) What is the tax on making an NFA firearm?

The tax is $200 for making any NFA firearm.

[26 U.S.C. 5821; 27 CFR 479.61]

(N4) What is the tax on the transfer of an NFA firearm?

The tax is $200 for the transfer of any firearm except a firearm classified as an "any other weapon" which is $5.

An unserviceable firearm may be transferred as a curio or ornament without payment of the transfer tax.

[26 U.S.C. 5811, 5852(e) and 5845(h); 27 CFR 479.11, 479.82 and 479.91]

(N5) How is this tax paid?

A check or money order made payable to the Bureau of ATF together with the application forms are to be mailed to: National Firearms Act Branch, Bureau of Alcohol, Tobacco, Firearms and Explosives, P.O. Box 530298, Atlanta, GA 30353–0298.

(N6) What should a State do when an unregistered NFA firearm is acquired through forfeiture or abandonment?

When a State wants to keep such NFA firearms for official use, the State must register the firearm by filing an ATF Form 10 Application for Registration of Firearms Acquired by Certain Governmental Entities. Since approval of the Form 10 is conditioned on an "official use only" basis, subsequent transfers will not be approved unless the transfer is to another government agency for official use.

[26 U.S.C. 5841; 27 CFR 479.104]

(N7) May a private citizen register a firearm not previously registered in the National Firearms Registration and Transfer Record?

No. The NFA permits only manufacturers, makers, importers, and certain governmental entities to register firearms.

[26 U.S.C. 5841(b) and 5861(d); 27 CFR 479.101(b) and 479.104]

(N8) What should a person do if he or she comes into possession of an unregistered NFA firearm?

Contact the nearest ATF office immediately. A listing of the telephone numbers can be found online at www.atf.gov.

(N9) Are there any exemptions from the making or transfer tax provisions of the NFA?

Yes. These are noted below, along with the required form number, if any, to apply for the exemption. Completed forms must be approved by the NFA Branch prior to the making or transfer:

(1) Tax–exempt transfer and registration of a firearm between special (occupational) taxpayers: ATF Form 3.

(2) Tax–exempt making of a firearm by a qualified manufacturer, or other than by a qualified manufacturer if the firearm is made on behalf of a Federal or State agency: ATF Form 1 and 2. Tax–exempt transfer and registration of the firearm on behalf of a Federal or State agency: ATF Form 5.

(3) Tax–exempt transfer and registration of an unserviceable firearm which is being transferred as a curio or ornament: ATF Form 5.

(4) Tax exempt transfer of a firearm to a lawful heir: ATF Form 5.

(5) Tax–exempt transfer by operation of law (e.g., court order).

[26 U.S.C. 5851–5853; 27 CFR 479.69, 479.70 and 479.88–91]

(N10) How does a person qualify to import, manufacture, or deal in NFA firearms?

The person must be licensed under the GCA and pay the required special (occupational) tax imposed by the NFA. After becoming licensed under the GCA, the licensee must file an ATF Form 5630.7, Special Tax Registration and Return National Firearms Act (NFA) with the appropriate tax payment with ATF. In addition, an importer (except importers of sporting shotguns and shotgun ammunition) must also be registered with ATF under the Arms Export Control Act of 1976, by filing an ATF Form 4587, Application to Register as an Importer of U.S. Munitions Import List Articles.

[26 U.S.C. 5801; 18 U.S.C. 923; 27 CFR 447.31, 478.41 and 479.34]

(N11) When must firearms special (occupational) taxes be paid and how much are the taxes?

These taxes must be paid in full on first engaging in business and thereafter on or before the first day of July. The current taxes are set out in the following table.

SPECIAL (OCCUPATIONAL) TAX RATES UNDER THE NFA

CLASS OF TAXPAYER	ANNUAL FEE
1 – Importer of Firearms (Including "Any Other Weapon")	$1000.
2 – Manufacturer of Firearms (Including "Any Other Weapon")	$1000.
3 — Dealer of Firearms (Including "Any Other Weapon")	$500.
1 — Importer of Firearms (Including "Any Other Weapon") REDUCED*	$500.
2 — Manufacturer of Firearms (Including "Any Other Weapon") REDUCED*	$500.

* REDUCED = Rates which apply to certain taxpayers whose total gross receipts in the last taxable year are less than $500,000.

A licensed manufacturer under contract to make NFA firearms for the U.S. Government may be granted an exemption from payment of the special (occupational) tax as a manufacturer of NFA firearms and an exemption from all other NFA provisions (except importation) with respect to the weapons made to fulfill the contract. Exemptions are obtained by writing the NFA Branch, stating the contract number(s) and the anticipated date of termination. This exemption must be renewed each year prior to July 1.

[26 U.S.C. 5801; 27 CFR 479.31, 479.32, 479.32a and 479.33]

(N12) Does a licensee that engages in both manufacturing and importing need to pay a separate special (occupational) tax payment for each activity?

Yes. A separate special (occupational) tax payment must be made for each of these activities. However, Class 1 (importer) and Class 2 (manufacturer) special (occupational) taxpayers are qualified to deal in NFA firearms without also having to pay special (occupational) tax as a Class 3 dealer.

[27 CFR 479.39]

(N13) May a licensed collector obtain NFA firearms in interstate commerce?

Yes, but only if the firearms are classified as curios or relics, are registered, and are transferred in accordance with the provisions of the NFA.

(N14) Are parts or kits which would convert a firearm into a machinegun subject to registration?

Yes.

[26 U.S.C. 5845(b); 27 CFR 479.11]

(N15) Who may qualify as a certifying official on an ATF Form 1 or ATF Form 4 for the making or transfer of an NFA firearm?

As provided by regulations, certifications by the local chief of police, sheriff of the county, head of the State police, or

State or local district attorney or prosecutor are acceptable. The regulations also provide that certifications of other officials are appropriate if found in a particular case to be acceptable to the Director. Examples of other officials who have been accepted in specific situations include State attorneys general and judges of State courts having authority to conduct jury trials in felony cases.

[27 CFR 479.63 and 479.85]

(N16) Is the chief law enforcement officer required to sign the law enforcement certification on an ATF Form 1 or ATF Form 4?

No. Federal law does not compel any official to sign the law enforcement certification. However, ATF will not approve an application to make or transfer a firearm on ATF Forms 1 or 4 unless the law enforcement certification is completed by an acceptable law enforcement official who has signed the certification in the space indicated on the form.

(N17) If the chief law enforcement official whose jurisdiction includes the proposed transferee's residence refuses to sign the law enforcement certification, will the signature of an official in another jurisdiction be acceptable?

No.

(N18) Does the registered possessor of a destructive device, machinegun, short–barreled shotgun, or short–barreled rifle need authorization to lawfully transport such items interstate?

Yes, unless the registered possessor is a qualified dealer, manufacturer or importer, or a licensed collector transporting only curios or relics. Prior approval must be obtained, even if the move is temporary. Approval is requested by either submitting a letter containing all necessary information, or by submitting ATF Form 5320.20, Application to Transport Interstate or to Temporarily Export Certain National Firearms Act (NFA) Firearms. This requirement does not apply to the lawful interstate transportation of silencers. Possession of the firearms also must comply with all State and local laws.

[18 U.S.C. 922(a)(4); 27 CFR 478.28]

(N19) If an individual is changing his or her State of residence and the individual's application to transport the NFA firearm cannot be approved because of a prohibition in the new State, what options does a lawful possessor have?

NFA firearms may be left in a safe deposit box in his or her former State of residence. Also, the firearm could be left or stored in the former State of residence at the house of a friend or relative in a locked room or container to which only the registered owner has a key. The friend or relative should be supplied with a copy of the registration forms and a letter from the owner authorizing storage of the firearm at that location.

The firearms may also be transferred in accordance with NFA regulations or abandoned to ATF.

(N20) May a transferor submit an application to transfer an NFA firearm prior to the date on which the transferor receives the weapon?

No.

(N21) If a person has a pistol and an attachable shoulder stock, does this constitute possession of an NFA firearm?

Yes, unless the barrel of the pistol is at least 16 inches in length (and the overall length of the firearm with stock attached is at least 26 inches). However, certain stocked handguns, such as original semiautomatic Mauser "Broomhandles" and Lugers, have been removed from the purview of the NFA as collectors' items.

[26 U.S.C. 5845; 27 CFR 479.11]

(N22) Does the possessor of an NFA firearm have to show proof of registration?

Yes. The approved application received from ATF serves as evidence of registration of the NFA firearm. This document must be made available upon request of any ATF officer. It is suggested that a photocopy of the approved application be carried by the possessor when the weapon is being transported.

[26 U.S.C. 5841(e); 27 CFR 478.101]

(N23) What is the status of unloaded or dummy grenades, artillery shell casings and similar devices?

Unloaded or dummy grenades, artillery shell casings, and similar devices, which are cut or drilled in an ATF approved manner so that they cannot be used as ammu- nition components for destructive devices are not considered NFA weapons.

(N24) Are muzzleloading cannons classified as destructive devices?

Generally, no. Muzzleloading cannons not capable of firing fixed ammunition and manufactured in or before 1898 and replicas thereof are antiques and not subject to the provisions of either the GCA or the NFA.

[26 U.S.C. 5845; 27 CFR 479.11]

(N25) Are grenade and rocket launcher attachments destructive devices?

Grenade and rocket launcher attachments for use on military type rifles generally do not come within the definition of destructive devices. However, the grenades and rockets used in these devices are generally within the definition.

[26 U.S.C. 5845; 27 CFR 479.11]

(N26) Are Paintball and/or Airgun Sound Suppressers NFA firearms?

The terms "firearm silencer" and "firearm muffler" mean any device for silencing, muffling, or diminishing the report of a portable firearm, including any combination of parts, designed or redesigned, and intended for use in assembling or fabricating a firearm silencer or firearm muffler, and any part intended only for use in such assembly or fabrication.

Numerous paintball and airgun silencers tested by ATF's Firearms Technology Branch have been determined to be, by nature of their design and function, firearm silencers. Because silencers are NFA weapons, an individual wishing to manufacture or transfer such a silencer must receive prior approval from ATF and pay the required tax.

[26 U.S.C. 5845; 27 CFR 479.11]

(N27) If I have any further questions as to the classification of a paintball or airgun silencer, who should I contact?

Please send a written request to ATF's Firearms Technology Branch, 244 Needy Road, Martinsburg, WV, 24505 or by fax at: (304) 616–4301.

[18 U.S.C. 921(a)(24); 26 U.S.C. 5845(a); 27 CFR 479.11]

O. MACHINEGUNS

(O1) May an unlicensed person make a machinegun?

Generally, no. However, applications to make and register machine guns on or after May 19, 1986 for the benefit of a Federal, State, or local government entity will be approved if documentation can be provided, along with the application to make a machinegun, which establishes that-the machine gun is particularly suitable for use by Federal, State or local governmental entities and is being made at the request and on behalf of such an entity

[18 U.S.C. 922(o)(2); 27 CFR 479.105(e)]

(O2) May machineguns be transferred from one registered possessor to another?

Yes. If the machinegun was lawfully registered and possessed before May 19, 1986, it may be transferred pursuant to an approved ATF Form 3 or Form 4, as applicable.

[18 U.S.C. 922(o)(2); 26 U.S.C. 5812]

P. NATIONAL INSTANT CRIMINAL BACKGROUND CHECK SYSTEM (NICS)

(P1) Who must comply with the requirements to conduct a NICS background check prior to transferring a firearm?

Licensed firearms importers, manufacturers, and dealers must conduct a NICS background check prior to the transfer of any firearm to a nonlicensed individual.

[18 U.S.C. 922(t); 27 CFR 478.102]

(P2) Are NICS checks conducted by ATF?

No. NICS checks are conducted by the Federal Bureau of Investigation (FBI).

[28 CFR 25.3]

(P3) Do all NICS checks go through the FBI's NICS Operations Center?

No. In many States, licensees initiate NICS checks through the State point of contact (POC) serving as an intermediary between a licensee and the federal databases checked by the NICS.

[28 CFR 25.2 and 25.6]

(P4) If the State is acting as a POC, does that mean that all NICS checks go through the POC rather than the FBI?

That depends on the State. In some States, the POC conducts background checks for all firearms transactions. In other States, licensees must contact the POC for handgun transactions and the FBI for long gun transactions. In some POC States, NICS checks for pawn redemptions are handled by the FBI.

(P5) How does a licensee know whether to contact the FBI or a POC in order to initiate a NICS check?

Your local office can advise you on the appropriate point of contact for NICS checks or you can check the ATF website at www.atf.gov or the NICS website www.fbi.gov.

(P6) Is there a charge for NICS checks?

The FBI does not charge a fee for conducting NICS checks. However, States that act as points of contact for NICS checks may charge a fee consistent with State law.

(P7) Must licensees enroll with the FBI to get access to NICS?

Licensees must be enrolled with the FBI before they can initiate NICS checks through the FBI's NICS Operations Center. Licensees who have not received an enrollment package from the FBI should call FBI NICS Customer Service at 1–877–FBI–NICS (324–6427) and ask that an enrollment package be sent to them. Licensees in States where a State agency is acting as a point of contact for NICS checks should contact the State for enrollment information.

(P8) Is a NICS background check required for the transfer of long guns as well as handguns?

Yes. Licensed manufacturers, importers, and dealers must conduct a background check for the transfer of all firearms subject to the GCA.

[18 U.S.C. 922(t); 27 CFR 478.102]

(P9) Are NICS background checks required for the transfer of antique firearms?

No. Because weapons that meet the definition of an "antique firearm" are not firearms subject to the GCA, licensees need not conduct a background check when transferring an antique firearm.

[18 U.S.C. 921(a)(3), 921(a)(16), 922(t); 27 CFR 478.11 and 478.102]

(P10) Must a licensee conduct a NICS background check for the transfer of firearms to another licensee?

No.

[18 U.S.C. 922(t); 27 CFR 478.102]

(P11) Must licensed collectors conduct a NICS background check prior to transferring a curio or relic firearm?

No. However, licensed manufacturers, importers and dealers who transfer a curio or relic firearm must conduct a NICS background check prior to transfer to a nonlicensee.

[18 U.S.C. 922(t); 27 CFR 478.102]

(P12) Is a NICS background check required for a licensee's loan or rental of a firearm to a nonlicensee?

If the firearm is loaned or rented for use on the licensee's premises, a background check is not required. However, if the firearm is loaned or rented for use off the premises, the licensee must conduct a background check prior to the transfer of the firearm.

[18 U.S.C. 922(t); 27 CFR 478.102 and 478.124]

(P13) Must licensees conduct a NICS background check for the sale of firearms to nonlicensees at gun shows?

Yes. A licensee conducting business temporarily at a gun show must comply with the background check provisions in the same manner as if the sale were taking place at the licensed premises.

[18 U.S.C. 922(t) and 923(j); 27 CFR 478.100 and 478.102]

(P14) Is the redemption of a pawned firearm subject to a NICS background check?

Yes. The redemption of a pawned firearm is a transfer subject to a NICS background check.

(P15) What should a licensed pawn-broker do when NICS denies the redemption of a pawned firearm?

Federal law prohibits the transfer of a firearm to a person who has been denied by NICS. Further guidance may be found in the ATF FFL Newsletter December 2002 Issue #1.

[18 U.S.C. 922(d) and (t)]

(P16) If an individual repeatedly pawns the same firearm, is the FFL required to do a NICS background check each time the firearm is redeemed?

Yes. The fact that the transferee has redeemed the firearm before does not excuse the pawnbroker from conducting a background check for each separate transaction.

[18 U.S.C. 922(t); 27 CFR 478.102]

(P17) Is a NICS background check required for the sale of a firearm to a law enforcement official for his or her personal use?

Yes. In such transactions, the law enforcement official is treated no differently from any other unlicensed transferee, and a NICS background check must be conducted.

[18 U.S.C. 922(t) and 925(a)(1)]

(P18) Are there transfers that are exempt from the NICS background check requirement?

Firearm transfers are exempt from the requirement for a NICS background check in three situations. These include transfers: (1) to transferees having a State permit that has been recognized by ATF as an alternative to a NICS check; (2) of National Firearms Act weapons to persons approved by ATF; and (3) certified by ATF as exempt because compliance with the NICS background check requirement is impracticable.

[18 U.S.C. 922(t); 27 CFR 478.102(d)]

(P19) What does it mean when a licensee receives a "cancelled" response from the FBI to a NICS background check?

On occasion, the FBI may provide the FFL with a "cancelled" response. Transactions will be cancelled if the FBI discovers that the NICS check was not initiated in accordance with ATF or FBI regulations, e.g., a NICS check was initiated for a non–authorized purpose or by a non–authorized individual. A licensee cannot transfer a firearm when a "cancelled" response is provided by NICS.

(P20) For purposes of the NICS background check requirements, what is meant by a "business day?"

A business day is defined as a day on which State offices are open in the State in which the proposed firearm transaction is to take place.

[18 U.S.C. 922(t)(1)(B)(ii); 28 CFR 25.2]

(P21) If a licensee contacts NICS on Thursday, December 2, and gets a "delayed" response, when may the licensee transfer the firearm if no further response is received from NICS?

The firearm may be transferred on Wednesday, December 8. Assuming State offices are open on Friday, Monday, and Tuesday, and closed on Saturday and Sunday, 3 business days would have elapsed at the end of Tuesday, December 7. Therefore, the licensee may transfer the firearm at the start of business on Wednesday, December 8. The 3 business days do not include the day the NICS check is initiated.

(P22) What should a licensee do if he or she gets a "denied" response from NICS or a State point of contact after 3 business days have elapsed, but prior to the transfer of the firearm?

If the licensee receives a "denied" response at any time prior to the transfer of the firearm, he or she may not transfer the firearm.

If the FBI provides a "denied" response after the third business day, and the firearm has already been transferred, the licensee should notify the NICS Section.

[18 U.S.C. 922(d); 27 CFR 478.99(c)]

(P23) What should a licensee tell a transferee who is denied by NICS?

The licensee should inform the transferee that the NICS check indicates that the transfer of the firearm should not be made, however the system does not provide a reason for the denial. The licensee should provide the transferee with the name and address of the denying agency and the NICS or State transaction number, and an appeals brochure created by the FBI or State POC outlining the transferee's appeal rights and responsibilities. An FBI appeal brochure is available through the FBI NICS section, or at www.fbi.gov/nics–appeals.

[28 CFR 25.10]

(P24) If a transferee receives a "denied" response from NICS, can the transferee find out why he or she or she was denied?

Yes. Although the licensee will not know the reason for the denial, the transferee may contact the FBI or the State point of contact in writing to request the reason for denial.

[28 CFR 25.10]

(P25) What NICS or State point of contact response information must a licensee record on ATF Form 4473?

Licensees must record any initial "proceed", "delayed", or "denied" response received, as well as any transaction number provided. If the initial NICS response was "delayed," a licensee must also record the date and response later provided, or that no resolution was provided within 3 business days if the licensee transfers the firearm after the 3 days. In addition, if a licensee receives a response from NICS after a firearm has been transferred, he or she or she must record this information.

[18 U.S.C. 922(t); 27 CFR 478.124(c)(3)(iv)]

(P26) Must licensees keep ATF Forms 4473 if the firearm transfer is denied or, for some other reason, not completed?

Licensees must keep each ATF Form 4473 for which a NICS check has been initiated, regardless of whether the transfer of a firearm is made. If the transfer is not made, the FFL must keep the Form 4473 for 5 years after the date of the NICS inquiry. If the transfer is made, the FFL must keep the Form 4473 for 20 years after the date of the sale or disposition. Forms 4473 with respect to a transfer that did not take place must be separately maintained.

[27 CFR 478.129(b)]

(P27) At what point during a transaction should a licensee contact NICS?

Licensees should contact NICS after the transferee has completed Section A of the ATF Form 4473.

[27 CFR 478.124]

(P28) For what period of time is a NICS check valid?

A NICS check is valid for 30 calendar days for any transaction. The 30 calendar day period is counted beginning on the day after NICS was initially contacted. Where more than 30 calendar days have passed since the licensee first contacted NICS, the licensee must initiate a new NICS check prior to transferring the firearm. It is not necessary to complete a new ATF Form 4473, but the results of the new NICS background check must be recorded on the form.

Example1: A NICS check is initiated on May15th. The licensee receives a "proceed" from NICS. The transferee does not return to pick up the firearm until June 22ⁿᵈ of the same year. The licensee must conduct another NICS check before transferring the firearm to the transferee.

Example 2: A NICS check is initiated on May 15th. The licensee receives a "delayed" response from NICS; no further response is received. The transferee does not return to pick up the firearm until June 16ᵗʰ of the same year. The licensee must conduct another NICS check before transferring the firearm to the transferee.

[27 CFR 478.102]

(P29) Can one NICS check be used for more than one firearms transaction?

No. A licensee must initiate a new NICS background check for each completed firearms transaction. However, a person may purchase or acquire several firearms in one transaction.

[27 CFR 478.102]

Example: A transferee completes an ATF Form 4473 for a single firearm on February 15. The licensee receives a "proceed" from NICS that day. The licensee signs the form, and the firearm is transferred. On February 20, the transferee returns to the licensee's premises and wishes to acquire a second firearm. The acquisition of the second firearm is a sep-

arate transaction. Therefore, a new NICS check must be initiated by the licensee.

Example: A transferee completes ATF Form 4473 for a single firearm on February 15. The licensee receives a "proceed" from NICS that day. The transferee does not return to pick up the firearm until February 20. Before the licensee completes the transfer of the first firearm, the transferee decides to acquire an additional firearm. The second firearm may be recorded on the same Form 4473. The acquisition of the 2 firearms is considered a single transaction. Therefore, the licensee is not required to conduct a new NICS check prior to transferring the second firearm.

(P30) Must a licensee always wait 3 business days before transferring a firearm to a transferee after receiving a "delayed" response?

A licensee may transfer a firearm to a transferee as soon as he or she receives a "proceed" from NICS (assuming that the transaction would be in compliance with State law). However, if the licensee does not receive a final "proceed" or "denied" response from NICS, he or she must wait until 3 business days have elapsed prior to transferring the firearm.

[18 U.S.C. 922(t); 27 CFR 478.102(a)(2)]

(P31) What happens if the transferee successfully appeals the NICS denial but more than 30 calendar days have elapsed since the initial background check was initiated?

The licensee must initiate another NICS check before the firearm may be transferred.

(P32) Does a permit qualify as an alternative to a NICS check if the purchaser is using it to purchase a type of firearm that is not covered by the permit?

Yes, assuming the transaction complies with State law.

Example: ATF recognizes the permit to purchase a handgun and the concealed weapons permit as alternatives to a NICS check in State A. Any purchaser who displays a permit to purchase a handgun or a concealed weapons permit in State A is not required to undergo a NICS check prior to purchasing a rifle, assuming the transaction complies with State law.

(P33) ATF has recognized the concealed weapons permits in State A and State B as valid alternatives to a NICS check. Can a resident of State A use a concealed weapons permit issued by State A to purchase a long gun in State B without undergoing a NICS check?

No. A permit qualifies as a NICS alternative only if it was issued by the State in which the transfer is to take place.

[18 U.S.C. 922(t); 27 CFR 478.102(d)]

(P34) Does a licensee who conducts a NICS check have to comply with State waiting periods before transferring a firearm?

Yes. Compliance with Federal background check provisions does not excuse a licensee from compliance with State law. A licensee must follow State law waiting periods.

Example: State X is acting as a point of contact for NICS checks. State law requires the licensee to wait 10 days, rather than 3 days, for a response to the background check prior to transferring a firearm. Because State law provides a 10 day period before a licensee may transfer a firearm, the licensee may not transfer the firearm until 10 days have elapsed since conducting the background check.

Example: State X is acting as a point of contact for NICS checks. State law allows a licensee to transfer a firearm 24 hours after conducting a NICS check and receiving a "delayed" response. Although State law would permit a licensee to transfer a firearm 24 hours after receiving a "delayed" response, the licensee must comply with Federal law which requires the licensee wait 3 business days prior to transferring a firearm where the licensee has not received notice the transfer would be prohibited.

Example: The law of State X provides for a 5 day waiting period before a handgun may be transferred by a licensee. An individual completes an ATF Form 4473 for the purchase of a handgun, a NICS check is conducted, and a "proceed" response is given by NICS. Although the licensee received a "proceed" response, the licensee must comply with the State waiting period, and the licensee may not transfer the firearm until the State 5 day waiting period has elapsed.

[18 U.S.C. 922(b)(2)]

(P35) Does an officer or employee of an entity that holds a Federal firearms license, such as a corporation, have to undergo a NICS check prior to acquiring a firearm for the personal collection of the officer or employee?

Yes. The exemption from completing an ATF Form 4473 and a NICS check applies only to the person holding the license, not to individual officers or employees of the person holding the license.

[18 U.S.C. 921(a)(1); 27 CFR 478.11]

(P36) How may a licensee participate in the raffling of firearms by an unlicensed organization?

Depending on how the organization arranges for the firearms to be transferred to the raffle winner, the licensee's participation may vary.

Example 1: A licensee transfers a firearm to the organization sponsoring the raffle. The representative acting on behalf of the organization must complete the ATF Form 4473 and undergo a NICS background check. When the buyer of a firearm is a corporation, association, or other organization, an officer or other representative authorized to act on behalf of the organization must complete the form with his or her personal information and attach a written statement, executed under penalties of perjury, stating that the firearm is being acquired for the use of the organization and the name and address of the organization. Once the firearm had been transferred to the organization, the organization can subsequently transfer the firearm to the raffle winner without an ATF Form 4473 being completed or a NICS check being conducted. This is because the organization is not a licensee. However, the organization cannot transfer the firearm to a person who is not a resident of the organization's State of residence nor can the organization knowingly transfer the firearm to a prohibited person.

Example 2: The licensee or a representative of the licensee brings a firearm to the raffle so that the firearm can be displayed. After the raffle, the firearm is returned to the licensee's premises. The licensee must complete an ATF Form 4473 and conduct a NICS background check prior to transferring the firearm to the winner of the raffle. If the firearm is a handgun, the winner of the raffle must be a resident of the State where the transfer takes place, or the firearm must be transferred through another licensee in the winner's State of residence. If the firearm is a rifle or shotgun, the licensee can lawfully transfer the firearm to the winner of the raffle as long as the transaction is over–the–counter and complies with the laws applicable at the place of sale and the State where the transferee resides.

Example 3: If the raffle meets the definition of a qualifying event, the licensee may conduct business at that event.

Please note, if the organization's practice of raffling firearms rises to the level of being engaged in the business of dealing in firearms, the organization must get its own Federal firearms license.

[18 U.S.C. 922(a)(5), 922(b)(3), 922(t), and 923(j); 27 CFR 478.99, 478.100, 478.102, and 478.124]

Q. MISDEMEANOR CRIME OF DOMESTIC VIOLENCE

(Q1) What is a "misdemeanor crime of domestic violence"?

A "misdemeanor crime of domestic violence" is an offense that:

(1) is a misdemeanor under Federal, State, or Tribal law;

(2) has, as an element, the use or attempted use of physical force, or the threatened use of a deadly weapon; and

(3) was committed by a current or former spouse, parent, or guardian of the victim, by a person with whom the victim shares a child in common, by a person who is cohabiting with or has cohabited with the victim as a spouse, parent, or guardian, or by a person similarly situated to a spouse, parent, or guardian of the victim.

However, a person is not considered to have been convicted of a misdemeanor crime of domestic violence unless:

(1) the person was represented by counsel in the case, or knowingly and intelligently waived the right of counsel in the case; and

(2) in the case of a prosecution for which a person was entitled to a jury trial in the jurisdiction in which the case was tried, either –

(a) the case was tried by a jury, or

(b) the person knowingly and intelligently waived the right to have the case tried by a jury, by guilty plea or otherwise.

In addition, a conviction would not be disabling if it has been expunged or set aside, or is an offense for which the person has been pardoned or has had civil rights restored (if the law of the jurisdiction in which the proceedings were held provides for the loss of civil rights upon conviction for such an offense) unless the pardon, expunction, or restoration of civil rights expressly provides that the person may not ship, transport, possess, or receive firearms, and the person is not otherwise prohibited by the law of the jurisdiction in which the proceedings were held from receiving or possessing firearms.

[18 U.S.C. 921(a)(33); 27 CFR 478.11]

(Q2) Must a misdemeanor crime of domestic violence (MCDV) be designated as a "domestic violence" offense?

No. A qualifying offense need not be designated as a domestic violence offense. For example, a conviction for assault may qualify as an MCDV even if the offense is not designated as a domestic violence assault.

[18 U.S.C. 921(a)(33) and 922(g)(9); 27 CFR 478.11 and 478.32(a)(9)]

(Q3) Does the prohibition on receipt or possession of firearms and ammunition apply if the person was convicted of an MCDV prior to the enactment of 18 U.S.C. 922(g)(9) on September 30, 1996?

Yes.

[18 U.S.C. 922(g)(9); 27 CFR 478.32(a)(9)]

(Q4) In determining whether a conviction in a State court is a "conviction" of a misdemeanor crime of domestic violence, does Federal, State or Tribal law apply?

The law of the jurisdiction determines whether a conviction has occurred. Therefore, if the law of the jurisdiction does not consider the person to be convicted, the person would not have the Federal disability.

[18 U.S.C. 921(a)(33); 27 CFR 478.11]

(Q5) What State and local offenses are "misdemeanors" for purposes of 18 U.S.C. 922(d)(9) and (g)(9)?

The definition of misdemeanor crime of domestic violence in the GCA includes any offense classified as a "misdemeanor" under Federal, State or Tribal law. In States that do not classify offenses as misdemeanors, the definition includes any State or local offense punishable by imprisonment for a term of 1 year or less or punishable by a fine.

[18 U.S.C. 921(a)(33); 27 CFR 478.11]

(Q6) Are local criminal ordinances "misdemeanors under State law" for purposes of sections 922(d)(9) and (g)(9)?

Yes, assuming a violation of the ordinance meets the definition of "misdemeanor crime of domestic violence" in all other respects.

(Q7) Is the relationship between the parties an element of an MCDV?

No. The "as an element" language in the definition of "misdemeanor crime of domestic violence" only applies to the use of force provision of the statute and not the relationship provision. However, to be disabling, the offense must have been committed by someone whose relationship to the victim meets the definition in the GCA.

[18 U.S.C. 921(a)(33); 27 CFR 478.11]

(Q8) What should an individual do if he or she has been convicted of a misdemeanor crime of domestic violence?

Individuals subject to this disability should immediately dispose of their firearms and ammunition, such as by abandonment to a law enforcement agency.

[18 U.S.C. 922(g)(9); 27 CFR 478.32]

(Q9) Does the disability apply to law enforcement officers?

Yes. The Gun Control Act was amended so that employees of government agencies convicted of misdemeanor crimes of domestic violence would not be exempt from disabilities with respect to their receipt or possession of firearms or ammunition. Thus, law enforcement officers and other government officials who have been convicted of a disqualifying misdemeanor may not lawfully possess or receive firearms or ammunition for any purpose, including performance of their official duties. The disability applies to firearms and ammunition issued by government agencies, purchased by government employees for use in performing their official duties, and personal firearms and ammunition possessed by such employees.

[18 U.S.C. 922(g)(9) and 925(a)(1); 27 CFR 478.32(a)(9) and 478.141]

(Q10) Is an individual who has been pardoned, or whose conviction was expunged or set aside, or whose civil rights have been restored, considered convicted of a misdemeanor crime of domestic violence?

No, as long as the pardon, expungement, or restoration does not expressly provide that the person may not ship, transport, possess, or receive firearms. A restoration of civil rights, however, is only effective to remove the Federal firearms disability if the law of the jurisdiction provides for the loss of civil rights for a conviction of such a misdemeanor.

[18 U.S.C. 921(a)(33); 27 CFR 478.11]

R. NONIMMIGRANT ALIENS

(R1) Who is a nonimmigrant alien?

A nonimmigrant alien is an alien in the United States in a nonimmigrant classification as defined by section 101(a)(15) of the Immigration and Nationality Act (8 U.S.C. 1101(a)(15). Generally, "nonimmigrant aliens" are tourists, students, business travelers, and temporary workers who enter the U.S. for fixed periods of time; they are lawfully admitted aliens who are not lawful permanent residents.

[27 CFR 478.11]

(R2) May a nonimmigrant alien who has been admitted to the United States under a nonimmigrant visa possess a firearm or ammunition in the United States?

An alien admitted to the United States under a nonimmigrant visa is prohibited from shipping, transporting, receiving, or possessing a firearm or ammunition unless the alien falls within one of the exceptions provided in 18 U.S.C. 922(y)(2), such as: a valid hunting license or permit, admitted for lawful hunting or sporting purposes, certain official rep-resentatives of a foreign government, or a foreign law enforcement officer of a friendly foreign government entering the United States on official law enforcement business.

[18 U.S.C. 922(g)(5)(B) and 922(y)(2); 27 CFR 478.11 and 478.32]

(R3) Can a nonimmigrant alien obtain a waiver from the prohibition on shipping, transporting, receiving, or possessing a firearm or ammunition?

Yes. An individual may apply to ATF for a waiver.

[18 U.S.C. 922(g)(5)(B) and 922(y)(3); 27 CFR 478.11 and 478.32]

(R4) May a nonimmigrant alien who has been admitted to the United States under a nonimmigrant visa and who falls within an exception, purchase a firearm or ammunition in the United States?

A nonimmigrant alien without residency in any State may not purchase and take possession of a firearm. A nonimmigrant alien may only purchase a firearm through a licensee where the licensee arranges to have the firearm directly exported. A nonimmigrant alien who falls within an exception may, however, purchase and take possession of ammunition.

A nonimmigrant alien who has established residency in a State may purchase and take possession of a firearm from an unlicensed person, provided the buyer and seller are residents of the same State, and no other State or local law prohibits the transaction. A nonimmigrant alien with residency in a State may purchase a firearm from a licensee, provided the sale complies with all applicable laws and regulations.

[18 U.S.C. 922(a)(9); 27 CFR 478.29a]

(R5) Does the prohibition on the receipt and possession of firearms and ammunition by aliens in nonimmigrant visa status apply to nonimmigrant aliens who lawfully enter the United States without a visa?

No. A nonimmigrant alien who is lawfully admitted to the United States without a visa (e.g. Visa Waiver Program), may acquire or possess a firearm in the United States.

(R6) What is an alien number or admission number?

An alien number is a unique 7, 8, or 9 digit number assigned to a noncitizen by the Department of Homeland Security upon the creation of a file.

An admission number is the number on a CBP Form I–94 or CBP Form I–94W, the arrival/departure form Customs and Border Protection (CBP) gives most nonimmigrant aliens when they arrive in the U.S. While most nonimmigrant aliens will automatically receive an admission number when they enter the U.S., Canadians will not. However, if a Canadian asks a CBP official for an admission number when he or she enters the United States, he or she will be given an admission number.

(R7) Is the possessor of a "green card" a nonimmigrant alien?

No. The possessor of a "green card" is a permanent resident and not in nonimmigrant status.

(R8) Is a valid hunting license or permit as an exception to the firearms prohibitions on nonimmigrant aliens only valid in the State in which it was issued?

No. A valid, unexpired hunting license or permit from any State within the United States satisfies the hunting license exception to the nonimmigrant alien prohibition. The hunting license or permit does not have to be from the State where the nonimmigrant alien is purchasing the firearm.

(R9) Does a Canadian citizen residing in the United States need an alien number or admission number to purchase a firearm?

Yes. All non–U.S. citizens need an alien number or admission number to purchase a firearm from a Federal firearms licensee (FFL). A FFL cannot complete the sale without an alien or admission number. This is the case even if you have a State permit that ATF has determined qualifies as a "NICS alternative" and therefore do not need to have a NICS background check.

(R10) Do nonimmigrant aliens need to obtain a permit to temporarily import firearms into the United States for hunting or other lawful sporting purposes?

Yes. All nonimmigrant aliens (with a few exceptions) must obtain an import permit from ATF to temporarily import firearms

and ammunition for hunting or other lawful sporting purposes. Please note this requirement applies to all nonimmigrant aliens, not all nonresidents (e.g., it does not apply to U.S. citizens residing abroad).

The form to be filed with ATF is an ATF Form 6 NIA (5330.3D) (Application/Permit for Temporary Importation of Firearms and Ammunition by Nonimmigrant Aliens) with documentation demonstrating that you fall within an exception to the nonimmigrant alien prohibition. The form is both the application and, once approved, the permit you present to the U.S. Customs and Border Protection when you enter the United States. The Form 6 NIA can be downloaded at www.atf.gov.

[27 CFR 478.115(d) and (e) and 478.120]

(R11) What documentation does an alien need to show U.S. Customs and Border Protection (CBP) when entering the United States with a firearm?

When entering the United States, an alien must show CBP both the approved Form 6 NIA (5330.3D) permit, and if admitted to the United States under a nonimmigrant visa, appropriate documentation demonstrating you fall within an exception to the nonimmigrant alien prohibition.

[27 CFR 478.120]

(R12) Do aliens that have an approved ATF Form 6 NIA need to provide Customs and Border Protection (CPB) with an ATF Form 6A?

No. Because you are only importing the firearms temporarily, you do not need to complete an ATF Form 6A or provide it to CBP.

[27 CFR 478.120]

(R13) How long is an ATF Form 6 NIA valid for?

The ATF Form 6 NIA is valid for one year from the date of approval. For conditions on importation see the instructions on the Form 6 NIA.

(R14) Is an ATF Form 6 NIA required to temporarily import an antique firearm?

No. Because antique firearms are not considered firearms for purposes of the GCA, none of the import regulations apply to the importation of antique firearms. Moreover, a nonimmigrant alien may pos-

sess antique firearms, even if the alien does not fall within an exception to the nonimmigrant alien prohibition.

[18 U.S.C. 921(a)(3) and (a)(16); 27 CFR 478.11]

(R15) Are there any restrictions on the types of firearms that can be temporarily imported into the United States for hunting or for a shooting competition?

Yes. National Firearms Act weapons (which include machineguns, short–barreled rifles and shotguns, and silencers); nonsporting firearms; U.S. government origin firearms or firearms that contain U.S. government origin manufactured parts or components; and firearms from certain proscribed countries may not be temporarily imported. If you think the firearm(s) you wish to import may be affected by these restrictions, please contact the Firearms and Explosives Imports Branch.

[18 U.S.C. 922(l) and 925(d); 26 U.S.C. 5844; 27 CFR 447.52 and 447.57; 27 CFR 478.111; 27 CFR 479.111]

(R16) May a firearm that was temporarily imported into the United States remain in the United States?

An alien that temporarily imports a firearm into the United States must remove the firearm from the country at the conclusion of the visit.

[27 CFR 478.115(d)(1)]

(R17) Does an alien with an approved Form 6 NIA import permit need to obtain an export permit to take the firearm(s) and/or remaining ammunition back out of the United States?

No.

(R18) Can an alien who enters the United States on a nonimmigrant alien visa rent a firearm for lawful hunting or sporting purposes while in the United States?

A nonimmigrant alien that possess a valid hunting license from a State within the United States or falls within any of the other exceptions or exemptions that allow nonimmigrant aliens to possess firearms may rent firearms to hunt or to use at a shooting range.

[18 U.S.C. 922(a)(5) and (9), 922(g)(5)(B) and 922(y); 27 CFR 478.99(a) and (c)(5)]

(R19) Does a United States citizen living abroad with no State of residence in the United States need an import permit to temporarily bring firearms into the United States for lawful purposes?

No. Unlike aliens, United States citizens are not required to obtain an import permit to temporarily bring firearms into the United States If the importation is other than a temporary importation, the citizen must file an ATF Form 6, Application and Permit for Importation of Firearms, Ammunition and Implements of War.

Contact the U.S. Customs and Border Protection for requirements regarding the temporary importation of firearms.

(R20) Would an import permit be required for a Canadian citizen driving through the United States from one part of Canada to another part of Canada?

No. An ATF Form 6 NIA import permit is not required because firearms in transit are not considered imported into the United States. However, the Canadian citizen must comply with the laws of each State which is being driving through. For information on State laws, contact the State's Attorney General Office for each State in which you are driving through.

[27 CFR 447.11, 27 CFR 478.11]

(R21) Does an alien residing in the United States who temporarily takes a firearm out of the United States need to obtain a Form 6 NIA to bring the firearm back into the country?

No Form 6 NIA is required when it can be established to the satisfaction U.S. Customs and Border Protection (CBP) official that the firearm was previously taken out of the United States. This can be satisfied by completing Customs Form 4457 prior to leaving the U.S. with the firearm. However, if you are a non-immigrant alien returning to the U.S. you must provide the CPB official with documentation demonstrating you are exempt from the general nonimmigrant alien prohibition on possessing firearms.

[18 U.S.C. 922(g)(5)(b); 27 CFR 478.99 and 478.115]

ATF POINTS OF CONTACT

Report Lost or Stolen Firearms, including National Firearms Act Weapons.. 888-930-9275 or 1-800-ATF-GUNS (1-800-283-4867)

Report Unlawful Firearms Activity 1-800-ATF-GUNS (1-800-283-4867)

Report Multiple Sales of Handguns

Please make sure to send a completed ATF Form 3310.4 to:
ATF National Tracing Center (NTC)
244 Needy Road, Martinsburg WV25405 304-260-1500

Rather than mailing this form, you may **FAX** it to the NTC at .... 877-283-0288 or
Email to .... MultipleHandgunSalesForms@atf.gov

(You also must send a copy of the Form 3310.4 to the
Chief Law Enforcement Office where your business is located
and attach a copy to the Form 4473 for the second handgun.)

Federal Firearms License Application or Renewal

Federal Firearms Licensing Center
P.O. Box 409567
Atlanta, GA 30384-9567
Toll-free 866-662-2750

ATF Forms and Publications

ATF Distribution Center .... (703) 870-7526 or
(703) 870-7528
Most forms and publications also are available at the ATF website at ....www.ATF.gov

Firearms Technology Branch

244 Needy Road
Martinsburg, WV 25405 .... (304) 616-4300
Email: ....fire_tech@atf.gov

Going Out-of-Business

Records should be shipped to:
ATF Out-of-Business Records Center
244 Needy Road, Martinsburg, WV 25405 800-788-7133

Questions About Federal Firearms Laws and Regulations

Contact your nearest ATF Field Office ....See listing starting on page 219

FOR QUESTIONS CONCERNING UNLAWFUL ACTIVITIES
CONTACT THE ATF CRIMINAL ENFORCEMENT FIELD DIVISIONS BELOW

Atlanta Field Division
2600 Century Parkway N.E.
Suite 300
Atlanta, GA 30345-3104
(404) 417-2600

Baltimore Field Division
31 Hopkins Plaza, 5th Floor
Balitmore, MD 21201
(443) 965-2000

Boston Field Division
10 Causeway Street
Room 791
Boston, MA 02222-1047
(617) 557-1200

Charlotte Field Division
6701 Carmel Road
Suite 200
Charlotte, NC 28226
(704) 716-1800

Chicago Field Division
525 West Van Buren Street
Suite 600
Chicago, IL 60607
(312) 846-7200

Columbus Field Division
230 West Street
Suite 400
Columbus, OH 43215
(614) 827-8400

Dallas Field Division
1114 Commerce Street
Room 303
Dallas, TX 75242
(469) 227-4300

Denver Field Division
950 17th Street, Suite 1800
Denver, CO 80202
(303) 575-7600

Detroit Field Division
1155 Brewery Park Blvd.
Suite 300
Detroit, MI 48207-2602
(313) 202-3400

Houston Field Division
5825 N. Sam Houston Parkway
Suite 300
Houston, TX 77086
(281) 716-8200

Kansas City Field Division
2600 Grand Blvd.
Suite 200
Kansas City, MO 64108
(816) 559-0700

Los Angeles Field Division
550 N. Brand Blvd.
Suite 800
Glendale, CA 91203
(818) 265-2500

Louisville Field Division
600 Dr. Martin Luther King Jr. Place
Suite 500
Louisville, KY 40202
(502) 753-3400

Miami Field Division
11410 NW 20th Street
Suite 200
Miami, FL 33172
(305) 597-4800

Nashville Field Division
5300 Maryland Way
Suite 200
Brentwood, TN 37027
(615) 565-1400

Newark Field Division
1 Garret Mountain Plaza
Suite 400
Woodland Park, NJ 07424
(973) 413-1179

New Orleans Field Division
One Galleria Blvd., Suite 1700
Metairie, LA 70001
(504) 841-7000

New York Field Division
32 Old Slip
Suite 3500
Manhattan, NY 10005
(646) 335-9000

Philadelphia Field Division
The Curtis Center
601 Walnut Street, Suite 1000 E
Philadelphia, PA 19106
(215) 446-7800

Phoenix Field Division
201 East Washington Street
Suite 940
Phoenix, AZ 85004
(602) 776-5400

San Francisco Field Division
5601 Arnold Road
Suite 400
Dublin, CA 94568-7724
(925) 557-2800

Seattle Field Division
Jackson Federal Building
915 2nd Avenue, Room 790
Seattle, WA 98174
(206) 204-3205

St. Paul Field Division
1870 MINN. World Trade Center
30 East Seventh Street
Suite 1900
St. Paul, MN 55101
(651) 726-0200

Tampa Field Division
400 North Tampa Street
Suite 2100
Tampa, FL 33602
(813) 202-7300

Washington Field Division
1401 H Street, NW
Suite 900
Washington, DC 20005
(202) 648-8010

**FOR QUESTIONS CONCERNING FEDERAL FIREARMS LAWS,
REGULATIONS, PROCEDURES OR POLICIES
CONTACT AN ATF INDUSTRY OPERATIONS FIELD OFFICE BELOW**

ALBUQUERQUE II FO (IO)
201 THIRD STREET SUITE 1550
ALBUQURQUE, NM 87102
(505) 346-6910

ATLANTA V FIELD OFFICE (IO)
2600 CENTURY PARKWAY, NE
ROOM 350
ATLANTA, GA 30345
(404) 417-2670

BALTIMORE V FIELD OFFICE (IO)
FALLON FEDERAL BUILDING
31 HOPKINS PLAZA, 5TH FLOOR
BALTIMORE, MD 21201
(443) 965-2120

BIRMINGHAM II FIELD OFFICE (IO)
920 18TH STREET NORTH
ROOM 237,
BIRMINGHAM, AL 35203
(205) 583-5950

BOSTON V FIELD OFFICE (IO)
10 CAUSEWAY STREET, ROOM 701
BOSTON, MA 02222
(617) 557-1250

BUFFALO II FIELD OFFICE (IO)
598 MAIN STREET, SUITE 201
BUFFALO, NY 14202
(716) 853-5160

CHARLESTON FIELD OFFICE (IO)
300 SUMMERS STREET, SUITE 1400
CHARLESTON, WV 25301
(304) 340-7820

CHARLOTTE III FIELD OFFICE (IO)
6701 CARMEL ROAD, SUITE 200
CHARLOTTE, NC 28226
(704) 716-1830

CINCINNATI II FIELD OFFICE (IO)
550 MAIN STREET
ROOM 1516
CINCINNATI, OH 45202
(513) 684-3351

CLEVELAND III FIELD OFFICE (IO)
5005 ROCKSIDE ROAD
SUITE 700
INDEPENDENCE, OH 44131
(216) 573-8140

COLUMBIA II FIELD OFFICE (IO)
1835 ASSEMBLY STREET, RM. 309
COLUMBIA, SC 29201
(803) 251-4640

DALLAS V FIELD OFFICE (IO)
1114 COMMERCE STREET
ROOM 303
DALLAS, TX 75242
(469) 227-4415

DENVER III FIELD OFFICE (IO)
950 17TH STREET, SUITE 1800
DENVER, CO 80202
(303) 575-7640

DETROIT V FIELD OFFICE (IO)
1155 BREWERY PARK BLVD
SUITE 300
DETROIT, MI 48207
(313) 202-3550

DOWNERS GROVE III FIELD OFFICE (IO)
3250 LACEY ROAD
SUITE 400
DOWNERS GROVE, IL 60515
(630) 725-5290

DUBLIN III FIELD OFFICE (IO)
5601 ARNOLD ROAD, SUITE 400
DUBLIN, CA 94568
(925) 557-2830

EL PASO II FIELD OFFICE (IO)
303 NORTH OREGON
EL PASO, TEXAS 79901
(915) 534-6449

FAIRVIEW HEIGHTS FIELD OFFICE (IO)
333 SALEM PLACE, SUITE 205
FAIRVIEW HEIGHTS, IL 62208
(618) 632-0704

FALLS CHURCH III FIELD OFFICE (IO)
7799 LEESBURG PIKE
SUITE 1050, N. TOWER
FALLS CHURCH, VA 22043-2413
(703) 287-1120

FORT PIERCE II FIELD OFFICE (IO)
1660 SW. ST. LUCIE WEST BLVD.
SUITE 400
PORT ST. LUCIE, FL 34986
(772) 924-2780

FORT WORTH II FIELD OFFICE (IO)
6000 WESTERN PLACE, SUITE 400
FT WORTH, TX 76107
(817) 862-2850

GLENDALE III FIELD OFFICE (IO)
550 N. BRAND BLVD., SUITE 800
GLENDALE, CA 91203
(818) 265-2540

GRAND RAPIDS II FIELD OFFICE (IO)
38 WEST FULTON, SUITE 200
GRAND RAPIDS, MI 49503
(616) 301-6100

GREENSBORO II FIELD OFFICE (IO)
1801 STANLEY ROAD
ROOM 300
GREENSBORO, NC 27407-2643
(336) 235-4950

HARRISBURG II FIELD OFFICE (IO)
MARKET SQUARE PLAZA
17 NORTH 2ND STREET, SUITE 1400
HARRISBURG, PA 17101
(717) 231-3400

HARTFORD FIELD OFFICE (IO)
450 MAIN STREET, ROOM 610
HARTFORD, CT 06103
(860) 240-3400

HELENA II FIELD OFFICE (IO)
10 WEST 15TH STREET
SUITE 2400
HELENA, MT 59626
(406) 441-3160

HOUSTON FIELD OFFICE VII (IO)
5825 N. SAM HOUSTON PARKWAY
SUITE 300
HOUSTON, TX 77086
(281) 716-8360

INDIANAPOLIS II FIELD OFFICE (IO)
151 N. DELAWARE STREET, SUITE 1000
INDIANAPOLIS, IN 46204
(317) 287-3500

JACKSON II FIELD OFFICE (IO)
100 WEST CAPITOL STREET,
SUITE 1403
JACKSON, MS 39269
(601) 863-0940

JACKSONVILLE III FIELD OFFICE (IO)
5210 BELFORT ROAD
SUITE 350
JACKSONVILLE, FL 32256
(904) 380-5580

KANSAS CITY III & VI FIELD OFFICE (IO)
2600 GRAND BOULEVARD, SUITE 200
KANSAS CITY, MO 64108
(816) 410-6000

LANSDALE FIELD OFFICE (IO)
100 WEST MAIN STREET
SUITE 300B
LANSDALE, PA 19446
(215) 362-1840

LAS VEGAS III FIELD OFFICE (IO)
8965 S. EASTERN AVENUE.
SUITE 200
LAS VEGAS, NV 89123
(702) 347-5930

LEXINGTON FIELD OFFICE (IO)
1040 MONARCH STEET
SUITE 250
LEXINGTON, KY 40513
(859) 219-4500

LITTLE ROCK II FIELD OFFICE (IO)
425 WEST CAPITOL, SUITE 775A
LITTLE ROCK, AR 72201
(501) 324-6457

LOUISVILLE II FIELD OFFICE (IO)
600 DR. MARTIN L KING JR PL.,
SUITE 354
LOUISVILLE, KY 40202
(502) 753-3500

LUBBOCK II FIELD OFFICE (IO)
SENTRY PLAZA III
5214 68TH STREET, SUITE 300
LUBBOCK, TX 79424
(806) 783-2750

MACON II FIELD OFFICE (IO)
1645 FOREST HILL ROAD
SUITE 200
MACON, GA 31210
(478) 405-2440

MC ALLEN II FIELD OFFICE (IO)
1100 EAST LAUREL AVENUE
SUITE 301
MCALLEN, TX 78504
(956) 661-7950

MERRILLVILLE II FIELD OFFICE (IO)
8420 INDIANA STREET 2ND FLOOR
MERRILLVILLE, IN 46410
(219) 755-6310

MIAMI VI FIELD OFFICE (IO)
11410 NW.
20TH STREET, SUITE 200
MIAMI, FL 33172
(305) 597-4960

MILWAUKEE II FIELD OFFICE (IO)
1000 NORTH WATER STREET
SUITE 1400
MILWAUKEE, WI 53202
(414) 727-6200

MOBILE II FIELD OFFICE (IO)
41 WEST INTERSTATE 65
SERVICE ROAD NORTH SUITE 350
MOBILE, AL 36608
(251) 706-8540

NASHVILLE II & IV FIELD OFFICE (IO)
5300 MARYLAND WAY, SUITE 200
BRENTWOOD, TN 37027
(615) 565-1412

NEW ORLEANS III FIELD OFFICE (IO)
ONE GALLERIA BLVD., SUITE 1700
METAIRIE, LA 70001
(504) 841-7120

NEW YORK VI FIELD OFFICE (IO)
32 OLD SLIP, SUITE 3500
NEW YORK, NY
(646) 335-9060

NEW JERSEY III FIELD OFFICE (IO)
3 GARRET MOUNTAIN PLAZA
SUITE 202
WOODLAND PARK, NJ 07424
(973) 247-3030

OKLAHOMA CITY I FIELD OFFICE (IO)
55 NORTH ROBINSON, ROOM 229
OKLAHOMA CITY, OK 73102
(405) 297-5073

OMAHA FIELD II OFFICE (IO)
17310 WRIGHT STREET, SUITE 204
OMAHA, NE 68130
(402) 952-2635

ORLANDO II FIELD OFFICE (IO)
3452 LAKE CYNDA DRIVE, SUITE 450
ORLANDO, FL 32817
(407) 384-2420

PHOENIX III FIELD OFFICE (IO)
201 EAST WASHINGTON STREET
SUITE 940
PHOENIX, AZ 85004
(602) 776-5480

PITTSBURGH III FIELD OFFICE (IO)
1000 LIBERTY AVENUE
SUITE 1414
PITTSBURGH, PA 15222
(412) 395-0540

PORTLAND III FIELD OFFICE (IO)
1201 N.E. LLOYD BLVD.
ROOM 720
PORTLAND, OR 97232
(503) 331-7820

RICHMOND II FIELD OFFICE (IO)
1011 BOULDER SPRINGS DRIVE
SUITE 300
RICHMOND, VA 23225
(804) 200-4141

ROANOKE II FIELD OFFICE (IO)
310 FIRST STREET, SW
SUITE 500
ROANAKE, VA 24011
(540) 983-6944

SACRAMENTO II FIELD OFFICE (IO)
1325 J STREET, SUITE 1530
SACRAMENTO, CA 95814
(916) 498-5095

SALT LAKE CITY II FIELD OFFICE (IO)
257 EAST 200 SOUTH, SUITE 475
SALT LAKE CITY, UT 84111
(801) 524-7000

SAN ANTONIO II FIELD OFFICE (IO)
8610 BROADWAY, SUITE 400
SAN ANTONIO, TX 78217
(210) 805-2777

SAN DIEGO III FIELD OFFICE (IO)
9449 BALBOA AVENUE, SUITE 200
SAN DIEGO, CA 92123
(858) 966-1030

SANTA ANA II FIELD OFFICE (IO)
34 CIVIC CENTER PLAZA
SUITE 6121
SANTA ANA, CA 92701
(714) 347-9150

SEATTLE II FIELD OFFICE (IO)
915 SECOND AVENUE
7TH FLOOR, ROOM 790
SEATTLE, WA 98174
(206) 204-7693

SPOKANE II FIELD OFFICE (IO)
1313 NORTH ATLANTIC STREET
SUITE 4100
SPOKANE, WA 99201
(509) 324-7866

SPRINGFIELD II FIELD OFFICE (IO)
3161 W. WHITE OAKS DRIVE
SUITE 200
SPRINGFIELD, IL 62704
(217) 547-3675

ST. LOUIS III FIELD OFFICE (IO)
ROBERT A. YOUNG FEDERAL BLDG.
1222 SPRUCE STREET, ROOM 6.205
ST. LOUIS, MO 63103
(314) 269-2250

ST. PAUL II FIELD OFFICE (IO)
30 EAST 7TH STREET, SUITE 1700
ST. PAUL, MN 55101
(651) 726-0220

SYRACUSE II FIELD OFFICE (IO)
100 SOUTH CLINTON STREET
ROOM 517
SYRACUSE, NY 13260
(315) 448-0898

TAMPA II FIELD OFFICE (IO)
923 U.S. HIGHWAY 301 SOUTH
TAMPA, FL 33619
(813) 612-2470

TUCSON III FIELD OFFICE (IO)
2255 WEST INA ROAD
3RD FLOOR
TUCSON, AZ 85741
(520) 297-2100

WILKES BARRE FIELD OFFICE (IO)
7 NORTH WILKES BARRE BLVD.
SUITE 271-M
WILKES-BARRE, PA 18702
(570) 820-2210

NON-ATF POINTS OF CONTACT

Directorate of Defense Trade Controls

U.S. Department of State
Directorate of Defense Trade Controls
Compliance & Registration Division
2401 E Street NW, SA-1, Room H1200
Washington, DC 20522-0112

Telephone 202-663-1282
Web site ....www.pmddtc.state.gov

Bureau of Industry and Security

Outreach and Educational Services Division
14th Street & Pennsylvania Ave., N.W. Washington, DC 20230

Telephone 202-482-4811
Web site ....www.bis.doc.gov

Federal Bureau of Investigation
Criminal Justice Information Services Division
NICS Section

Appeal Services Team, Module A-1
Post Office Box 4278
Clarksburg, WV 26302-4278

Telephone 877-444-6427
Web site ....www.fbi.gov/about-us/cjis/nics/nics

Canada Firearms Centre

Royal Canadian Mounted Police
Canadian Firearms Program
Ottawa, ON K1A 0R2

Telephone 800-731-4000
Web site ....www.rcmp-grc.gc.ca/cfp-pcaf/index-eng.htm
E-mail ....cfp-pcaf@rcmp-grc.gc.ca

Questions about State Laws or Local Ordinances

Contact your State Police, local law enforcement authority or State Attorney General's Office

See listing of State Attorney Generals starting on page 224

STATE ATTORNEYS GENERAL

Alabama
Office of the Attorney General
501 Washington Avenue
Montgomery, AL 36104
(334) 242-7300

Alaska
Office of the Attorney General
P.O. Box 110300
Diamond Courthouse, 4th Floor
Juneau, AK 99811-0300
(907) 465-2133

American Samoa
Office of the Attorney General
American Samoa
Government Executive Office Bldg.
Pago Pago, AS 96799
011 (684) 633-4163

Arizona
Office of the Attorney General
1275 West Washington Street
Phoenix, AZ 85007
(602) 542-4266

Arkansas
Office of the Attorney General
200 Tower Building
323 Center Street
Little Rock, AR 72201-2610
(800) 482-8982

California
Office of the Attorney General
1300 I Street, Suite 1740
Sacramento, CA 95814
(916) 445-9555

Colorado
Office of the Attorney General
Ralph L. Carr Colorado Judicial Center
1300 Broadway, 10th Floor
Denver, Colorado 80203
720-508-6000

Connecticut
Office of the Attorney General
55 Elm Street
Hartford, CT 06141-0120
(860) 808-5318

Delaware
Office of the Attorney General
Carvel State Office Building
820 North French Street
Wilmington, DE 19801
(302) 577-8500

District of Columbia
Office of the Corporation Counsel
441 4th Street, NW,
Washington DC 20001
(202) 727-3400

Florida
Office of the Attorney General
The Capitol, PL 01
Tallahassee, FL 32399-1050
(850) 414-3300

Georgia
Office of the Attorney General
40 Capitol Square
Atlanta, GA 30334-1300
(404) 656-3300

Guam
Office of the Attorney General
590 S. Marine Corps Drive
ITC Bldg., Suite 706
Tamuning, Guam 96913
(671) 475-3324

Hawaii
Office of the Attorney General
425 Queen Street
Honolulu, HI 96813
(808) 586-1500

Idaho
Office of the Attorney General
700 W. Jefferson Street
P.O. Box 83720
Boise, ID 83720-1000
(208) 334-2400

Illinois
Office of the Attorney General
James R. Thompson Center South
100 West Randolph Street
12th Floor
Chicago, IL 60601
(312) 814-3000

Indiana
Office of the Attorney General
Indiana Government Center
302 West Washington Street
5th Floor
Indianapolis, IN 46204
(317) 232-6201

Iowa
Office of the Attorney General
Hoover State Office Building
1305 East Walnut
Des Moines, IA 50319
(515) 281-5164

Kansas
Office of the Attorney General
120 S.W. 10th Avenue, 2nd Floor
Topeka, KS 66612-1597
(785) 296-2215

Kentucky
Office of the Attorney General
State Capitol, Room 118
700 Capitol Avenue Frankfort, KY 40601
(502) 696-5300

Louisiana
Office of the Attorney General
Department of Justice
P.O. Box 94005
Baton Rouge, LA 70804
(225) 326-6079

Maine
Office of the Attorney General
6 State House Station
Augusta, ME 04333-0006
(207) 626-8800

Maryland
Office of the Attorney General
200 St. Paul Place
Baltimore, MD 21202-2202
(410) 576-6300

Massachusetts
Office of the Attorney General
One Ashburton Place Boston, MA
02108-1698
(617) 727-2200

Michigan
Office of the Attorney General
P.O. Box 30212
525 West Ottawa Street Lansing, MI
48909-0212
(517) 373-1110

Minnesota
Office of the Attorney General
1400 Bremer Tower
445 Minnesota Street
St. Paul, MN 55101
(651) 296-3353

Mississippi
Office of the Attorney General
Department of Justice
Post Office Box 220
Jackson, MS 39205-0220
(601) 359-3680

Missouri
Office of the Attorney General
Supreme Court Building
207 West High Street
Jefferson City, MO 65101
(573) 751-3321

Montana
Office of the Attorney General
Justice Building
215 North Sanders Helena, MT 59620-1401
(406) 444-2026

Nebraska
Office of the Attorney General
2115 State Capitol
Lincoln, NE 68509
(402) 471-2682

Nevada
Office of the Attorney General
Old Supreme Court Building
100 North Carson Street
Carson City, NV 89701
(775) 684-1100

New Hampshire
Office of the Attorney General
State House Annex
33 Capitol Street
Concord, NH 03301-6397
(603) 271-3658

New Jersey
Office of the Attorney General
Richard J. Hughes Justice Complex
25 Market St., CN 080
Trenton, NJ 08625
(609) 292-8740

New Mexico
Office of the Attorney General
Post Office Drawer 1508
Santa Fe, NM 87504-1508
(505) 827-6000

New York
Office of the Attorney General
Department of Law -The Capitol
2nd Floor
Albany, NY 12224
1-800-771-7755

North Carolina
Office of the Attorney General
Department of Justice
9001 Mail Service Center
Raleigh, NC 27699-9001
(919) 716-6400

North Dakota
Office of the Attorney General
State Capitol
600 East Boulevard Avenue
Bismarck, ND 58505-0040
(701) 328-2210

No. Mariana Islands
Office of the Attorney General
Administration Building
P.O. Box 10007
Saipan, MP 96950-8907
(670) 664-2341

Ohio
Office of the Attorney General
State Office Tower
30 East Broad Street, 14th Floor
Columbus, OH 43266-0410
(614) 466-4986

Oklahoma
Office of the Attorney General
State Capitol, Room 112
313 NE 21st Street
Oklahoma City, OK 73105
(405) 521-3921

Oregon
Office of the Attorney General
Justice Building
1162 Court Street
NE Salem, OR 97301-4096
(503) 378-4400

Pennsylvania
Office of the Attorney General
1600 Strawberry Square
Harrisburg, PA 17120
(717) 787-3391

Puerto Rico
Office of the Attorney General
P.O. Box 9020192
San Juan, PR 00902-0192
(787) 721-2900

Rhode Island
Office of the Attorney General
150 South Main Street
Providence, RI 02903
(401) 274-4400

South Carolina
Office of the Attorney General
Rembert C. Dennis Office Bldg.
P.O. Box 11549
Columbia, SC 29211-1549
(803) 734-3970

South Dakota
Office of the Attorney General
1302 E. Highway 14, Suite 1
Pierre, SD 57501-8501
(605) 773-3215

Tennessee
Office of the Attorney General
P.O. Box 20207
Nashville, TN 37202-0207
(615) 741-3491

Texas
Office of the Attorney General
Capitol Station
P.O. Box 12548
Austin, TX 78711-2548
(512) 463-2100

Utah
Office of the Attorney General
Utah State Capitol Complex
350 North State Street Suite 230
Salt Lake City UT 84114-2320
(801) 538-9600

Vermont
Office of the Attorney General
109 State Street
Montpelier, VT 05609-1001
(802) 828-3171

Virgin Islands
Office of the Attorney General
Department of Justice
34-38 Kronprindsens
Gade GERS Building,
2nd Floor St. Thomas,
Virgin Islands 00802
(340) 774-5666

Virginia
Office of the Attorney General
900 East Main Street
Richmond, VA 23219
(804) 786-2071

Washington
Office of the Attorney General
1125 Washington Street, SE
P.O. Box 40100
Olympia, WA 98504-0100
(360) 753-6200

West Virginia
Office of the Attorney General
State Capitol, Room E26
1900 Kanawha Boulevard East
Charleston, WV 25305
(304) 558-2021

Wisconsin
Office of the Attorney General
114 East State Capitol
P.O. Box 7857
Madison, WI 53707-7857
(608) 266-1221

Wyoming
Office of the Attorney General
State Capitol Building, Room 123
200 W. 24th Street
(307) 777-7841

TO: FEDERAL FIREARMS LICENSEES (FFL)

On November 30, 1998, the permanent provisions of the Brady Handgun Violence Prevention Act of 1993 (Brady Act) went into effect. The Brady Act required the U.S. Attorney General to establish a National Instant Criminal Background Check System (NICS) that any FFL could contact by telephone, or other electronic means, for information to be supplied immediately, on whether the receipt of the firearm by a prospective transferee would violate state or federal law.

In order to request a NICS background check, the FFL must first enroll in the NICS with the FBI. To obtain an enrollment form and additional NICS information, access the NICS FFL Web site at <www.fbi.gov/nics-ffl>. If you are an FFL without Internet access, you may contact the FBI Criminal Justice Information Services (CJIS) Division's NICS Section Customer Service at 1-877-FBI-NICS (324-6427) to retrieve copies of the information. The site includes but is not limited to the following:

- **Contact the NICS FFL Liaison Specialist**—E-mail the NICS Liaison Specialist with questions, comments, and/or new enrollments.

- **NICS Resolution Cards for Your Denied/Delayed Customers**—Order these NICS Resolution Cards online (50 card limit per request).

- **Update your FBI NICS FFL Contact Information**—If the NICS Section must contact your business, do we have the correct contact information on file for you? Now the FFL can update their contact information online by submitting a form to the NICS Section. Remember, if your main contact information has changed, the FFL must contact the Bureau of Alcohol, Tobacco, Firearms and Explosives (ATF) Licensing Center in writing by e-mail to <nlc@atf.gov>, fax 1-866-257-2749, or via U.S. mail to 244 Needy Road, Martinsburg, West Virginia 25405.

- **Receive NICS Messages**—FFLs and their employees can receive e-mails pertaining to NICS system changes/outages/general updates, etc. The FFL will have the option to add their e-mail address, request the e-mail address on file be replaced with a new e-mail address, or ask to be removed from the e-mail list all together. There is no limit to how many store employees can request updates. The FFL number is a requirement for any request. Once added to the list, e-mails will be received from this address <nicsfflupdates@leo.gov>.

- **The FFL Enrollment Form**—Requires you pick a code word. Each time you contact the NICS Section telephonically, you will be asked to supply your code word and your FFL number for verification of your identity. This code word is used for authentication of your identity and should be treated as **confidential information**. Additionally, you can request to conduct your NICS transaction online by using the NICS E-Check 2.0. This requires the FFL to complete sections 12-14 on the enrollment form. The FFL will be provided a user name and establish a password that will be used for authentication of your identity. The Enrollment Form Instructions are located on the backside of the enrollment form and should be read completely prior to completing the form.

- **The FFL Acknowledgment of Responsibilities under the NICS**—Outline the NICS-related responsibilities of each FFL. These responsibilities also apply to any FFLs, employees, agents, or other representatives. Any employee who will perform NICS background checks must read and comply with the policies and procedures outlined in the acknowledgment. Two signature pages are included in the agreement. The first signature page should be retained by the FFL. The second signature page is included on the enrollment form. This signature area must be signed by the licensee, witnessed, and returned to the FBI.

- **The NICS FFL User Manual**—Explains the NICS process.

- **The NICS Appeal Information**—For customers who are denied the purchase of a firearm based on a NICS background check. The appellant can file an appeal online or print the appeal brochure and fingerprint card to mail the appeal. Appeal packets are also available by calling the NICS Customer Service at 1-877-FBI-NICS (324-6427). The appellant begins the process via the Web site at <www.fbi.gov/nics-appeals>.

- **How to Obtain and Challenge your FBI Identification Record Brochure**—For customers with a criminal history interested in obtaining their FBI record.

- **The NICS E-Check 2.0 Information**—Provides links to the NICS E-Check 2.0 Web site. It also illustrates the functionality, security, and benefits of the new and improved NICS E-Check 2.0.

- **The FFL Quick Reference Guide**—Provides a quick overview of the steps required in conducting a NICS background check.

- **The NICS Information Sheet**—Explains a delay response. This sheet should be distributed to prospective gun purchasers whose transactions have been **delayed**.

- **The NICS Delay Instruction Sheet**—Provides instructions for the FFL on the processing of the delayed firearm transaction.

The FFL has two possible ways to conduct a NICS transaction:

1. Internet—Conduct your NICS transactions online by utilizing the NICS E-Check 2.0. Aren't having two options to conduct your NICS transactions better than one? To enroll with the NICS E-Check 2.0, you must complete sections 12-14 on the enrollment form in addition to registering to use the NICS E-Check on the dedicated Web site at <www.nicsezcheckfbi.gov>.

2. Telephonically—Once the NICS Section receives/completes your enrollment form, you will are automatically enrolled to conduct NICS transactions telephonically. Please be advised that calls may be monitored and recorded for any authorized purpose.

The NICS Section is dedicated to providing services to FFLs that enable your business to operate in an efficient and organized manner. One of the valuable tools that you can use to maximize efficiency and organization is the NICS E-Check. Please understand that we try to handle all incoming transactions as efficiently as possible; however, during peak times, it is not always possible to avoid hold times on the telephone. Even during times of high transaction volume, the NICS E-Check was still running efficiently with an **average wait time of less than two minutes once the transaction was initiated**.

The NICS E-Check 2.0 allows you to access the system by using any browser on any device with Internet access, for example, computer, laptop, tablet, smart phone, etc. The FFL and/or manager receives administrative controls, which includes the ability to create and modify employee NICS E-Check accounts. You simply log in to the NICS E-Check 2.0 using a user name and password, similar to how many of us manage online accounts.

The NICS E-Check 2.0 offers many benefits which are listed below:

- You will have administrative controls and can create accounts for all employees you wish to have access to the NICS E-Check.

- You will be able to lock or suspend accounts for employees that have left the company or are on extended leave.

- You will be able to reset passwords for employees that have forgotten their password.

- You will be able to access the NICS E-Check using any computer or browser.

- Ability to retrieve background checks 24/7 regardless of whether initiated on NICS E-Check 2.0 or at the NICS Contracted Call Center.

- No hold times for the NICS Contracted Call Center.

- No hold times for an FBI NICS representative to review a transaction.

- Ability to print completed NICS background check search requests.

- The availability of messages regarding NICS operational status.

- Added customer protection against identity theft.

- Added safeguard against theft of license number and code word.

In addition to this letter, you have been provided instructions on how to register to use the NICS E-Check 2.0 and the features and testimonials of the NICS E-Check. Please use these documents to maximize the services offered to you by the NICS Section.

Please complete the FFL Enrollment Form and return it to the FBI within two weeks of receipt. You may return this document via facsimile at (304) 625-0897, e-mail at <fbinicsteamcc@leo.gov>, or mail to:

Federal Bureau of Investigation
NICS Section
Post Office Box 4278
Clarksburg, WV 26302-9951

You may also enroll by calling the NICS Customer Service at 1-877-FBI-NICS (324-6427), press option 2, and then option 4. However, if you enroll through the NICS Customer Service, you will only be enrolled to conduct your NICS transactions telephonically. To enroll for NICS E-Check 2.0 follow the enclosed instructions. All enrollment forms **must** be returned to the NICS Section containing the signature of the FFL and a witness. Your ability to perform the background checks required under the Brady Act is dependent on completion and return of the FFL Enrollment Form. **Upon utilization of your code word for the first time, you are acknowledging you have read and understand the obligations and responsibilities under the NICS as an FFL.**

We look forward to working with you. If you have questions regarding the materials online, please call the NICS Customer Service at 1-877-FBI-NICS (324-6427).

230

What is a Business Day?

A business day is a 24-hour day (**beginning at 12:01 a.m.**) on which state offices are open. A business day **does not include Saturday, Sunday, or holidays.** The count for the three business days begins the day after the check was initiated.

If the NICS is contacted on:	The firearm can be legally transferred on*
Sunday	Thursday
Monday	Friday
Tuesday	Saturday
Wednesday	Tuesday
Thursday	Wednesday
Friday	Thursday
Saturday	Thursday

If there is a state or federal holiday during the time period, the answer may be delayed one additional day. Additional state laws may also apply.

What Do I Give My Delayed/Denied Customers?

NICS Resolution Card

The FBI Criminal Justice Information Services (CJIS) Division's National Instant Criminal Background Check System (NICS) Section is pleased to offer a tool for firearm dealers and their customers. The NICS Section has noticed an increased propensity for a denied or delayed customer to be unsure about the best avenue to pursue when receiving a delayed or deny response from their NICS background check. In response, the NICS Resolution Card has been created. The purpose of this card is two-fold: (1) to define the most appropriate action for the customer to pursue and (2) to educate and encourage customers to utilize the NICS Web site. Customers who initiate their inquiry through the NICS Web site will experience an improved user-friendly, streamlined process.

Federal Firearms Licensee (FFL): We are recommending the following practice be employed when utilizing the card.

DENY side of the card:

If a prospective transferee has been *denied* the transfer of a firearm, the FFL should give the individual a NICS Resolution Card, **circle** the word "DENY" on the top of the card, and **provide the individual with the NICS Transaction Number (NTN)**. The FFL should write the NTN on the line provided on the bottom of the card. The Deny side of the card provides the customer with information on how to appeal this decision on-line or via the U.S. Mail.

DELAY side of the card:

If a prospective transferee gets **extended and/or experiences continuous delays**, the FFL may provide a NICS Resolution Card, **circle** the word "DELAY" on the top of the card, and **provide the individual with the NTN**. The FFL should write the NTN on the line provided on the bottom of the card. The Delay side of the card provides the customer with information on how to appeal this decision on-line or via the U.S. mail.

The specific reason for delay and deny cannot be provided over the phone due to the Privacy Act of 1974.

Both options can be found at <www.fbi.gov/nics-appeals>. For customers without Internet access, the NICS Resolution Card also provides the mailing address and basic instructions for sending requests to the NICS Section.

To request a shipment of the NICS Resolution Cards, visit the NICS FFL Web site at <www.fbi.gov/nics-ffl>, or contact the NICS Customer Service at 1-877-FBI-NICS (324-6427), option 2.

Deny

The National Instant Criminal Background Check System (NICS) is a descriptor-based name search. If you believe you have been erroneously denied you may submit a request to appeal the deny decision.

The agency which processed your NICS transaction is required to have an appeal process. You may make application first to the *denying agency*, i.e., either the state or local law enforcement agency that processed your transaction. As an alternative to appealing directly through the denying agency, you may elect to submit your appeal request to the FBI Criminal Justice Information Services Division's NICS Section.

The NICS Appeal Web site was established to help guide NICS *denied* appellants through the appeal process.

http://www.fbi.gov/nics-appeals

For customers without Internet access, you may mail an appeal request to the address listed below:

FBI NICS Section
Attention: Appeal Services Team
P.O. Box 4278
Clarksburg, WV 26302-9922

The appeal submission must include:
- ✓ Request the reason for your denial
- ✓ NICS Transaction Number (NTN)
- ✓ Your full name
- ✓ Your address
- ✓ Your phone number

You will be notified via U.S. mail once the status of your appeal is determined.

Positive proof of identification is needed to make a final determination on many appeal requests for a denied decision. Providing a set of your rolled fingerprint impressions prepared by law enforcement or an authorized fingerprinting agency is highly recommended to help expedite your appeal of a deny decision.

The NICS Section cannot release the reason for a deny determination over the phone.

NICS Transaction Number (NTN):

_____
(NTN Required to Appeal)   June 2014

Delay

The National Instant Criminal Background Check System (NICS) is a descriptor-based name search. If you are experiencing an extended delay, you may submit a request to appeal the delay decision.

The NICS Appeal Web site was established to help guide NICS *delayed* appellants through the appeal process.

http://www.fbi.gov/nics-appeals

You will be notified via U.S. Mail once the status of your appeal is determined.

For customers with no Internet access, you may mail the appeal request to:

FBI NICS Section
Attention: Appeal Services Team
P.O. Box 4278
Clarksburg, WV 26302-9922

The appeal submission must include:
- ✓ A request for the reason for your delay
- ✓ NICS Transaction Number (NTN)
- ✓ Your full name
- ✓ Your address
- ✓ Your phone number
- ✓ Positive proof of your identity

Positive proof of your identity requires that you provide a set of your rolled fingerprint impressions prepared by a law enforcement agency or other authorized fingerprinting agency. A fingerprint card can be downloaded from the Web site or will be mailed to you if you have no Internet access.

The NICS Section cannot release the reason for a delay determination over the phone.

The NICS Section suggests that you wait at least 30 days prior to filing an appeal on a delay to give the NICS Section's staff time to complete the transaction. If your background check is completed, the NICS Section will notify the Federal Firearm Licensee with a final status.

NICS Transaction Number (NTN):

_____
(NTN Required to Appeal)   June 2014

www.ingramcontent.com/pod-product-compliance
Lightning Source LLC
Chambersburg PA
CBHW081358270326
41930CB00015B/3344

9 781939 473585